INTRODUCTION TO BIOLOGY

By the same author

Introduction to Biology: Third Tropical Edition
Introduction to Biology: West African Edition
Introduction to Human and Social Biology (with Brian Jones)
Introduction to Biology: Colour Edition
Introduction to Genetics
Experimental Work in Biology
   1 Food Tests
   2 Enzymes
   3 Soil
   4 Photosynthesis
   5 Germination and Tropisms
   6 Diffusion and Osmosis
   7 Respiration and Gaseous Exchange
   8 Human Senses
(Teachers' Notes are available for each book)
Experimental Work in Biology
Combined Edition
Class Experiments in Biology
(with C. J. and P. C. Worsley)
Life Study

Visual material (for further details see page 211)

Slides: six sets, each of 20 colour slides, with notes
Transparencies: 84 b/w unmounted overhead projector transparencies on Human Physiology, 23 on Insects, 12 on Pollination and Fertilization

Foreign editions

Original edition

Holland   *Inleiding tot de Biologie*, Wolters-Noordhoff NV, Groningen
Italy   *Biologia*, Casa Editrice Scode, Milan

Tropical edition

India   *Introduction to Biology*, Allied Publishers Private Ltd, Bombay
Singapore   *Introduction to Biology* and *Introduction to Biology* (Chinese edition), Eastern Universities Press, Singapore

© D. G. Mackean 1962, 1963, 1965, 1969, 1973

First edition 1962
Second edition 1963
Third edition 1965
Fourth edition 1969
Fifth edition 1973
Reprinted with corrections 1976, 1977, 1978, 1980, 1981, 1982, 1984

Printed in Great Britain by Jarrold and Sons Ltd, Norwich

0 7195 2830 5 boards
0 7195 2831 3 limp

D0255897

JOHN MURRAY · LONDON

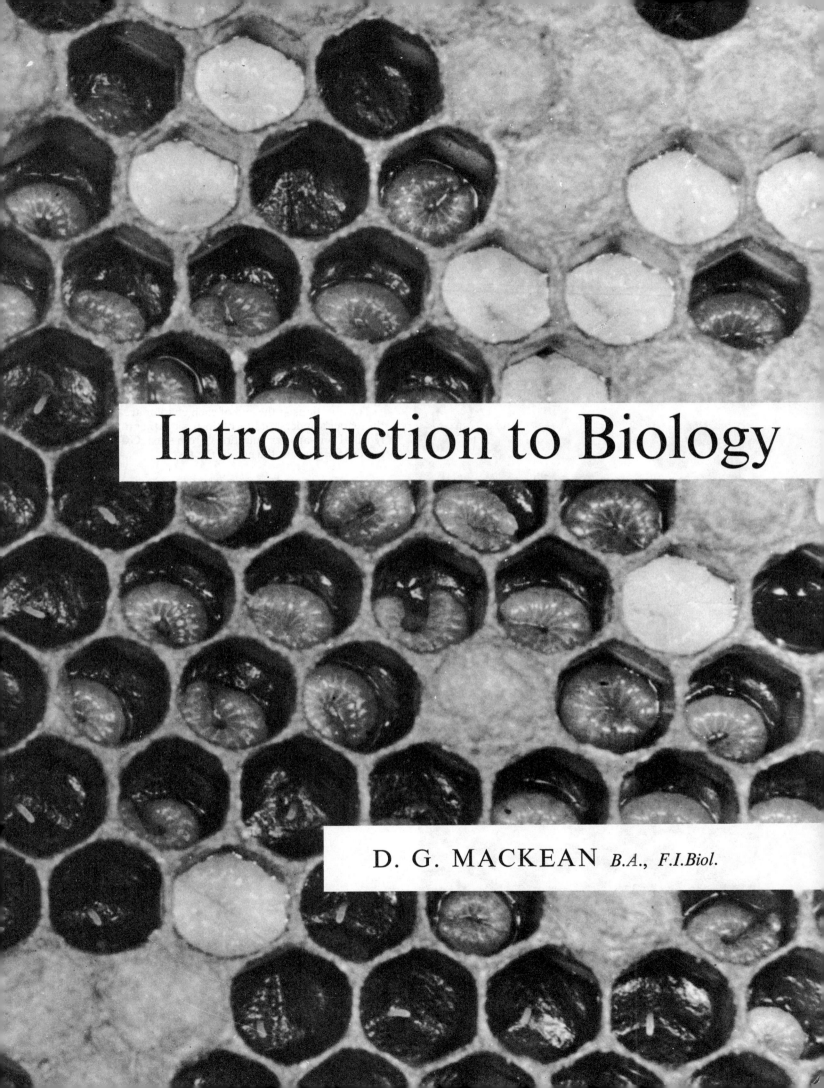

Introduction to Biology

D. G. MACKEAN B.A., F.I.Biol.

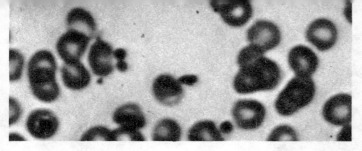

Preface

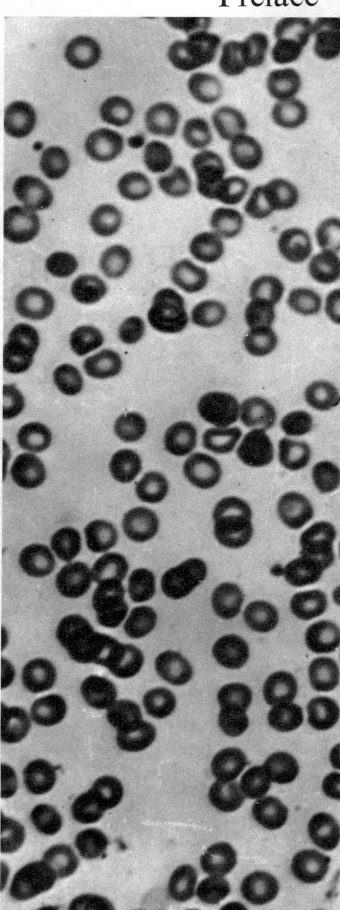

Introduction to Biology is written for O level and CSE students and the contents represent a body of biological knowledge common to syllabuses of the major GCE boards in the United Kingdom.

The book does not reflect a particular attitude or offer a novel approach to the subject but is meant as a reference book with clearly explained and fully illustrated information from which students can readily find the facts they need. It is hoped that teachers will refer students to the text at appropriate points in their own programmes of teaching rather than treat it as a course to be followed chapter by chapter.

The experimental work described in the text is intended mainly for revision or for students who have been unable to carry out the experiments for themselves. Comprehensive instructions for these and many other experiments, and information about their preparation, are given in *Experimental Work in Biology*.

The questions at the end of each chapter are intended to test the reader's understanding of the topic rather than invite simple recall. Long-answer questions from past GCE examination papers are given at the end of the text.

DGM

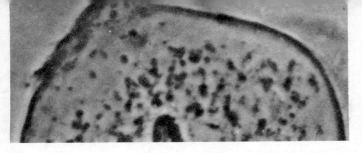

Contents

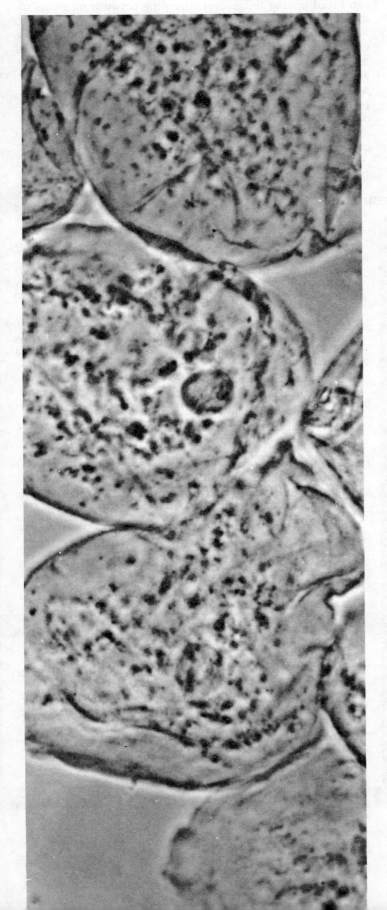

1	The Varieties of Living Organisms	6
2	Characteristics of Living Organisms	9
3	Cells	10
4	Structure of the Flowering Plant	14
5	Storage Organs and Vegetative Reproduction	22
6	Buds and Twigs	27
7	Sexual Reproduction in Flowering Plants	30
8	Seeds, Germination and Tropisms	39
9	Respiration	47
10	Photosynthesis and the Nutrition of Green Plants	50
11	Food Chains and the "Balance of Nature"	57
12	Diffusion and Osmosis	62
13	Translocation and Transpiration	68
14	Bacteria	72
15	Fungi	75
16	Soil	78
17	Food and Diet in Man	82
18	The Digestion, Absorption and Metabolism of Food	85
19	Blood, its Composition, Function and Circulation	93
20	Breathing	101
21	Excretion	106
22	Skin and Temperature Control	109
23	Sexual Reproduction	112
24	The Skeleton, Muscles and Movement	119
25	Teeth	125
26	The Sensory Organs	128
27	Co-ordination	137
28	Insects	146
29	Fish	159
30	Frogs and Tadpoles	163
31	Birds	168
32	Simple Plants and Animals	174
33	Chromosomes and Heredity	179
34	Heredity and Genetics	194
35	Evolution and the Theory of Natural Selection	203
	Books for Further Reading	210
	Visual Material	211
	Reagents	211
	Acknowledgements	211
	Examination Questions	212
	Glossary	214
	Index	220

The diagrams are by the author

1 The Varieties of Living Organisms

THE Earth is populated by enormous numbers of different plants and animals. When these organisms are studied and described, it becomes apparent that they can be divided into groups. In each group of plants or animals, the members resemble each other strongly. These similarities are not always obvious but become apparent when the characteristics of the group are known. Bees and butterflies, for example, though differing considerably in appearance, size and habits, belong to the same group, insects, because they both have hard outer skeletons, six legs and two pairs of wings.

The first division of organisms is into plant and animal kingdoms. The next division is into *phyla* (singular: *phylum*), for example, the first nine groups of animals listed in the table. The smallest natural group of animals or plants is the *species*.

For example, birds are not a species, but robins are. Generally, all members of a species are alike in all important respects and can breed among themselves. Breeding between members of different species does not happen very frequently in Nature.

Placing an organism in a particular category is not always easy. Certain single-celled creatures are not definitely animals or plants but possess characteristics of both. Fungi and bacteria are sometimes placed in the plant kingdom although they do not contain chlorophyll and differ considerably from the green plants in their methods of obtaining food. The table of living organisms given below is not complete and does not conform to any strict biological classification but offers a simplified and convenient scheme.

Living Organisms

ANIMALS

A—ANIMALS WITHOUT VERTEBRAL COLUMNS: INVERTEBRATES (Fig. 1.1 *a–h*)

1 **Single-celled animals** can usually be seen only with a microscope and all live in water where they occur in vast numbers.
2 **Coelenterates.** Examples are sea anemones, jelly-fish, corals and Hydra.
3 **Flatworms.** Mostly small fresh-water animals found under stones and floating leaves in streams. The group includes parasitic tape-worms and liver flukes.
4 **(True) Worms.** This phylum includes the earthworm, many little worms that live in ponds, lugworms, ragworms and bristle-worms of the sandy coasts.
5 **Crustacea.** Familiar crustaceans are crabs, lobsters, crayfish, prawns, shrimps, many small fresh-water creatures such as the fresh-water shrimp, water-flea and water-louse.
6 **Insects** possess six legs and usually wings. Examples are: butterflies, ants, bees, grasshoppers, flies, mosquitoes and beetles.
7 **Arachnids** have eight walking legs. Examples are spiders, scorpions, ticks and mites.
8 **Molluscs.** In this phylum are snails, slugs, whelks, oysters, and other "shellfish"; squid and octopus.
9 **Echinoderms** include the starfish and sea urchins.

B—ANIMALS WITH VERTEBRAL COLUMNS: VERTEBRATES (Fig. 1.2 *a–e*)

I POIKILOTHERMIC ("variable body temperature")

1 **Fish** breathe by means of gills and have bodies covered with scales. Examples: shark, herring, pike, stickleback.
2 **Amphibia** (e.g. frogs, newts, toads) have no scales on their bodies; but spend much of their lives on land.
3 **Reptiles** are land-dwelling animals with scaly bodies. Examples: lizards, snakes, tortoises, crocodiles.

II HOMOIOTHERMIC ("constant body temperature")

4 **Birds** have bodies that are covered with feathers. Examples: sparrow, duck, penguin.
5 **Mammals** have bodies that are covered with fur; their young are born alive and suckled with milk. Examples: cows, dogs, cats, whales, seals, apes, man.

PLANTS

A—PLANTS WHICH DO NOT HAVE FLOWERS

1 **Single-celled plants** are similar to single-celled animals, but they are green and obtain their food differently. When in great numbers they often make pond-water look green. They also occur as a green powdery dust on tree trunks.
2 The **algae** include the green slimy filaments on ponds and sea-weeds.
3 **Liverworts** are small, flat, green, leaf-like plants found in clusters in damp places, stream banks, and in cellars and caves to which light has access.
4 **Mosses** are small green plants growing in dense colonies.
5 **Ferns** are of many kinds and include bracken.
6 **Coniferous trees.** Examples are larch, spruce, pine, cypress.
7 **Fungi** include moulds, toadstools, mushrooms, bracket fungi. (Fungi are not always classified as plants.)

B—FLOWERING PLANTS

1 **Monocotyledons.** Narrow-leaved plants with only one cotyledon in their seeds. Examples: grasses, reeds, rushes, cereals, iris.
2 **Dicotyledons.** Broad-leaved plants with two cotyledons in their seeds.
   (i) *Herbaceous plants.* Examples: daisy, buttercup, dandelion.
   (ii) *Shrubs.* Woody, bushy plants. Examples: blackthorn, or sloe, box, privet.
   (iii) *Deciduous trees.* Examples: oak, ash, hazel, beech.

EXERCISE

Try to classify the following animals and plants according to the lists given above:

Python, chestnut tree, antelope, mussel, cowslip, lugworm, tadpole, Douglas fir, whale, caterpillar, turtle, puffball, jackdaw, alligator, flea, trout, bluebell, mildew, earwig, cedar, otter, barley.

Initially it is good practice to set out the answers as follows:

Cow. Animal; vertebrate; homoiothermic; mammal.

Toadstool. Plant; non-flowering; fungus.

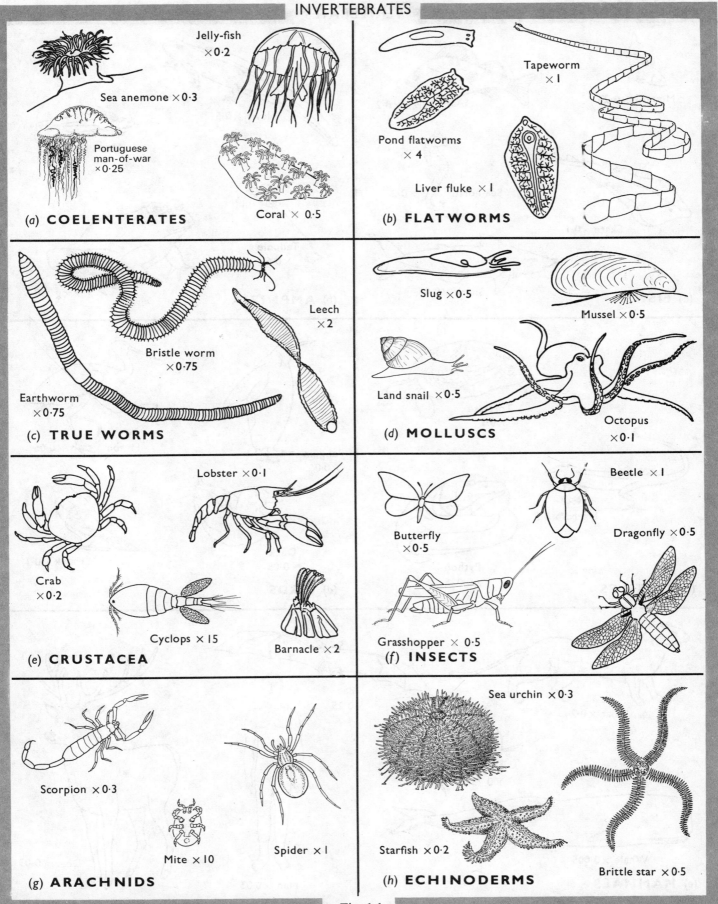

(a) **COELENTERATES**
Sea anemone ×0·3
Jelly-fish ×0·2
Portuguese man-of-war ×0·25
Coral × 0·5

(b) **FLATWORMS**
Pond flatworms × 4
Liver fluke × I
Tapeworm × I

(c) **TRUE WORMS**
Earthworm ×0·75
Bristle worm ×0·75
Leech ×2

(d) **MOLLUSCS**
Slug ×0·5
Mussel ×0·5
Land snail ×0·5
Octopus ×0·1

(e) **CRUSTACEA**
Crab ×0·2
Lobster ×0·1
Cyclops × 15
Barnacle ×2

(f) **INSECTS**
Butterfly ×0·5
Beetle × I
Dragonfly ×0·5
Grasshopper × 0·5

(g) **ARACHNIDS**
Scorpion ×0·3
Mite × 10
Spider × I

(h) **ECHINODERMS**
Sea urchin ×0·3
Starfish ×0·2
Brittle star ×0·5

Fig. 1.1

VERTEBRATES

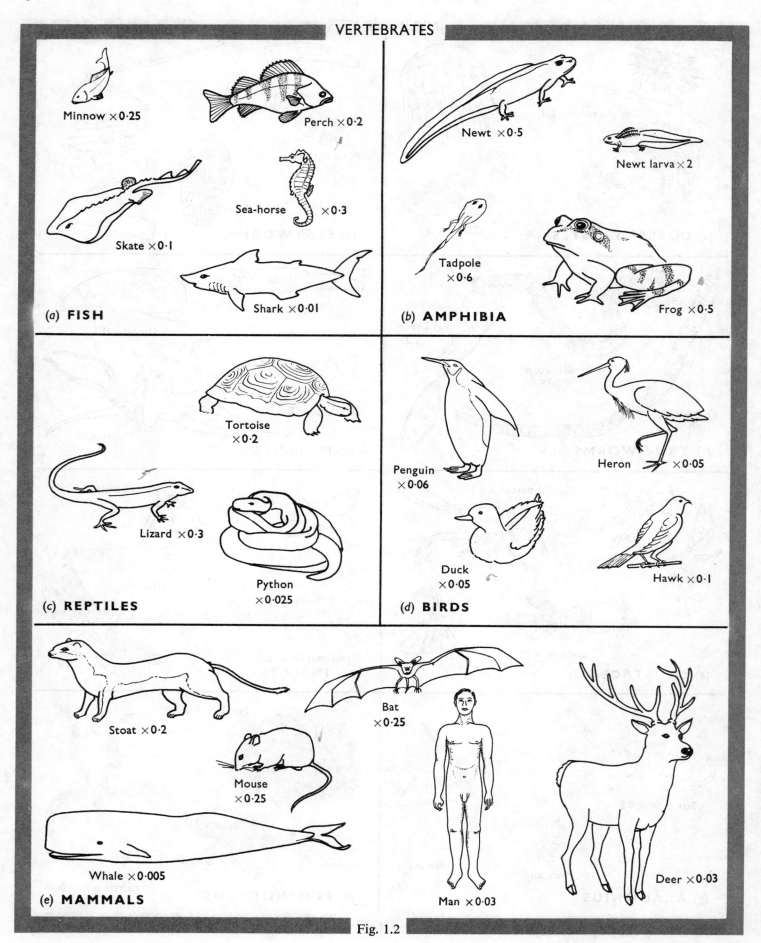

(a) **FISH**
Minnow ×0·25
Perch ×0·2
Sea-horse ×0·3
Skate ×0·1
Shark ×0·01

(b) **AMPHIBIA**
Newt ×0·5
Newt larva ×2
Tadpole ×0·6
Frog ×0·5

(c) **REPTILES**
Tortoise ×0·2
Lizard ×0·3
Python ×0·025

(d) **BIRDS**
Penguin ×0·06
Heron ×0·05
Duck ×0·05
Hawk ×0·1

(e) **MAMMALS**
Stoat ×0·2
Mouse ×0·25
Bat ×0·25
Man ×0·03
Deer ×0·03
Whale ×0·005

Fig. 1.2

2 Characteristics of Living Organisms

BIOLOGY is the study of life (Greek *bios*=life, *logos*=knowledge) which, in practice, means the study of living things.

In most animals, the characteristics by which we know they are alive are self-evident: they move about, they feed, they have young, and they respond to changes in their surroundings.

These features are less obvious in plants and certain small animals; and when dealing with organisms like bacteria and viruses the distinctions between living and non-living can often be drawn only by a trained scientist with the appropriate apparatus and techniques at his disposal. The main differences between living organisms and non-living objects can be summarized as follows:

1. **Respiration.** This is the process by which energy is made available as a result of chemical changes within the organism. The commonest of these is the chemical decomposition of food as a result of its combination with oxygen. This is not a particularly obvious occurrence in plants and animals; but it is fairly easy to demonstrate that living creatures take in air, remove some of the oxygen from it and increase the volume of carbon dioxide in it. More simply expressed it can be said that living organisms take in oxygen and give out carbon dioxide. Sometimes this takes place with obvious breathing movements. Respiration also results in a rise of temperature, which is more easily detectable in animals than in plants.

2. **Feeding.** This is an essential preliminary to respiration, since energy comes ultimately from food. The production of food in the leaves of a tree is less obvious than the feeding of an animal, which moves actively in search of food. Feeding also results in growth.

3. **Excretion.** Living involves a vast number of chemical processes, including respiration, many of which produce substances that are poisonous when moderately concentrated. The elimination of these from the body of the organism is called excretion.

4. **Growth.** Strictly, growth is simply an increase in size, but it usually implies also that the organism is becoming more complicated and more efficient. An illustration of this is an animal which changes its form from larva to adult, for example, a frog or a butterfly.

5. **Movement.** An animal can generally move its whole body, whereas the movements of the higher plants are usually restricted to certain parts such as the opening and closing of petals, or to movements of parts as a result of growth.

6. **Reproduction.** No organisms have a limitless life, but although individuals must die sooner or later their life is handed on to new individuals by reproduction, resulting in the continued existence of the species.

7. **Sensitivity** (Irritability). Sensitivity is the ability to respond to a stimulus. Obvious signs of sensitivity are the movements made by animals as a result of noises, on being touched, or on seeing an enemy.

Differences between animals and plants

Plants and animals have in common, to a greater or lesser extent, all the features listed above, but there are some fundamental differences between them, of which one of the most important is the method of feeding.

1. **Method of Feeding.** Animals take in food that is chemically very complicated (i.e. composed of large molecules); it consists either of plant products or of other animals. This food is reduced to simpler material by the process of digestion, and in this form it can be taken up by the body.

Plants, in general, take in very simple substances that are composed of small molecules, namely carbon dioxide from the air, and water and dissolved salts from the soil. In their leaves they combine this carbon dioxide and water into sugar, using sunlight as a source of energy. From the sugar so produced, and the salts taken in from the soil, green plants can make all of the substances needed for their existence.

Feeding in animals thus involves a breaking-down process while feeding in plants is a building-up, or synthesis.

2. **Chlorophyll.** The green colour found in most plants is important for the absorption of sunlight and is due to chlorophyll, which is not present in any animal. This difference is, in effect, one indication of the fundamental difference in feeding. (It should be noted, however, that fungi do not possess chlorophyll and are often classified as a separate kingdom.)

3. **Cellulose.** In their structures, notably their cell walls, plants have a large quantity of a substance called cellulose, which is never present in animal structures.

4. **Movement.** Unlike many animals, most of the familiar plants do not move about as complete organisms, but certain microscopic plants move as actively as microscopic animals.

5. **Sensitivity.** Although both plants and animals respond to stimuli, the response of an animal usually follows almost immediately after the application of even a very brief stimulus. In plants, on the other hand, a response may take place over a matter of hours or days, and then only if the stimulus persists for a relatively long time.

QUESTIONS

1. A motor car moves, takes in oxygen and gives out carbon dioxide, consumes fuel but nevertheless is not a living creature. In what ways does it not "qualify" as a living organism?

2. A sponge-like organism is found adhering to a rock in a marine pool. How would a microscopic examination help to decide whether it was a plant or an animal?

3 Cells

NEARLY all plants and animals have one important characteristic in common: they are made up of cells. If any structures from plants or animals are examined microscopically they will be seen to consist of more or less distinct units—cells—which, although too small to be seen with the naked eye, in their vast numbers make up the structures or organs.

Since the cells of any organ are usually specially developed in their size, shape and chemistry to carry out one particular function (e.g. muscle cells for contracting) there is, strictly speaking, no such thing as a "typical" cell of plants or animals. Nevertheless, certain of the features common to most cells can be illustrated diagrammatically.

Parts of the cell (see Figs. 3.1–3.6)

Cell membrane. All cells are bounded by a very thin flexible membrane which retains the cell contents and controls the substances entering and leaving the cells.

Cell wall. In plant cells only, there is a wall outside the cell membrane. It confers shape and, to some extent, rigidity on the cell. While the cell is growing the cell wall is fairly plastic and extensible, but once the cell has reached full size, the wall becomes tough and resists stretching.

Unless impregnated with chemicals, as in the cells of corky tree bark, the cell wall is freely permeable to gases and water, i.e. it allows them to pass through in either direction. The cell wall is made by the cytoplasm and is non-living, being made of a transparent substance called cellulose.

The *middle lamella* is the layer which first forms between cells after a plant cell has divided (Fig. 3.5c) and may remain visible between mature cells in microscopical preparations.

Protoplasm is the material inside the cell which is truly alive. There are two principal kinds of protoplasm; the protoplasm which constitutes the nucleus (*see* below) is called *nucleoplasm*. All other forms of protoplasm are referred to as *cytoplasm*.

Cytoplasm is jelly-like and transparent, fluid or semi-solid and may contain particles such as chloroplasts or starch grains. In some cells it is able to flow about. In the cytoplasm the chemical processes essential to life are carried on. The cell membrane is selectively permeable, allowing some substances to pass through and preventing others from doing so. This selection helps to maintain the best conditions for chemical reactions in the protoplasm.

The **nucleus** consists of nucleoplasm bounded by a nuclear membrane. It is always embedded in the cytoplasm, is frequently ovoid in shape and lighter in colour than the cytoplasm. In diagrams it is often shaded darker because most microscopical preparations are stained with dyes to show it up clearly. It is less easily seen in the unstained cell. The nucleus is thought to be a centre of chemical activity, playing a part in determining the shape, size and function of the cell and controlling most of the physiological processes within it.

Without the nucleus the cell is not capable of its normal functions or of division, although it may continue to live for a time. When cell division occurs, the nucleus initiates and controls the process (Fig. 3.5).

Vacuole. In animal cells there may be small droplets of fluid in the cytoplasm, variable in size and position. In plant cells the vacuole is usually a large, permanent, fluid-filled cavity occupying the greater part of the cell. In plants, this fluid is called cell sap and may contain salts, sugar and pigments dissolved in water. The outward pressure of the vacuole on the cell wall makes the plant cells firm, giving strength and resilience to the tissues.

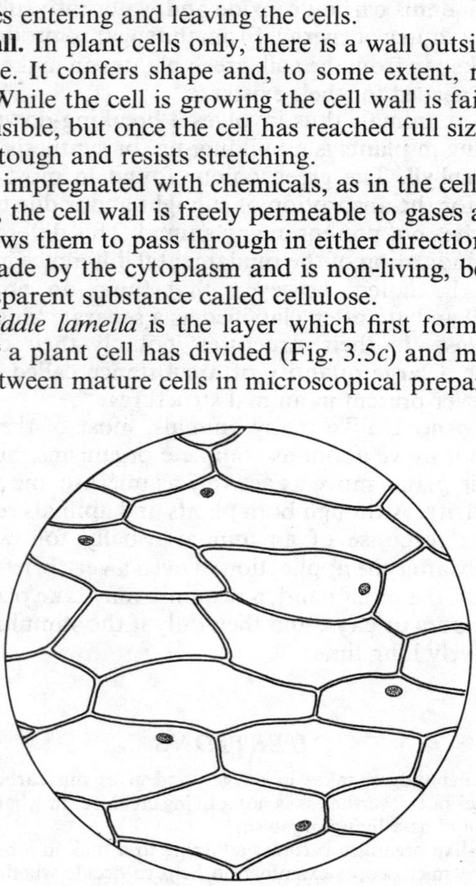

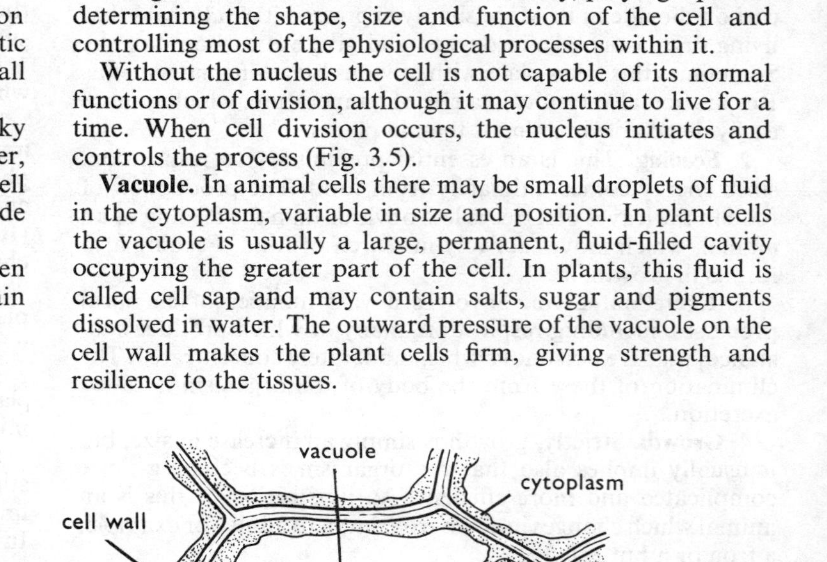

Fig. 3.1   Epidermis from onion scale seen under the microscope

The epidermis is one cell thick, so that under the microscope the transparent cells can be seen. The shape of each cell is partly determined by the pressure of the other cells round it. If a cell were isolated it would be rounded or oval.

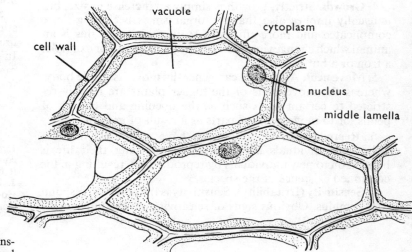

Fig. 3.2   A group of similar cells highly magnified to show cell structures

Cell division

Some, but not all, cells are able to divide and produce new cells as shown in Fig. 3.5. Cell division and subsequent cell enlargement taking place in many cells results in the growth of organisms.

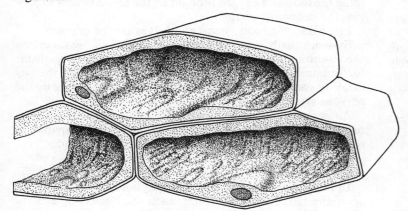

Fig. 3.3   Stereogram of plant cells

It is important to remember that, although cells look flat in sections or thin strips of tissue, they are in fact three-dimensional and may seem to have different shapes according to the direction in which the section is cut.

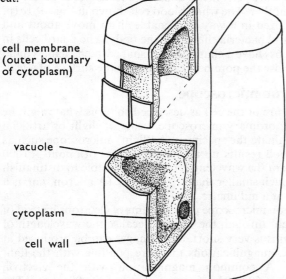

cell membrane
(outer boundary
of cytoplasm)

vacuole

cytoplasm

cell wall

Fig. 3.4(a)   Stereogram of one plant cell

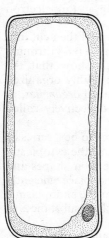

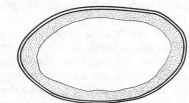

Fig. 3.4(b)   Transverse section

If the cell at (a) is cut across, it will look like (b) under the microscope; if cut longitudinally it will look like (c).

Fig. 3.4(c)   Longitudinal section

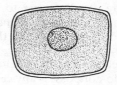

(a) A plant cell about to divide has a large nucleus and no vacuole.

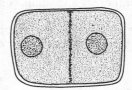

(b) The nucleus divides first.

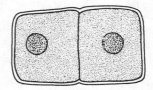

(c) The middle lamella develops and separates the two cells.

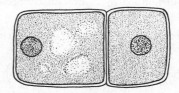

(d) The cytoplasm lays down a primary wall and layers of cellulose on each side of the middle lamella.

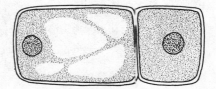

(e) One of the cells develops a vacuole and enlarges. The other cell retains the ability to divide again.

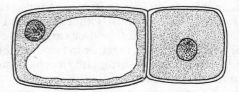

(f) Although each cell has its own primary wall, these often give the appearance of a single, intermediate wall between cells.

Fig. 3.5

Differences between plant and animal cells

1. Although there is a very wide range of variation, the cells of plant tissues are usually easier to demonstrate under the microscope than are the cells of animal tissues. This is partly because the plant cells are larger and their cell walls give them a distinctive outline.

2. Plants have cell walls made of cellulose. Animal cells have no cell walls and do not possess any cellulose.

3. Mature plant cells have only a thin lining of cytoplasm, with a large central vacuole. Animal cells consist almost entirely of cytoplasm (Fig. 3.7 and Plate 1). If any vacuoles are present they are usually temporary and small, concerned with excretion or secretion.

4. Animal cells never contain chloroplasts (p. 18) whereas these are present in a great many plant cells.

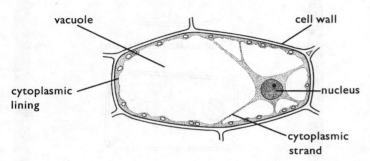

Fig. 3.6   A plant cell

Sometimes the nucleus appears in the centre of the cell, but it is still surrounded by cytoplasm connected by strands to that lining the wall.

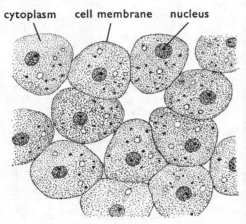

Fig. 3.7   A group of animal cells

Relation of cells to the organism as a whole

Although each cell can carry on the vital chemistry of living, it is not capable of existence on its own. A muscle cell could not obtain its own food or oxygen. Other specialized cells, present in a tissue or organ, collect food or carry oxygen. Unless individual cells are grouped together in large numbers and made to work together by the co-ordinating mechanisms of the body, they cannot exist for long.

Tissue. A tissue such as bone, nerve or muscle in animals, and epidermis, phloem or pith in plants, is made up of many hundreds of cells of one or a few types, each type having a more or less identical structure and function so that the tissue can also be said to have a specific function, e.g. nerves conduct impulses, phloem carries food.

Organs consist of several tissues grouped together making a functional unit: for example, a muscle is an organ containing long muscle cells held together with connective tissue and permeated with blood vessels and nerve fibres. The arrival of a nerve impulse causes the muscle fibres to contract, using the food and oxygen brought by the blood vessels to provide the necessary energy.

In a plant, the roots, stems and leaves are the organs.

System usually refers to a series of organs whose functions are co-ordinated to produce effective action in the organism: for example, the heart and blood vessels constitute the circulatory system; the brain, spinal cord and nerves make up the nervous system.

An organism results from the efficient co-ordination of the organs and systems to produce an individual capable of separate existence and able to perpetuate its own kind.

Specialization

In both their structure and their physiology, cells are often specialized to perform particular functions. For example, nerve cells (p. 138) are very long and are able to conduct electrical impulses rapidly. This makes them highly suitable for sending information from one part of the body to another. The cells forming the sieve tubes in the veins of plants (see Fig. 4.3b, p. 15) are specialized for conducting food. They are elongated and their end walls are perforated, allowing substances to pass from one cell to another. The white blood cell shown in Fig. 19.1c (p. 93) is specialized in a way that enables it to move about and engulf harmful bacteria that enter the body. The guard cells in Fig. 12.6 (p. 64) have a characteristic shape which enables them to open or close the pore between them, in the surface of a leaf.

The electron microscope

The structure of the cell as described so far is what might be seen with an ordinary microscope, using daylight or artificial light to illuminate the specimen. A "light" microscope such as this gives good results up to magnifications of about × 1000 but, because of the wavelength of light, is unable to distinguish structures much smaller than 0·3 microns. (A micron, μm, is a thousandth of a millimetre.)

The electron microscope passes beams of electrons instead of beams of light through the object. Because the wavelength of electron beams is very short, clear images can be produced at much greater magnifications than are possible with the light microscope. A common magnification with the electron microscope is × 50,000 but in some cases × 500,000 is used. The image can be seen by projecting it on to a fluorescent screen, as in a television tube, or by photographing it (an electron micrograph, Plate 1b).

The specimen is embedded in clear plastic and then cut into slices thin enough to allow the electron beam to pass through. Electron micrographs taken by this means show that the cytoplasm is not a structureless jelly, but a highly organized material, containing specialised structures called organelles, as well as food reserves and other granules, collectively called inclusions (Fig. 3.8).

Organelles. Endoplasmic reticulum. This seems to be a series of tubes and flattened sacs extending throughout the cytoplasm. The channels are not permanent but change their shapes and connections. Enzyme reactions (p. 85) can take place efficiently on the surface of the membranes which form the endoplasmic reticulum, and the system of channels might allow substances to move rapidly about in the cell.

Ribosomes. Some of the endoplasmic reticulum membranes

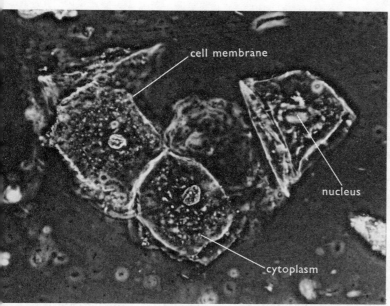

(*Brian Bracegirdle*)

Plate 1a. CELLS FROM CHEEK LINING AS SEEN UNDER LIGHT MICROSCOPE (×1000)

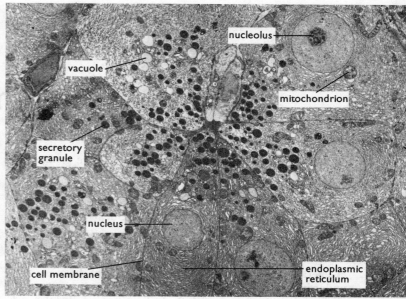

(*J. H. Kugler*)

Plate 1b. ELECTRON MICROGRAPH OF CELLS IN THE PANCREAS (×22 000)

have particles attached to them. These are called ribosomes and are known to be the sites where proteins (p. 82) are built up. The process of assembling amino acids to make proteins as described on p. 186, takes place on the ribosomes.

Mitochondria (singular = mitochondrion). Mitochondria are present in the majority of cells. Their shape varies from sausage-like to spherical and they can move about in the cells, often collecting at sites where rapid chemical activity is taking place or where oxygen is abundant. As shown in Fig. 3.8, their internal membranes appear to be thrown into folds which greatly increase their surface area. On these internal membranes there may be enzymes which control the process of respiration (p. 47). In this process, molecules of food such as glucose are broken down with the aid of oxygen and enzymes, to release energy. The mitochondria, therefore, are the main energy converters of the cell.

Nuclear Membrane. Electron micrographs show the nucleus to have a membrane round it. In this membrane there appear to be pores and it is thought that substances pass through the pores into the cytoplasm where they influence the reactions taking place.

Nucleolus. The nucleolus is thought to be the region in the nucleus where the substance RNA (ribo-nucleic acid) is made. This substance has properties similar to the DNA described on p. 186 and is probably the chemical which passes out of the nuclear pores to control the protein synthesis on the ribosomes.

PRACTICAL WORK

Plant cells. The epidermis can be stripped fairly easily from the inside of an onion scale (Fig. 3.1) or the lower surface of many leaves such as rhubarb. Since the epidermis here is one cell thick, the cells can be seen in transparency if a small piece of the tissue is placed flat on a slide, covered with a drop of water, and examined under the low power of the microscope. A little iodine solution may stain the nuclei light brown, and any starch grains present will turn dark blue. Cells and chloroplasts can be seen in a moss leaf if it is mounted flat on a slide with a drop of water.

Animal cells. If a finger is run round the inside of the cheek, and the fluid so collected is smeared on a slide, examination under the low

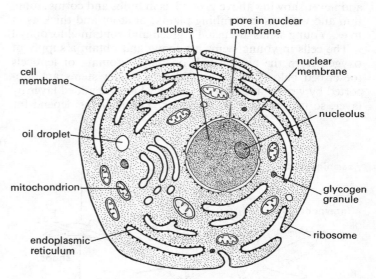

Fig. 3.8 General diagram of animal cell as seen by the electron microscope. The cell has been dehydrated, stained and sectioned. It is assumed that the simplified features shown do represent structures in the living cell.

power of the microscope will show numbers of epithelial cells scraped from the mouth lining (Plate 1a). The nuclei can be seen without staining but will show up more clearly if a drop of methylene blue solution is placed on the slide for about one minute. The slide is tilted to let the stain run off. The cytoplasm and nuclei will both take up the stain but the nuclei will be darker.

QUESTIONS

1. What features are (*a*) possessed by both plant and animal cells, (*b*) possessed by plant cells only?
2. With what materials must cells be supplied if they are to survive?
3. In what ways would you say that the white blood cell (Fig. 19.1*c*) is less specialized than the nerve cell (Fig. 27.3*a*)?
4. In many microscopical preparations of animal tissues, it is difficult to make out the cell boundaries and yet the disposition and numbers of cells can usually be determined. Which cell structure makes this possible?

4 Structure of the Flowering Plant

THE flowering plant consists of a portion above ground, the shoot, and a portion below ground, the root, although this does not imply that any part of a plant below ground must be a root.

The shoot is usually made up of a stem, bearing leaves, buds and flowers.

Stem

General characteristics (Fig. 4.1). A stem has leaves at regular intervals and a terminal bud at the growing point. The region of the stem from which the leaf springs is called the *node*, and the length of stem between the nodes, the *internode*.

Commonly, the stem is erect, but it may be horizontal as in strawberry runners; underground as in rhizomes; very short and never showing above ground as in bulbs and corms; long, thin and weak as in climbing plants; or stout and thick as in trees. Young stems are usually green and contain chlorophyll

The cells in young stems are living and obtain a supply of oxygen from the air through openings, stomata or lenticels (described below), in their epidermis. Older stems are supported by woody and fibrous tissues which are added layer by layer, so increasing their thickness. Young stems depend for

their rigidity on the turgidity of their cells, the cylindrical distribution of their conducting tissues and the opposing stresses of the pith and epidermis. Running through the stem are tubes which conduct water from the soil up to the leaves and food from the leaves to various parts of the plant.

Functions of the stem. It (*a*) supports the structures of the shoot; (*b*) spaces out the leaves so that they receive adequate air and sunlight; (*c*) allows conduction of water from soil to leaves, and food from leaves to other parts of the plant; (*d*) holds flowers above ground, thus assisting pollination by insects or wind. (*e*) If the stem is green, photosynthesis (Chapter 10) may occur in it.

Detailed structure (Figs. 4.2, 4.3 *a* and *b*, Plates 2 and 3). A fairly typical stem, such as that of a sunflower, is in the form of a cylinder. The outer layer of cells forms a skin, the *epidermis*, the inner cells make up the *cortex* and *pith*. Between the cortex and pith lie a number of *vascular bundles* containing specialized cells which carry food and water.

EPIDERMIS. The single layer of closely fitting cells is effective in holding the inner cells in shape, preventing loss of water, affording protection from damage and preventing the entry of fungi, bacteria and dust. This layer is relatively impermeable to

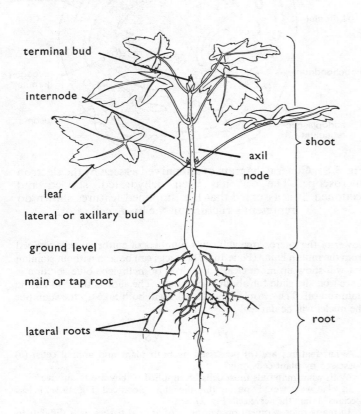

Fig. 4.1   Structure of a typical flowering plant

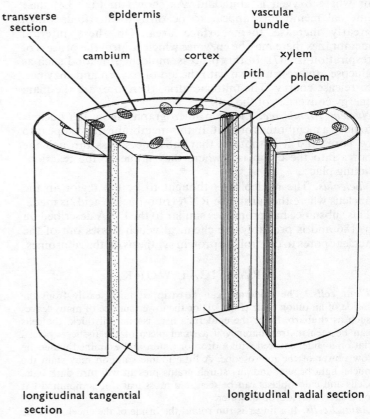

Fig. 4.2   Stereogram of plant stem

14

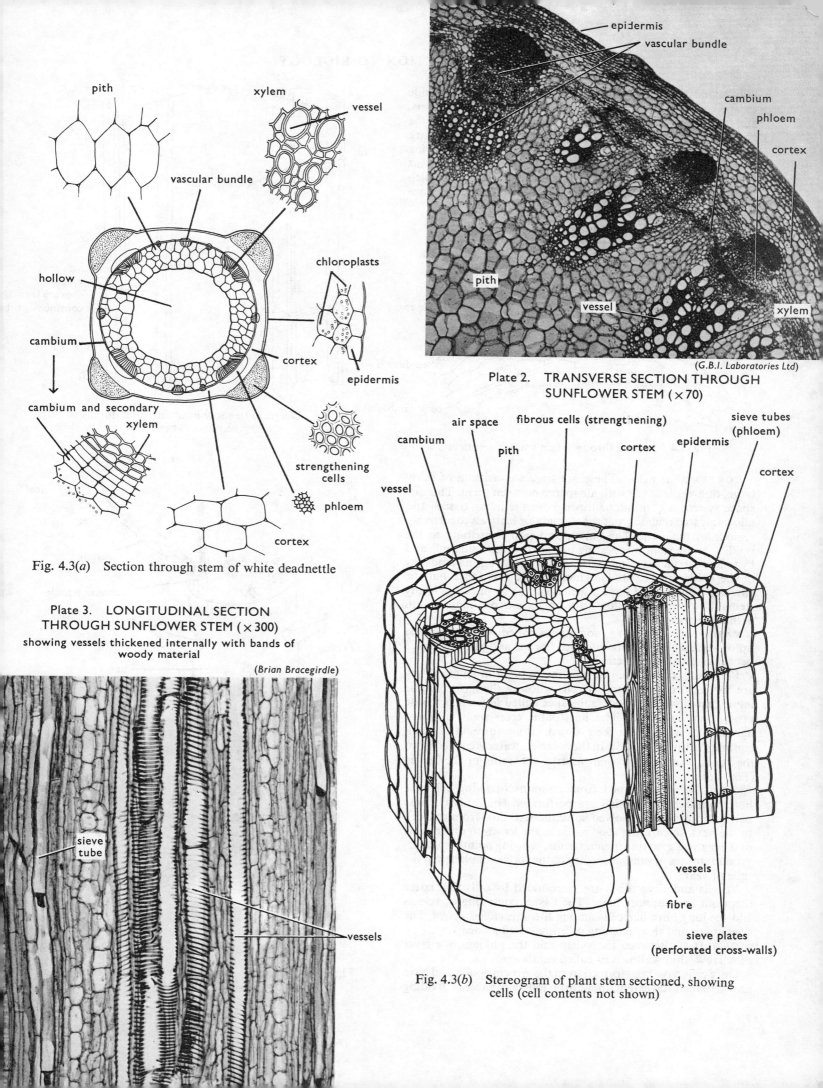

pith

xylem

vessel

vascular bundle

chloroplasts

hollow

cambium

cambium and secondary
xylem

cortex

epidermis

strengthening
cells

phloem

cortex

Fig. 4.3(a)   Section through stem of white deadnettle

Plate 3.   LONGITUDINAL SECTION
THROUGH SUNFLOWER STEM (×300)
showing vessels thickened internally with bands of
woody material

(Brian Bracegirdle)

sieve
tube

vessels

epidermis

vascular bundle

cambium

phloem

cortex

pith

vessel

xylem

(G.B.I. Laboratories Ltd)

Plate 2.   TRANSVERSE SECTION THROUGH
SUNFLOWER STEM (×70)

air space   fibrous cells (strengthening)

cambium

pith

vessel

cortex   epidermis

sieve tubes
(phloem)

cortex

vessels

fibre

sieve plates
(perforated cross-walls)

Fig. 4.3(b)   Stereogram of plant stem sectioned, showing
cells (cell contents not shown)

liquids and gases, and oxygen can enter, and carbon dioxide escape, only through stomata (described below) in young stems, and lenticels in older stems. The *lenticels* are small gaps in the bark, usually circular or oval and slightly raised on the bark surface. In them, the cells of the bark fit loosely, leaving air gaps which communicate with the air spaces in the cortex (Fig. 4.4). The epidermis is usually in a state of strain, in which it tends to shrink along its length. This shrinking effect contributes to the rigidity of the stem.

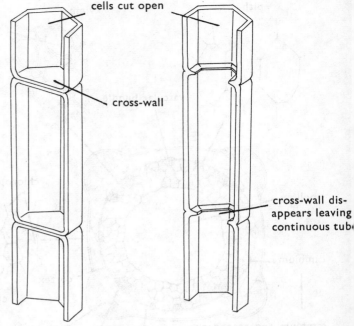

Fig. 4.5　Diagram to show how vertical columns of cells give rise to vessels

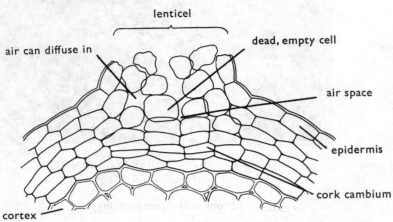

Fig. 4.4　Section through stem showing lenticel

CORTEX AND PITH. These are tissues consisting of fairly large, thin-walled cells with air spaces between them. This air-space system is continuous throughout the living tissues and allows air to circulate from the stomata or lenticels to all living regions of the stem. The cortex and pith contribute to the rigidity of the stem by pressing out against the epidermis and by tending to increase in length against the shrinking tendency of the epidermis (Expt. 1). These tissues also space out the vascular bundles and have a general value as packing. Many stems, however, are hollow with only a narrow band of pith within the cortex.

VASCULAR BUNDLES, sometimes called veins, are made up of vessels and sieve tubes, with fibrous and packing tissue between and around them.

Vessels consist of long tubes a metre or so in length. They are formed from columns of cells whose walls have become impregnated with a woody substance called *lignin* and whose protoplasm has died. The horizontal cross-walls of these lignified cells have broken down, thus forming a long continuous tube (Fig. 4.5). In these vessels water is carried from the roots, through the stem and to the veins in the leaves (Fig. 4.6).

Sieve tubes are formed from columns of living cells the horizontal walls of which are perforated (Fig. 3.8c). These perforations allow dissolved substances to flow from one cell to the next, so carrying food made in the leaves to other parts of the plant, e.g. to the ripening fruits, growing points or underground storage organs, according to the species of plant and the time of year.

Vessels and sieve tubes are surrounded by cells that space them out and support them. The tissue, consisting of vessels and the long fibre-like cells among them, is called *xylem*. The sieve tubes and their packing cells are called *phloem*.

CAMBIUM. Between the xylem and the phloem is a layer of narrow, thin-walled cells called cambium.

Once cells have been formed from the growing point and have grown to their full extent, they are no longer capable of dividing

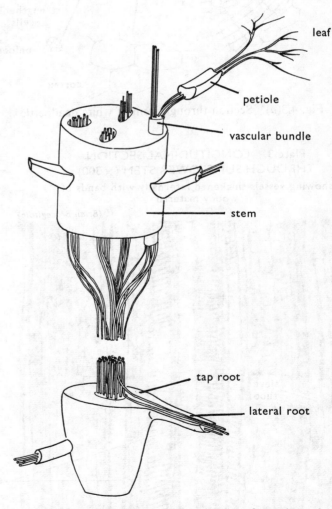

Fig. 4.6　To show distribution of veins from lateral root to leaf

to make new cells. They may have become changed in structure and specialized to a particular function, as has, for example, a sieve tube. The cells in the cambium, however, do not lose their ability to divide and are able to multiply and make new cells.

Although at first the cambium is restricted to the vascular bundles, it later forms a continuous cylinder within the stem between the cortex and pith. Its cells divide in such a way as to make new xylem cells, called secondary xylem, on the inside and new phloem cells, secondary phloem, on the outside. In woody plants like trees, this continues throughout their life-time, and as the cambium continues to divide and add new cells the stem increases in thickness, a process called *secondary thickening*. In such woody stems the epidermis is often replaced by a dead, corky layer, bark, which itself is made by a separate layer of cork cambium just beneath the epidermis. The phloem becomes a thin layer of living cells between the bark and the woody core of xylem (Fig. 4.7).

Strength of stems. Vertical stems are likely to experience sideways forces when the wind blows against them. The turgor (*see* Chapter 12) of the cells, the opposing forces of the epidermis tending to shrink, and the pith tending to extend, all contribute to the stem's resilience. The vascular bundles usually contain the toughest structures in the stem, the woody vessels and, often, long stringy fibrous cells running alongside them. When the vascular bundles are arranged in a cylinder near the outside of the stem they add to its strength, a cylindrical structure being much more resistant to bending than a solid structure of the same weight. In many other stems the strengthening tissue is distributed in such a way as to make the stem resistant to bending stresses; for example, the "square-sectioned" stem of the deadnettle family with the vascular bundles and strengthening strands of cells in the corners (*see* Fig. 4.3a).

Leaf

General structure (Fig. 4.8 *a* and *b*). A leaf is a flat, green *lamina* or blade made from a soft tissue of thin-walled cells, supported by a stronger network of veins. Leaves are sometimes joined to the stem by a stalk, *petiole*, which continues into its *midrib* (or the main vein). Sometimes there is no leaf stalk.

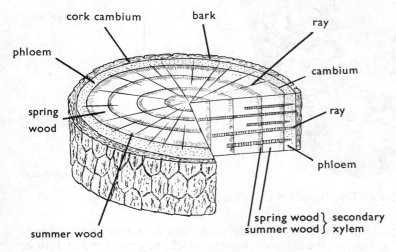

Fig. 4.7 Diagrammatic section through a woody stem showing four seasons' growth

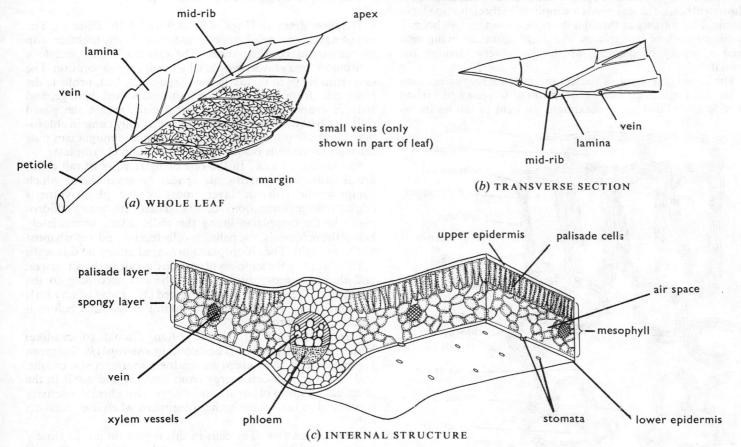

(a) WHOLE LEAF

(b) TRANSVERSE SECTION

(c) INTERNAL STRUCTURE

Fig. 4.8 Leaf structure

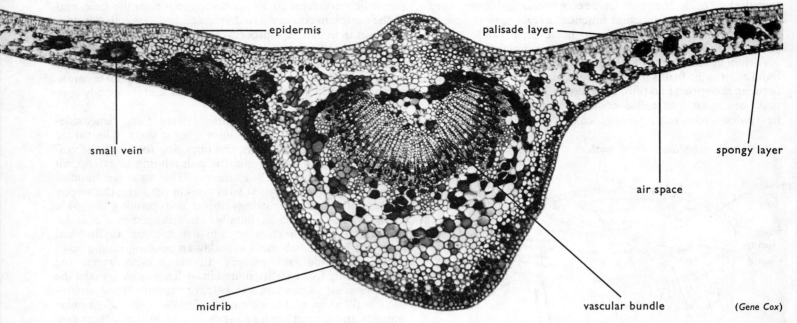

Plate 4.   TRANSVERSE SECTION THROUGH A LEAF (×30)

Functions. The important function of leaves is to make food, in the form of carbohydrates, by *photosynthesis*. The water necessary for this process is conveyed through the vessels which run in the vascular bundles branching from the stem, and through the petiole and midrib, dividing repeatedly to form a network of tiny veins throughout the lamina. In addition, for photosynthesis, the leaf needs a supply of carbon dioxide from the air. This diffuses in through the pores, stomata (*see* below), in one or both of its surfaces. For respiration, all living cells need a supply of oxygen, which also enters through the stomata.

The broad, flat shape of the leaf presents a large surface area to the air, facilitating rapid absorption of oxygen and carbon dioxide and allowing the maximum sunlight to fall on its ex-posed surface. Most leaves are thin in section and, in consequence, the distance through which the gases have to diffuse, from the atmosphere to the cells inside, is small, and gaseous exchange can be fairly rapid. The permeability to gases and the large surface area of the leaf are also characteristics which encourage rapid evaporation of water vapour.

Detailed structure (Figs. 4.8c, 4.9 and 4.10, Plate 4). The EPIDERMIS is a single layer of cells fitting closely together with no air spaces between them. The epidermis may secrete a continuous waxy layer, *cuticle*, which reduces evaporation. The epidermis helps to maintain the shape of the leaf, protects the inner cells from bacteria, fungi and mechanical damage, and reduces evaporation. The epidermal cells, except the guard cells (*see* below) of the stomata, do not usually contain chloroplasts and are transparent. In consequence, sunlight can pass through to the cells below, which do contain chloroplasts.

PALISADE LAYER. In the one or more rows of tall cylindrical cells, with narrow air spaces between them, which comprise the palisade layer, most of the photosynthesis (carbohydrate formation) occurs. There are many chloroplasts in the cytoplasm lining the walls. Lying immediately below the epidermis, the palisade cells receive and absorb most of the sunlight. The chloroplasts arranged along the side walls are not far from the supplies of carbon dioxide in the air spaces, and they can move up or down the cell according to the intensity of the sunlight. The elongated cells result in very little sunlight being absorbed by horizontal cross-walls before it reaches the chloroplasts.

CHLOROPLASTS are small, often discoid (discus-like) bodies made of protein. They contain *chlorophyll*, the green pigment which gives green plants their characteristic colour, and which can absorb energy from sunlight and use it in the chemical build-up of sugars and starch. This chemical activity is believed to take place in the chloroplast when it is receiving light.

SPONGY LAYER. The cells in this region do not fit closely together, and large air spaces are left between them. The air spaces communicate with each other and, through the stomata,

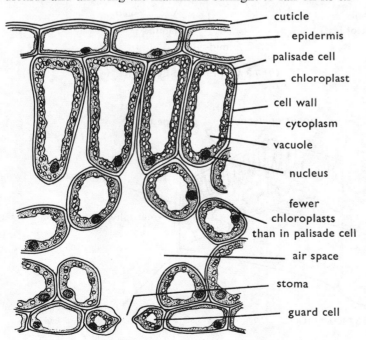

Fig. 4.9   Section through leaf to show cell structure

with the atmosphere, thus allowing air to circulate in them and reach most of the internal cells of the leaf. The cells of the spongy layer can photosynthesize, but they receive less sunlight than do the palisade cells, and contain fewer chloroplasts. The palisade and spongy layers are known collectively as *mesophyll*.

STOMATA (Fig. 4.11, Plate 5). Usually more abundant on the lower side of the leaf, stomata are openings in the epidermis. They are formed between two *guard cells* which, according to their internal pressure, or *turgor*, can increase or reduce the size of the stoma or close it completely. The conditions which affect the opening or closure of the stomata are thought to be connected principally with light intensity and, in some cases, with the loss of water. The mechanism by which they open is a chain of events leading to an increase in the concentration of sugars in the cell sap in the vacuoles of the guard cells. When this happens, the osmotic potential (p. 63) of the cell sap decreases and the guard cells withdraw water from their neighbours. This increases the turgor pressure in the guard cell, which tends to swell. The wall of the cell is particularly thick along its inner border so that it does not readily stretch. The stretching of the outer walls, however, causes the guard cells to curve away from each other and so increases the gap between them.

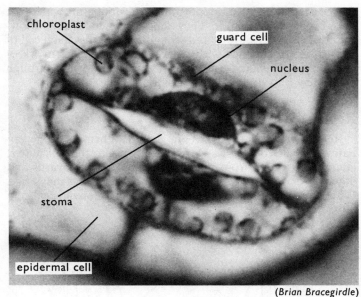

(Brian Bracegirdle)

Plate 5.   A STOMA OF A LEAF (×800)

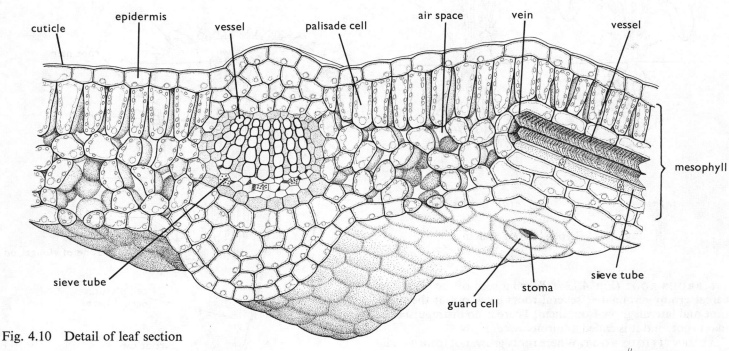

Fig. 4.10   Detail of leaf section

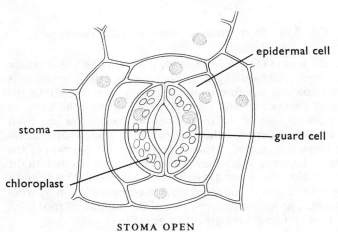

STOMA OPEN

Fig. 4.11

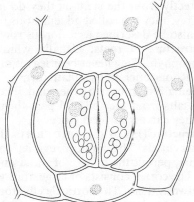

STOMA CLOSED

MIDRIB AND VEINS support the leaf, conducting water into it and food away from it. They contain vascular bundles surrounded by other fibrous and strengthening cells. Each cell of the leaf is not supplied with its own vein, but the network of veins is very fine, and water has to pass from a vein through only a few cells to reach, say, a palisade cell.

Roots

Root systems. TAP ROOT (Fig. 4.12a). When a seed germinates, a single root grows vertically down into the soil. Later, lateral roots grow from this at an acute angle outwards and downwards, and from these laterals other branches may arise. Where a main root is recognizable the arrangement is called a tap-root system.

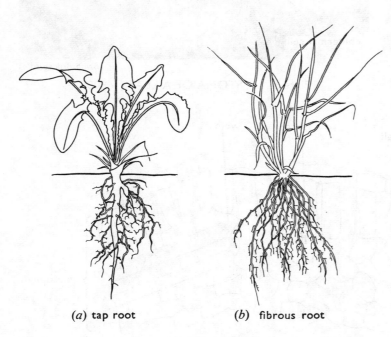

(a) tap root   (b) fibrous root

Fig. 4.12   Types of root system

FIBROUS ROOT (Fig. 4.12b). When a seed of the grass and cereal group germinates, several roots grow out at the same time and laterals grow from them. There is no distinguishable main root, and it is called a fibrous system.

ADVENTITIOUS ROOT. Where roots grow, not from a main root, but directly from the stem as they do in bulbs, corms, rhizomes or ivy, they are called adventitious roots, but such a system may also be described as a fibrous rooting system.

General structure of roots. Usually white, roots cannot develop chlorophyll. They never bear leaves or axillary buds.

Function. Roots anchor the plant firmly in the soil and prevent its being blown over by the wind. They absorb water and mineral salts from the soil and pass them into the stem. Frequently they can act as food stores.

Detailed structure (Fig. 4.13). EPIDERMIS. This is a layer of cells without a cuticle. The younger regions, particularly those with root hairs, permit the uptake of water and solutes.

CORTEX. The cortex consists of large, thin-walled cells with air spaces between them. The cortical cells store food material and the innermost layer of cells may regulate the inward passage of water and dissolved substances.

VASCULAR TISSUE. This is in the centre of the root, and initially the phloem strands lie between the radial arms of the central xylem. The branches that form the lateral roots grow from this region and force their way through the cortex, bursting through the outer layer to reach the soil. The centrally placed vascular tissue well-adapts the root to the strain that it is likely to experience along its length while holding the plant firmly in the soil when the shoot is being blown sideways by the wind. This makes an interesting comparison with the cylindrical distribution of vascular tissue in the stem and the lateral strain to which it might be subjected in the same conditions.

GROWING POINT (Fig. 4.14). At the root tip (Plate 6) is a region where the cells are dividing rapidly. Behind the root tip, the new cells produced by the dividing region absorb water and develop vacuoles, the intake of water causing the cells to elongate, the cell walls still being relatively plastic. This area behind the root tip is thus the region of extension and, since the upper part of the root is firmly anchored, it pushes the root tip down, or sideways, between the soil particles. The root tip is protected from damage by the *root cap*, layers of cells which are continually produced by the dividing region and replaced from the inside as fast as the outer ones are worn away by the abrasion of the soil particles.

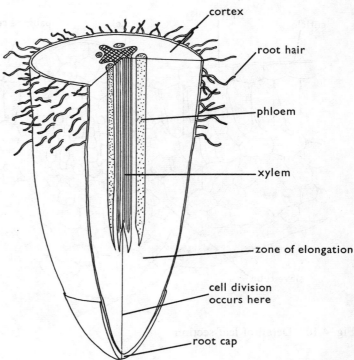

Fig. 4.13   Stereogram of root (vertically shortened)

ROOT HAIRS (Fig. 4.14) provide the main absorbing region of the root. They are tiny, finger-like outgrowths from the cells of the epidermis before it dies. They appear just above the zone of elongation, and there are none at the root tip or in the older regions of the root. The root hairs grow out from the cells and between the soil particles, their shape therefore being determined, to some extent, by the position of the particles between which they grow. Their cell walls stick to the soil particles, which cannot easily be washed off. This helps to keep the soil firm round the roots and reduces erosion by wind and rain. The total absorbing area of the millions of root hairs in a root system is very great (*see* also Fig. 12.8).

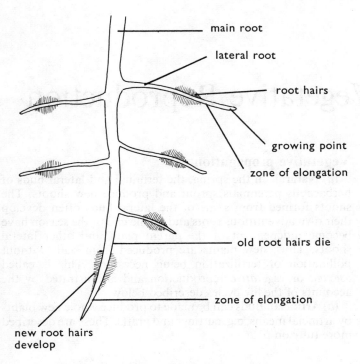

Fig. 4.14 Regions of a root system

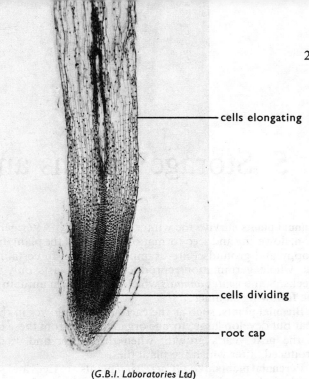

(G.B.I. Laboratories Ltd)

Plate 6. LONGITUDINAL SECTION OF
THE ROOT TIP OF AN ONION (×20)

PRACTICAL WORK

1. *To show the tensions in stems.* A strip of epidermis from a rhubarb petiole is partially removed as illustrated in Fig. 4.15a. When replaced in position it will have become too short, showing the shrinking stress that exists in the epidermis.

A cork-borer is pushed into the pith and removed, but without extracting any tissue. The cylinder of pith so formed, freed from the constraint of the epidermis, expands and protrudes slightly. This shows the elongating tendency of the inner tissues (Fig. 4.15b).

2. *Cells and vessels.* Some celery or rhubarb stalks are cut up into pieces about 2 cm long and left for a few days in a macerating fluid of 10 per cent chromic and nitric acids. This will break down the intercellular material so that the cells can be teased apart with mounted needles. The macerating acids should be washed off or repeatedly diluted before the material is handled. A little of such material, torn into pieces which are as small as possible, put on a slide with a little water will show individual cells and vessels under the lower power of the microscope.

3. *Stomata.* Stomata can be seen if a piece of the lower epidermis of a rhubarb leaf is stripped off and placed on a slide under the microscope. The stomata can be made to close by putting a little strong salt or sugar solution on the tissue. This will withdraw water by osmosis and cause the guard cells to lose their turgor.

[*Note*: this experiment is described more fully in the laboratory manual *Diffusion and Osmosis—see* p. 1.]

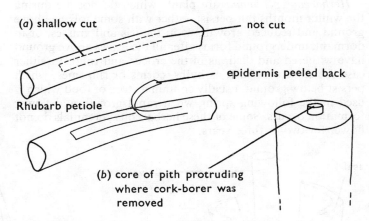

Fig. 4.15 To show opposite tensions in pith and epidermis

QUESTIONS

1. What are the main functions of (a) stems, (b) roots, (c) leaves?
2. What are the differences in (a) structure, (b) function, between vessels and sieve tubes?
3. How do roots, stems and leaves obtain supplies of oxygen for respiration?
4. What are the advantages of having a network of veins in a leaf (e.g. Fig. 4.8a)?
5. By what means does a rooting system achieve a large absorbing surface?
6. How is lateral stress resisted by (a) a young stem, (b) an old stem?
7. In what ways does the broad, thin structure of dicotyledonous leaves adapt them to their functions?

5 Storage Organs and Vegetative Reproduction

Annual plants survive the winter as seeds only. After germination, flowering and seed formation, the rest of the plant dies off. Poppy and groundsel are examples of these. In certain cases the whole germination-reproduction cycle lasts only a few weeks. Some *winter annuals* which germinate in autumn die in the following summer.

Biennial plants, such as the carrot, do not flower in the first year but develop large storage organs. Food from these is used in the next year's growth, when the flower and seeds are produced, after which the plant dies.

Perennial plants. *Woody perennial* plants are trees and shrubs in which the trunk and branches persist and grow from year to year. Those which shed all their leaves in autumn are called *deciduous*; while *evergreens*, such as the conifers and holly, shed their leaves throughout the year.

Herbaceous perennials are plants which do not die during the winter months but persist, either with some foliage above ground and reduced growth, as in the iris and grasses, or in dormant, underground forms after all the leaves above ground have withered and died, as in the crocus and tulip. In either case the structures such as bulbs, corms or rhizomes, which persist below ground, usually contain a store of food which is used in the following spring when rapid growth and flower formation occurs. Some of these herbaceous perennials do not flower for two or three years.

Vegetative propagation

(*a*) **Natural.** In the spring, the terminal and lateral buds of herbaceous perennials sprout and produce new shoots. The shoots formed from some of the lateral buds often develop their own adventitious roots and by the end of the season have become independent of the parent plant and other lateral shoots. Thus, new plants are produced from buds without pollination or fertilization being necessary. This is called *asexual* or *vegetative* reproduction and is illustrated by the accounts of the life cycles described below.

(*b*) **Artificial.** Buds can be made to produce whole new plants by artificial means, e.g. cuttings and grafts. These are described more fully on p. 26.

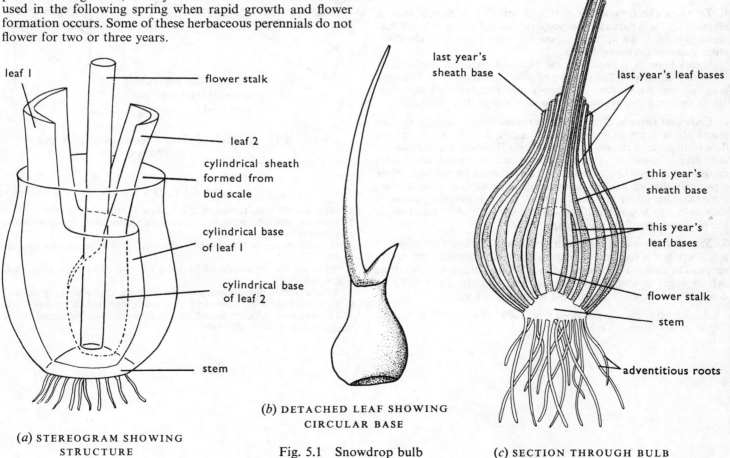

(*a*) STEREOGRAM SHOWING STRUCTURE

(*b*) DETACHED LEAF SHOWING CIRCULAR BASE

Fig. 5.1   Snowdrop bulb

(*c*) SECTION THROUGH BULB

22

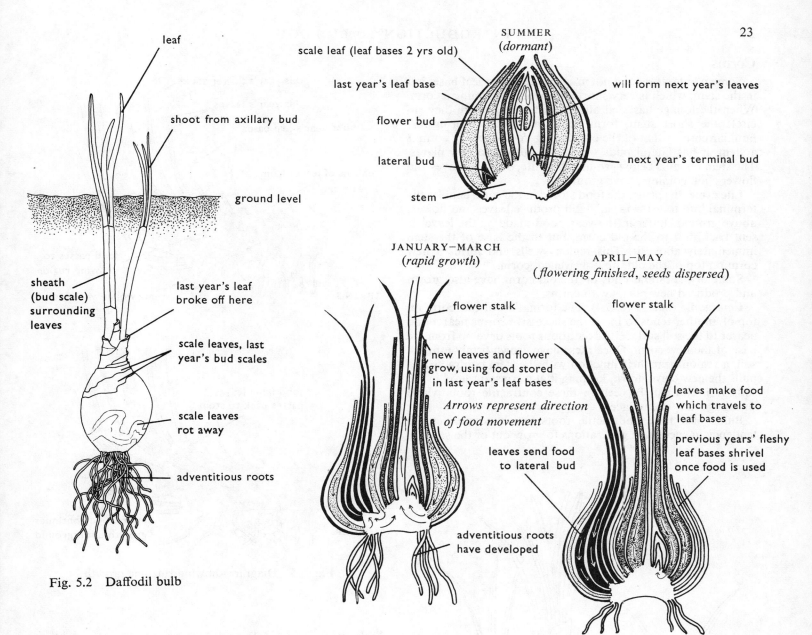

leaf

shoot from axillary bud

ground level

sheath (bud scale) surrounding leaves

last year's leaf broke off here

scale leaves, last year's bud scales

scale leaves rot away

adventitious roots

Fig. 5.2   Daffodil bulb

SUMMER (*dormant*)

scale leaf (leaf bases 2 yrs old)

last year's leaf base

flower bud

lateral bud

stem

will form next year's leaves

next year's terminal bud

JANUARY–MARCH (*rapid growth*)

flower stalk

new leaves and flower grow, using food stored in last year's leaf bases

Arrows represent direction of food movement

leaves send food to lateral bud

adventitious roots have developed

APRIL–MAY (*flowering finished, seeds dispersed*)

flower stalk

leaves make food which travels to leaf bases

previous years' fleshy leaf bases shrivel once food is used

Bulbs

Bulbs are condensed shoots with fleshy leaves. The stem is very short and never grows above ground. The internodes are short; the leaves are very close together and they overlap. The outer leaves are scaly and dry and protect the inner ones (Fig. 5.2) which are thick and fleshy with stored food. In the snow-drop and daffodil, the bulb is formed by the bases of the leaves which completely encircle the stem (Fig. 5.1) and it is to these cylindrical leaf bases that the food is sent by the rest of the leaf above ground. In the leaf axils of the bulb are lateral buds which can develop into new bulbs and shoots.

Life cycle (Fig. 5.3). In winter, the adventitious roots grow out of the stem and the terminal bud begins to grow above the ground, making use of the stored food in the fleshy leaves which consequently shrivel. During the spring some of the food made in the leaves of the daffodil is sent to the leaf bases, which swell and form a new bulb inside the old one. Food is also sent to one or more lateral buds which consequently grow to form daughter bulbs. The shrivelled storage leaves of the old bulb become the dry scaly leaves which surround the newly formed daughter bulbs. The new bulb and the daughter bulbs will produce independent plants next season. Thus bulb formation is one form of vegetative reproduction.

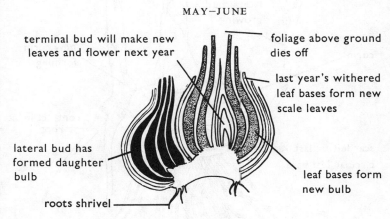

MAY–JUNE

terminal bud will make new leaves and flower next year

lateral bud has formed daughter bulb

roots shrivel

foliage above ground dies off

last year's withered leaf bases form new scale leaves

leaf bases form new bulb

Fig. 5.3   Annual cycle in a daffodil type of bulb

Corms

Some plants store food not in special leaves or leaf bases but in the stem, which is very short and swollen, forming a corm. When the foliage has died off, the leaf bases, where they encircle the short stem, form protective scaly coverings. A familiar corm is that of the crocus, Fig. 5.4. Since the corm is a stem, it has lateral buds which can grow into new plants. The stem remains below ground all its life, only the leaves and flower stalk coming above ground.

Life cycle. In spring, the food stored in the corm enables the terminal bud to grow rapidly and produce leaves and flowers above ground. Later in the year, food made by the leaves is sent back, not to the old corm, but to the base of the stem immediately above it. This region swells and forms a new corm on top of the old, now shrivelled, corm.

Some of the lateral buds on the old corm have also grown and produced new plants with corms.

Contractile roots (Fig. 5.4). The formation of one corm on top of another tends to bring the successive corms nearer and nearer to the soil surface. Adventitious roots develop from the base of the new corm. Once these have grown firmly into the soil, a region near their junction with the stem contracts and pulls the new corm down, keeping it at a constant level in the soil. Wrinkles can be seen in these contractile roots where shrinkage has taken place.

Bulbs also have contractile roots which counteract the tendency in successive generations to grow out of the soil.

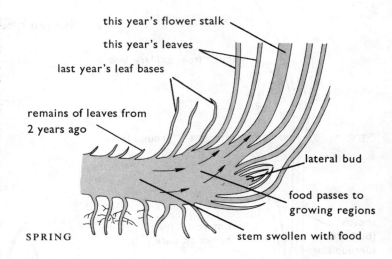

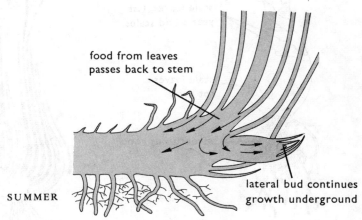

Fig. 5.5   Diagram showing rhizome growth

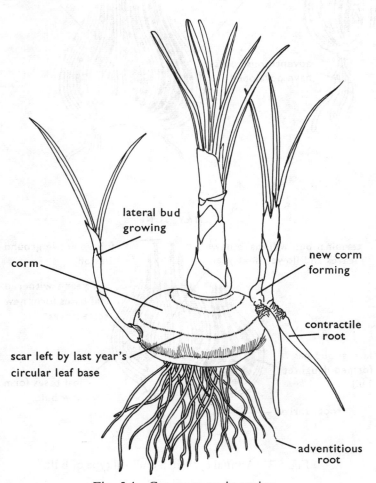

Fig. 5.4   Crocus corm in spring

Rhizomes (Figs. 5.5 and 5.7)

In plants with rhizomes, the stem remains below ground but continues growth horizontally. The old part of the stem does not die away as in bulbs and corms, but lasts for several years. In the iris, the terminal bud turns up and produces leaves and flowers above ground. The old leaf bases form circular scales round the rhizome, which is swollen with food reserves.

Life cycle (Fig. 5.5). The annual cycle of a rhizome is similar to that of a corm. In summer, food from the leaves passes back to the rhizome, and a lateral bud uses it, grows horizontally underground, and so continues the rhizome. Other lateral buds produce new rhizomes which branch from the parent stem. The terminal buds of these branches curve upwards and produce new leafy shoots and flowers. Adventitious roots grow from the nodes of the underground stem.

Stem tubers (Fig. 5.6)

In the potato plant, lateral buds at the base of the stem produce shoots which grow laterally at first and then down into the ground. These are comparable to rhizomes, as they are underground stems with tiny scale leaves and lateral buds. Unlike rhizomes, however, they do not swell evenly along their length with stored food.

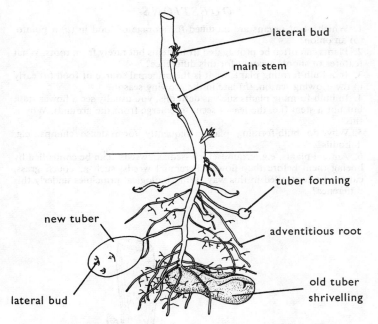

Fig. 5.6(a) Stem tubers growing on a potato plant

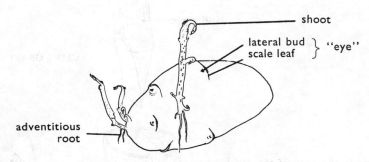

Fig. 5.6(b) Potato tuber sprouting

Annual cycle. Food made in the leaves passes to the ends of these rhizomes, which swell and form the tubers we call potatoes. Since the potato tuber is a stem, it has leaves and axillary buds; these are the familiar "eyes". Each one of these can produce a new shoot in the following year, using the food stored in the tuber (Fig. 5.6b). The old tubers shrivel and rot away at the end of the season.

Advantages of food storage

Food in the storage organs enables very rapid growth in the spring. A great many of the spring and early summer plants have bulbs, corms, rhizomes or tubers: e.g. daffodil, snowdrop and bluebell, crocus and cuckoo pint, iris and lily of the valley and lesser celandine.

Early growth enables the plant to flower and produce seeds before competition with other plants, for water, mineral salts and light, reaches its maximum. This must be particularly important in woods, where in summer the leaf canopy prevents much light from reaching the ground and the tree roots tend to drain the soil of moisture over a great area.

In agriculture, man has exploited many of these types of plant and bred them for bigger and more nutritious storage organs for his own consumption.

Advantages of vegetative reproduction

Since food stores are available throughout the year and the parent plant with its root system can absorb water from quite a wide area, two of the hazards which beset seed germination are reduced. Whereas buds are produced in an environment where the parent is able to flourish, many seeds dispersed from plants never reach a suitable situation for effective germination. Vegetative reproduction does not usually result in rapid and widespread distribution of offspring in the same way as seed dispersal, but tends to produce a dense clump of plants with little room for competitors between them. Such groups of plants are very persistent and, because of their underground food stores and buds, can still grow after their foliage has been destroyed by insects, fire, or man's cultivation. Those of them in the "weed" category are difficult to eradicate mechanically, since even a small piece of rhizome bearing a bud can give rise to a new colony. However, selective weed killers reduce this difficulty.

Gardeners make use of vegetative propagation when they divide up the rhizomes, tubers or rootstocks at the end of the flowering season. Each part so divided will grow in the following year to make a separate plant.

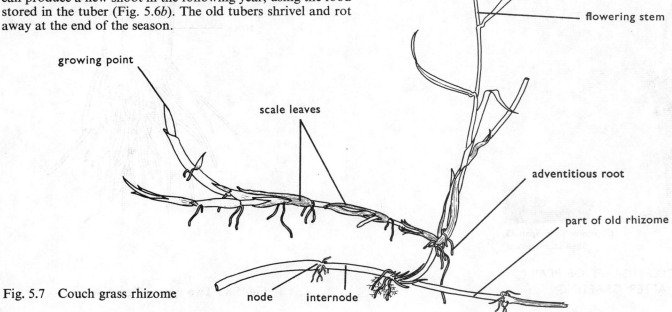

Fig. 5.7 Couch grass rhizome

Artificial propagation

(a) **Cuttings.** It is possible to produce new individuals from certain plants by putting the cut end of a shoot into water or moist earth. Adventitious roots grow from the base of the stem into the soil while the shoot continues to grow and produce leaves.

In practice, the cut end of the stem is treated with a rooting hormone (*see* p. 44) to promote root growth, and evaporation from the shoot is reduced by covering it with polythene or a glass jar. Carnations, geraniums and chrysanthemums are commonly propagated from cuttings.

(b) **Grafting.** A bud or shoot from one plant is inserted under the bark on the stem of another, closely related, variety so that the cambium layers of both are in contact. The rooted portion is called the *stock* and the bud or shoot being grafted is the *scion* (Fig. 5.8).

Rose plants grown from seed would be extremely variable. To obtain consistent characteristics, a bud from the desired variety is grafted on to the stem of a plant grown from seed. The stock is then cut away, above the grafted bud and the bud grows, using the water and nutrients supplied by the stock. When flowers are produced they are of the cultivated type and colour.

By cutting and grafting, the inbred characteristics of the plant are preserved e.g. an apple tree produced by germinating a seed from a good eating apple would yield a variety of apples, many of them being small, sour "crab-apples", but the fruit from the graft would retain their size and flavour.

1. Which plant organs are modified for storage of food in (*a*) a potato, (*b*) an onion?
2. Plants can often be propagated from stems but rarely from roots. What features of shoots account for this difference?
3. In a bulb-forming plant, what is the principal source of food (*a*) early in the growing season, (*b*) late in the growing season?
4. In bulb-forming plants such as daffodils, you usually see a flower stalk but not a stem (i.e. the leaves seem to emerge from the ground). Why is this?
5. Why do bulb-forming plants frequently form dense clumps, e.g. daffodils?
6. Annual plants, e.g. groundsel, classed as "weeds" can be controlled by hoeing them before they flower. Perennial weeds, such as couch grass, cannot be controlled in this way. What biological principles underly this difference?

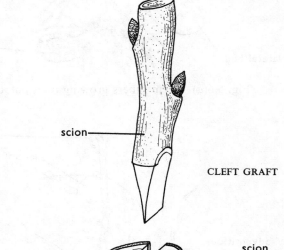

scion

CLEFT GRAFT

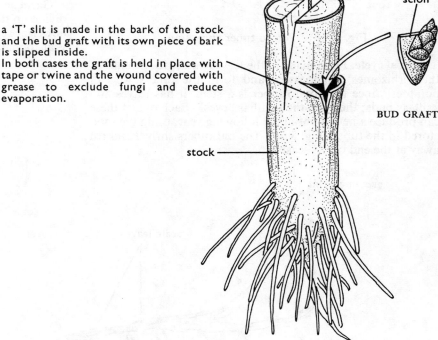

scion

a 'T' slit is made in the bark of the stock and the bud graft with its own piece of bark is slipped inside.
In both cases the graft is held in place with tape or twine and the wound covered with grease to exclude fungi and reduce evaporation.

stock

BUD GRAFT

(*Reprinted from* Span **15**,
Shell International)

Plate 7. CLEFT GRAFT OF PEAR
6 WEEKS AFTER GRAFTING

Fig. 5.8   Two types of grafting

6 Buds and Twigs

BUDS

Structure (Fig. 6.1*a*)

A bud is a "condensed" shoot. Its stem is very short and its leaves are so close that they overlap, each one wrapping round the next above it. The inner leaves are crinkled and folded, since a large surface area is packed into a small space. The outermost leaves are often thicker and tougher, and sometimes black or brown. These are the *bud scales* and they protect the more delicate, inner, foliage leaves from drying up, from mechanical damage by birds, insects, etc., and, to some extent, from extremes of temperature. At the end of the bud's short stem is either a flower, or a growing point where rapid cell division will take place later on when the next bud is forming.

Function

The bud-forming habit gives the plant the advantages of being able to present a photosynthesizing surface to the atmosphere very rapidly after seasonal conditions become favourable. The leaves, already formed in the bud, are available almost at once, whereas it would take many weeks for them to grow from the single cells at the growing point.

In most plants the buds are formed when conditions are favourable for plant growth, e.g. in the summer months, and the close packing of the leaves and the thick bud scales protect the leaves from desiccation and low temperatures during the winter.

Types of bud. Terminal buds are formed at the ends of main shoots or branches during the season's growth. There are also buds in the axils of the leaves. These are called *axillary* or *lateral* buds. If they do not grow in the following year, they are also described as dormant buds.

Terminal buds when they grow, continue growth in length, whereas lateral buds make new branches. Either type may produce a flower instead of, or in addition to, a leafy shoot. If this happens in a terminal bud, growth is continued in the following season by one or more lateral buds, since the flower or inflorescence drops away leaving no growing point.

Growth Fig. 6.1 (*a–c*). In the spring, the stem of the bud begins to elongate and the bud scales are pushed apart. As the stem grows in length it spaces out the leaves, which unfold and spread out their surface (Fig. 6.2). The bud scales often curl back and in a few weeks fall off. On exposure to light the chlorophyll in the leaves develops fully, and photosynthesis begins soon after.

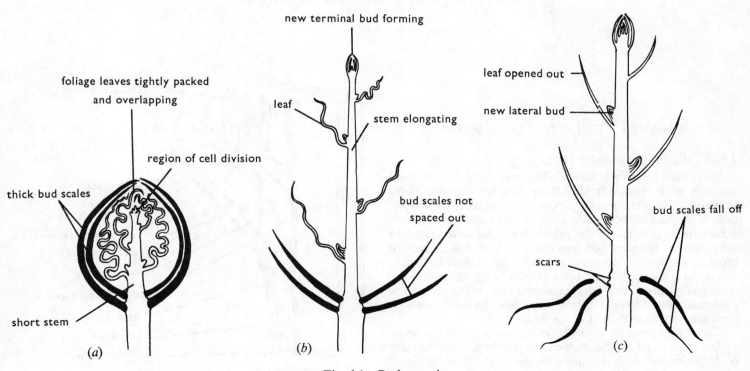

Fig. 6.1   Bud growth

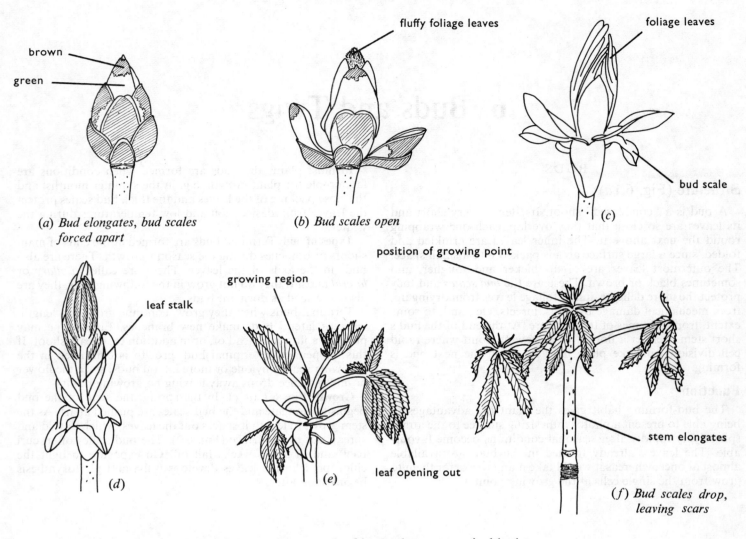

(a) *Bud elongates, bud scales forced apart*

(b) *Bud scales open* — fluffy foliage leaves

foliage leaves — bud scale — (c)

leaf stalk — (d)

growing region — position of growing point — leaf opening out — (e)

stem elongates — (f) *Bud scales drop, leaving scars*

Fig. 6.2   Stages in growth of horse-chestnut terminal bud

CHARACTERISTICS OF TWIGS

Leaf fall, or abscission (Fig. 6.3)

Most trees shed their leaves. Deciduous trees do so in the autumn, while evergreens shed theirs in small numbers all the year round. In certain trees, abscission takes place as follows: cells at the base of the leaf-stalk divide and form layers of cells across the region where the petiole joins the stem. The new tissue nearest the stem becomes corky and the vessels become blocked so that the leaf is deprived of water (Plate 8). Before this occurs, the contents of the leaf cells begin to break down chemically, producing the characteristic red and yellow autumnal tints. The cells' contents are digested and the soluble products absorbed back into the tree.

The cells beyond the corky layer degenerate, and the dried-up leaf falls, leaving a scar on the stem protected from the entry of bacteria and fungi by impermeable cork. The details of abscission vary in different species.

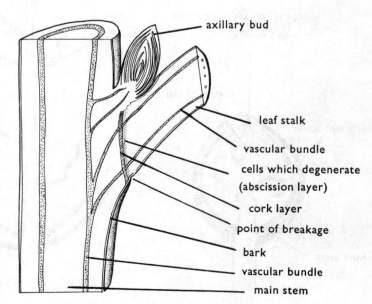

axillary bud — leaf stalk — vascular bundle — cells which degenerate (abscission layer) — cork layer — point of breakage — bark — vascular bundle — main stem

Fig. 6.3   Diagram to show leaf fall

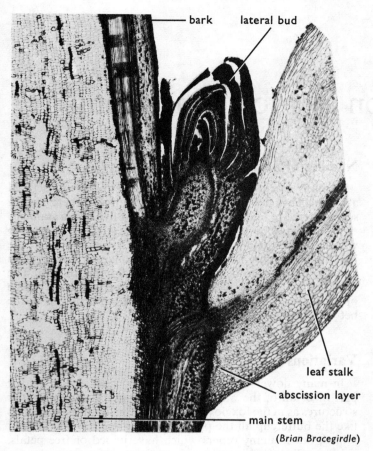

Plate 8. LEAF FALL IN SYCAMORE.
LONGITUDINAL SECTION (×10)

(Brian Bracegirdle)

Earlier in the year, when the terminal bud was sprouting, the bud scales, unlike the foliage leaves, were not spaced out on the stem, and when they fell off they left narrow scars, close together, extending from a quarter to half-way round the stem. These are commonly called girdle scars and, since they mark the position of each year's terminal bud, the length of stem between each set of girdle scars represents one year's growth (Figs 6.4–6.6).

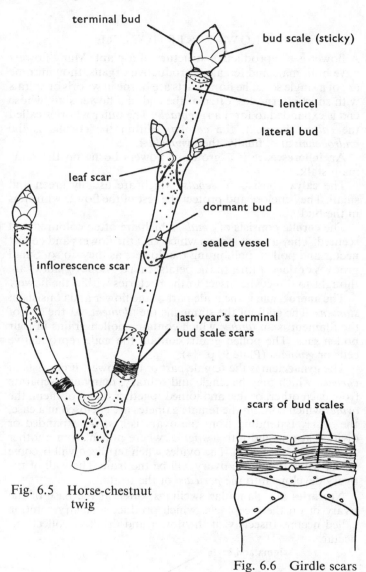

Fig. 6.5 Horse-chestnut twig

Fig. 6.6 Girdle scars of horse-chestnut

Winter twigs

The leaf scars on twigs are a characteristic shape for each species, the sealed vascular bundles making a pattern of dots in them. Since each leaf usually has a bud in its axil, above each leaf scar there should be a lateral bud.

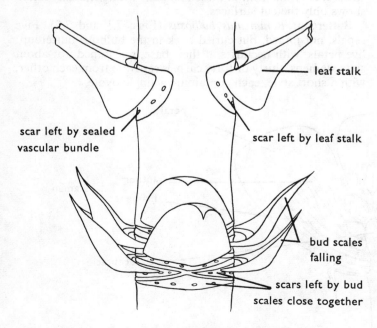

Fig. 6.4 The formation of "girdle" scars and leaf scars

QUESTIONS

1. Which plant organs are present in a bud?
2. Bud scales are considered to be modified leaves. In what ways do they differ from the ordinary foliage leaves?
3. When a bud sprouts, what change in form takes place in (a) the stem, (b) the leaves?
4. What could be the stimuli which cause the buds to open in the spring?
5. The distance between groups of terminal bud scale scars can be used to determine the age of a twig. What is the connexion between these "girdle scars" and the seasonal growth?
6. From what unfavourable seasonal conditions is autumn leaf-fall supposed to protect a deciduous tree? Suggest reasons why most conifers can survive the winter without shedding their leaves.

7 Sexual Reproduction in Flowering Plants

FLOWER STRUCTURE

A flower is a reproductive structure of a plant. Many flowers have both male and female reproductive organs, though some are of a single sex. The floral parts are borne in whorls or spirals with short internodes, often at the end of a flower stalk whose end is expanded to form a *receptacle*. The outer whorl is called the *calyx*, the next, the *corolla*. Within the corolla is the *androecium* and, finally, the *gynaecium*.

An **inflorescence** is a group of flowers borne on the same main stalk.

The **calyx** consists of *sepals*, which are usually green and small. They enclose and protect the rest of the flower while it is in the bud.

The **corolla** consists of *petals*, which are often coloured and scented. They attract insects which visit the flowers and collect nectar and pollen, pollinating the flowers as they do so. Small grooves or darker lines in the petals, called "honey guides" are thought to direct the insect to the nectaries within the flower.

The **androecium** is the male part of the flower and consists of *stamens*. The stalk of the stamen is the *filament*. At the end of the filament is an *anther* which contains pollen grains in four pollen sacs. The pollen grains contain the male reproductive cells or *gametes* (Plate 9, p. 34).

The **gynaecium** is the female part of the flower. It consists of *carpels*, which may be single and solitary, many and separate from each other, or few and joined together. In all of them, the *ovules* which contain the female gametes are enclosed in a case, the *ovary*. Extending from the ovary is a *style*, expanded or divided at one end into a *stigma*, where pollen from another flower will be received. The ovules when fertilized will become seeds, while the whole ovary will be the fruit. The wall of the ovary develops into the *pericarp* of the fruit.

Nectaries are glandular swellings, often at the base of the ovary or on the receptacle, which produce a sugary solution called nectar. Insects visit the flower and drink or collect this nectar.

Number of parts

In many species of flowering plant, the structures described above occur in definite numbers. For example, if there are five sepals there are likely to be five petals and five or ten stamens.

Whorls may be repeated; for example, there may be two whorls of five petals or two whorls of five stamens. In the buttercup and rose families there are numerous stamens and carpels, and the numbers vary from one plant to another. The floral parts usually alternate so that petals do not come opposite sepals but between them, and stamens are borne between petals, and so on.

Variations

In many flowers, petals or sepals are joined, or fused, for part of or all of the way along their length, forming tubular structures as in the foxglove and deadnettle families. In flowers like the buttercup all the petals are the same size and are not joined, but in many others which have joined or free petals, some petals differ in size and shape from others, as in the pea family and deadnettles.

The half-flower

A drawing of a half-flower is a convenient method of representing flower structure. The flower is cut in halves with a razor blade, the outline of the cut surfaces drawn, and the structures visible behind these filled in. A longitudinal section shows only the cut surfaces.

Buttercup: *Ranunculus bulbosus* (Figs. 7.1 and 7.2). Five sepals, not joined, but curled back in the bulbous buttercup: five petals, with nectaries at their base, not joined up: about sixty stamens, thirty or forty carpels separate from each other, with a short style, each containing a single ovule.

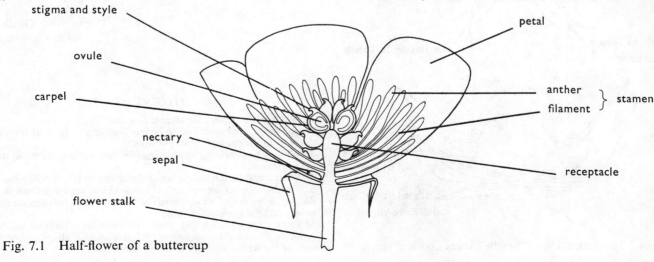

Fig. 7.1   Half-flower of a buttercup

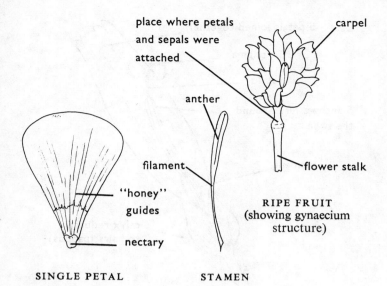

place where petals and sepals were attached

carpel

anther

filament

"honey" guides

nectary

flower stalk

SINGLE PETAL

STAMEN

RIPE FRUIT
(showing gynaecium structure)

Fig. 7.2 Floral parts of buttercup

White deadnettle: *Lamium album* (Fig. 7.3). Five spiky sepals, joined at the base to make a cup, surrounding a tube of five fused petals of which the uppermost is particularly well-developed. Only four stamens, with their black anthers close together under the top petal and their filaments joined to the petal tube. A forked stigma, with a long style joined to the petals for part of its length and leading to a four-lobed ovary.

Lupin (Figs. 7.4 and 7.5). Five fused sepals, with a two-lobed appearance; five petals, not all joined but of different shapes and sizes. The uppermost petal is called the standard, and the two partly joined petals at the side are the wings. Within the wings are two partly joined petals forming a boat-shaped keel. Inside the keel are ten stamens the filaments of which are fused to form a sheath round the ovary. The ovary is long, narrow and pod-shaped, and consists of one carpel with about ten ovules. The style ends in a stigma, just within the pointed end of the keel.

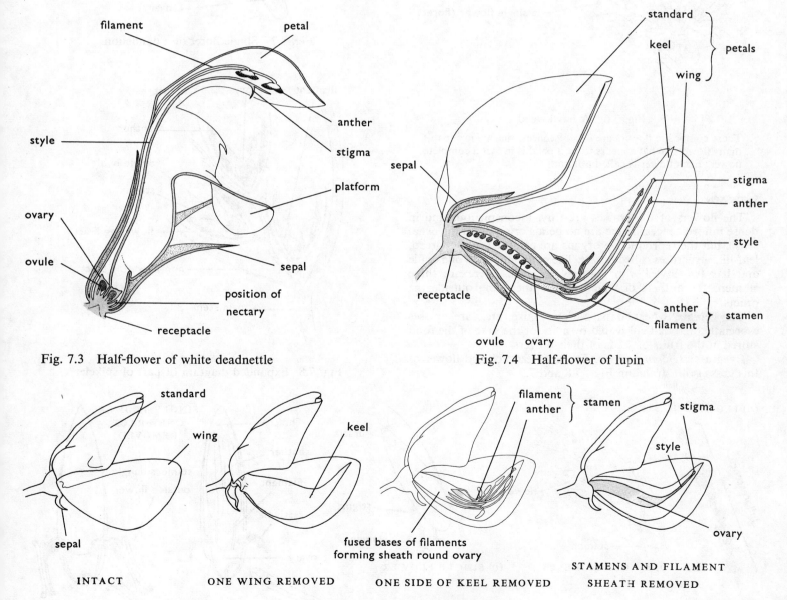

Fig. 7.3 Half-flower of white deadnettle

Fig. 7.4 Half-flower of lupin

Fig. 7.5 Lupin flower dissected

Composite flowers

The flowers of the *Compositae* family, daisies, dandelions, hawkweeds, etc., are arranged in dense inflorescences (Fig. 7.6). What at first appears to be a petal in the flower head is actually a complete flower, often called a *floret* (Fig. 7.7).

In some Compositae, the outer florets with conspicuous petals have no reproductive organs and are called sterile: the inner florets with corollas of tiny, fused petals carry the reproductive organs. The various daisies are examples of this type of inflorescence.

Dandelion: *Taraxacum officinale.* In the outer florets of the dandelion inflorescence there are five petals fused together to form a tube at the base but flattening out at the top. The sepals are reduced to a whorl of fine hairs, the five fused anthers are grouped round the style, which has a forked stigma. There is a single ovule in the ovary.

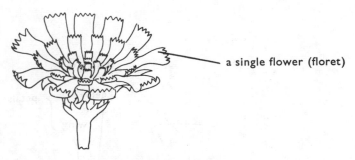

Fig. 7.6 A hawkweed

These composite flowers are inflorescences, that is, groups of many flowers. What appears to be a petal is in fact a complete flower. Each flower is called a floret.

Grasses

The flowers of the grasses are tiny, inconspicuous and in dense inflorescences. There are no petals or sepals in the usual sense, but the reproductive organs are enclosed in two green, leaf-like structures called *bracts*. The ovary contains one ovule and has two styles with feathery stigmas. There are three stamens the anthers of which, when ripe, hang outside the bracts.

The cereals, wheat, oats, barley, maize, etc., are grasses especially bred and cultivated by man for the sake of the food stored in the fruits or seeds of their flowers.

Ryegrass: *Lolium perenne.* The inflorescence and flower of this grass is illustrated in Figs. 7.8 and 7.9.

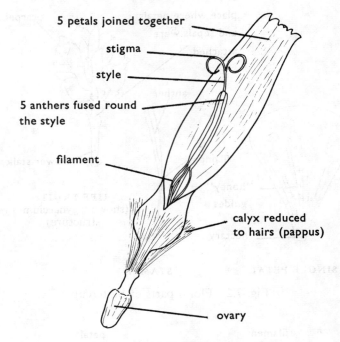

Fig. 7.7 Single floret of a dandelion

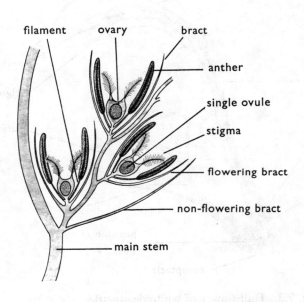

Fig. 7.8 Expanded diagram of part of spikelet

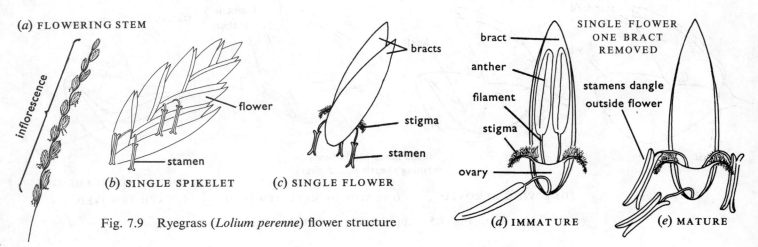

Fig. 7.9 Ryegrass (*Lolium perenne*) flower structure

POLLINATION

The transfer of pollen from anthers to stigma is called pollination. Cross-pollination is the transfer of pollen from the anthers of one flower to the stigma of another flower of the same species. In some species self-pollination occurs either regularly or when cross-pollination has failed to take place, as in the willow herb. In cross-pollination pollen is usually transferred on the bodies of insects which enter the flowers, or by chance air-currents carrying the pollen from one flower to the next. The structures of many flowers are closely adapted to the method of insect or wind pollination. The Table below gives the main differences between these two kinds of flower.

Insect pollination mechanisms

Essentially, these mechanisms involve an insect's visiting one flower, becoming dusted with pollen from the ripe stamens, and then visiting another flower where some of the pollen on its body adheres to the stigma.

When ripe, the pollen sacs of the anther split open and expose the pollen, which can then be dislodged (Fig. 7.11 *a* and *b*).

In many flowers the anthers and stigma do not ripen at the same time, so that a visiting insect is not likely to effect self-pollination. Before the stigma is ripe the lobes may not be expanded or the chemicals needed for the growth of the pollen tube may not be present.

Fig. 7.10

air bladders increase surface area

(*a*) Pollen of insect-pollinated flower of Hollyhock   (*b*) Pollen of Pine (wind pollinated)

The buttercup. Some flowers are elaborately adapted to pollination by a particular group of insects like moths or bees. In the buttercup, however, it seems likely that the general wanderings of various insects, e.g. bees, beetles and ants, round the petals and nectaries bring the insects' bodies into contact with the reproductive organs resulting in the transfer of pollen from anthers to stigmas, either in the same flower (self-pollination), or from one flower to another (cross-pollination).

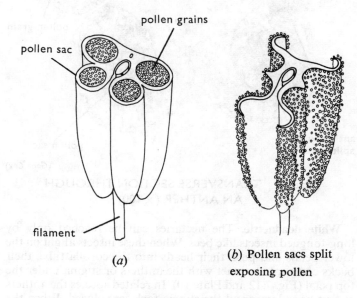

pollen grains

pollen sac

filament

(*a*)   (*b*) Pollen sacs split exposing pollen

Fig. 7.11   Structure of anther (top cut off)

Wind and Insect Pollinated Flowers Compared

WIND POLLINATED	INSECT POLLINATED
1 Small, inconspicuous flowers; petals often green. No scent or nectar.	*1* Relatively large flowers or conspicuous inflorescences. Petals brightly coloured and scented; mostly with nectaries.

Insects respond to the stimulus of colour and scent and are "attracted" to the flowers. When in the flower, they collect or eat the nectar from the nectaries, or pollen from the anthers.

2 Anthers large and loosely attached to filament so that the slightest air movement shakes them. The whole inflorescence often dangles loosely (hazel male catkins), and the stamens hang out of the flower exposed to the wind (Fig. 7.9 *c* and *e*).	*2* Anthers not so large and firmly attached to the filament. They are not usually carried outside the flower but are in a position within the petals where insects are likely to brush against them.

The wind is more likely to dislodge pollen from exposed, dangling anthers than from those enclosed in petals.

3 Large quantities of smooth, light pollen grains produced by the anthers.	*3* Smaller quantities of pollen produced. The grains often have spiky patterns or stick together in clumps (Fig. 7.10).

With wind pollination, only a very small proportion of pollen grains is likely to land on a ripe stigma. If large quantities of pollen are not shed, the chances of successful pollination become very poor. Smooth, light grains are readily carried in air currents and do not stick together.
In pollination by insects, fewer of the pollen grains will be wasted. The patterned or sticky pollen grains are more likely to adhere to the body of the insect.

4 Feathery stigmas hanging outside the flower (Fig. 7.9 *c* and *d*).	*4* Flat or lobed, sticky stigmas inside the flower.

The feathery stigmas of grasses form a "net" of relatively large area in which flying pollen grains may be trapped.

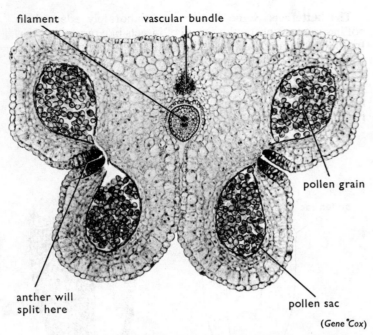

Plate 9. TRANSVERSE SECTION THROUGH
           AN ANTHER (×100)

(Gene Cox)

White deadnettle. The nectaries can be reached only by long-tongued insects like bees. When these insects alight on the lower petals and push their heads into the corolla tube, their backs come into contact with the anthers or stigma under the top petal (Fig. 7.12 and Plate 10). In related species the anthers at first hang lower and the stigma forks are closed. When the anthers have shed their pollen the stigma opens and bends lower than the anthers.

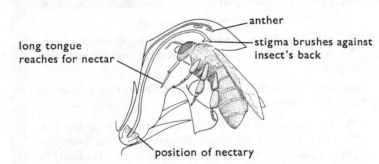

Fig. 7.12   Pollination of white deadnettle

Lupins. Lupins have no nectar, and the bees which visit them collect only pollen from the flowers (Fig. 7.13). Other members of this family, e.g. the clovers, do produce nectar.

The weight of the insect when it alights on the "wings" of the flower depresses them. Near their base, the wings are linked to the petals of the keel so that these too are forced down, and the stamens and stigma protrude from a hole at the end of the keel and touch the underside of the insect's body. In the lupin, the anthers push out the pollen which has collected in the pointed tip of the keel, and it emerges rather like tooth-paste from a tube, much of it adhering to the insect's body. When the insect alights on an older flower which has shed all its pollen, the style and stigma protrude from the keel and pollen from the insect's body will stick to the stigma.

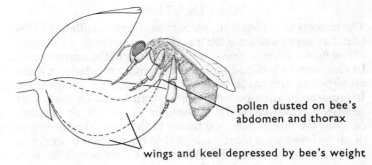

Fig. 7.13   Pollination of lupin or sweet pea

Grasses

At first, the feathery stigmas protrude from the flower, and pollen grains floating in the air are trapped by them. Later, the anthers hang outside the flower, the pollen sacs split, and the wind blows the pollen away. This sequence varies with species.

If the branches of a hazel tree with ripe male catkins, or the flowers of the ornamental pampas grass, are shaken, a shower of pollen can easily be seen (Plate 11).

Incompatibility

In both wind- and insect-pollinated flowers, pollen from a certain species may reach the stigma of a different species. Usually the chemicals present in the cells of the stigma prevent further development of the "foreign" pollen grains.

Importance of pollination to agriculture

After fertilization the ovary of a plant develops into a fruit. Fruit formation therefore depends on fertilization, which can follow only after pollination.

Farmers and fruit growers are well aware that a good yield of fruit will occur only if most of the available flowers have been pollinated. Many of the cereals are self-pollinated or wind-pollinated, and the fact that they are growing close together makes the latter effective. Insect pollination of a field of clover or an orchard of apples, however, needs a fairly dense population of insects, particularly bees.

FERTILIZATION

The following four generalizations apply to both plants and animals.

1. A *gamete* is a reproductive cell. A male gamete is usually small with a nucleus and little cytoplasm; it is the gamete which leaves the male organ and moves about, either by its own power or by external agencies like wind or insects.

2. The female gamete is larger, with a nucleus and more protoplasm than the male; it sometimes contains food reserves. Often it never leaves the female organ or body, where it is produced, until after it is fertilized; even when it does do so it is usually immobile. The male gamete in a flowering plant is a nucleus in the pollen grain; in most animals it is the sperm. The female gamete in plants is a large egg-cell in the ovule, while in animals it is the ovum.

3. The product of the fusion of male and female gametes is called a *zygote*.

4. *Fertilization* is the fusion (joining together) of the nuclei of male and female gametes to form a zygote. After fertilization the zygote undergoes cell division and growth, developing into a new individual or a preliminary form, an embryo, a seed, or a larva.

Plate 10. BEE VISITING WHITE DEADNETTLE

Plate 11. HAZEL CATKINS SHEDDING POLLEN

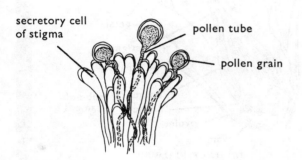

Fig. 7.14 Pollen grains growing on stigma
of crocus

Fertilization in plants

Fertilization follows pollination, but the interval of time between the two events varies in different species from sixteen hours to twelve months. The pollen grain absorbs nutriment secreted by the stigma, and the cytoplasm in the grain grows out as a tube. This tube grows down through the style between the cells (Fig. 7.14) absorbing a nutritive fluid from them. On reaching the ovary it grows to one of the ovules and enters it through a hole, the micropyle (Fig. 7.15). The tip of the pollen tube breaks open in the ovule, and the male nucleus, which has been passing down the tube, enters the ovule and fuses with the female nucleus there (Plate 12).

Each egg-cell of an ovule can be fertilized only by a male nucleus from a separate pollen grain.

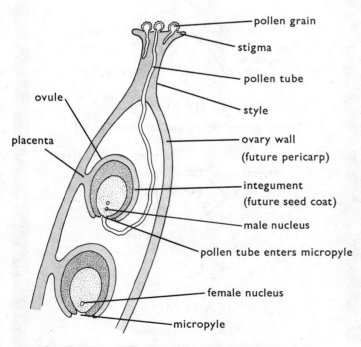

Fig. 7.15 Diagram of fertilization

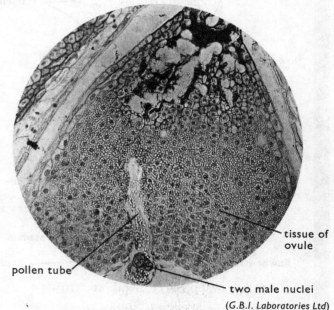

Plate 12. POLLEN TUBE PENETRATING
PINE OVULE (×800)

Result of fertilization

Fruit and seed formation. After fertilization the petals, stamens, style and stigma wither and usually fall off (Fig. 7.16).

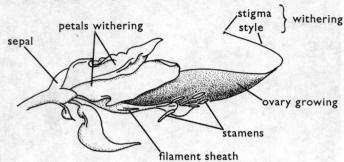

Fig. 7.16   Lupin flower after fertilization

The sepals may persist in a dried and shrivelled form. Food made in the leaves reaches the fertilized ovules and the ovary, which grow rapidly. Inside the ovule cell division and growth produces a seed containing a potential plant or embryo. The embryo consists of a miniature root or radicle, a small shoot or plumule, and one or two leaves, the cotyledons, which sometimes contain food reserves. The integuments of the ovules become thicker and harder, forming the testas of the seeds, and finally, water is withdrawn from the seeds, making them dry and hard. In this condition they are best able to withstand extremes of temperature and other adverse conditions.

The ovary wall may become dry and hard, forming a capsule or pod as in the poppy and lupin, or it may become succulent and fleshy as in tomatoes, gooseberries and plums.

Fruits. The whole ovary after fertilization is called a fruit. In the strawberry (Fig. 7.19) and rose hip (Fig. 7.21), the pips are the fruits and the fleshy part is the receptacle. In the apple (Fig. 7.20) and pear the swollen receptacle is fused to the outside of the ovary wall. In either case, the whole structure is often called a "false fruit".

In agriculture such fruits have been specially bred and selected for their large, edible receptacles.

The distinction between fruit and vegetable at the green-grocer's shop does not correspond to the biological definition of fruit; for example, runner beans, french beans, cucumber, marrow and tomato are all fruits.

Fruit formation. Examples of the formation of fruits from fertilized flowers are shown in Figs. 7.17–7.22.

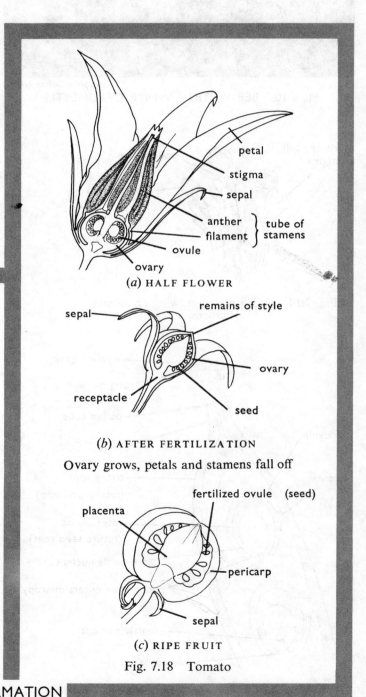

(a) HALF FLOWER

(b) AFTER FERTILIZATION
Ovary grows, petals and stamens fall off

(c) RIPE FRUIT
Fig. 7.18   Tomato

(a) PETALS HAVE FALLEN

(b) LONGITUDINAL SECTION
Fig. 7.17   Raspberry

FRUIT FORMATION

FRUIT FORMATION

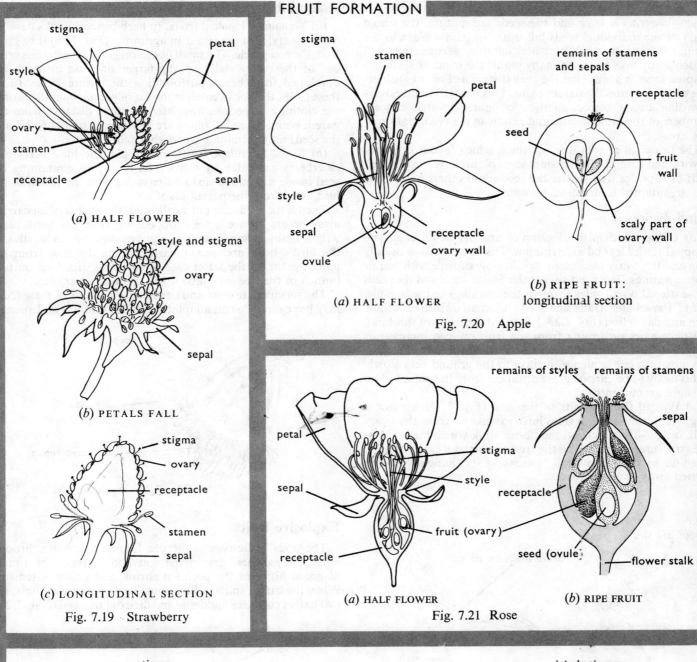

(a) HALF FLOWER

(b) PETALS FALL

(c) LONGITUDINAL SECTION

Fig. 7.19 Strawberry

(a) HALF FLOWER

Fig. 7.20 Apple

(b) RIPE FRUIT:
longitudinal section

(a) HALF FLOWER

(b) RIPE FRUIT

Fig. 7.21 Rose

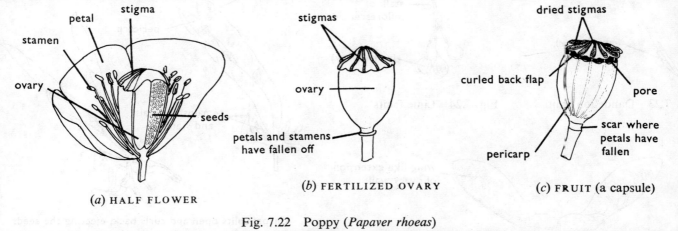

(a) HALF FLOWER

(b) FERTILIZED OVARY

(c) FRUIT (a capsule)

Fig. 7.22 Poppy (*Papaver rhoeas*)

DISPERSAL OF FRUITS AND SEEDS

When flowering is over and the seeds are mature, the whole ovary or the individual seeds fall from the parent plant to the ground, where, if conditions are suitable, germination will subsequently take place. In many plants the fruits or seeds are adapted in such a way that they are distributed over considerable distances from the parent plant. This helps to reduce overcrowding among, and competition for light and water between, members of the same species and results in the colonization of new areas.

The principal adaptations are those which favour dispersal by wind and animals. In addition, some plants have "explosive" pods or capsules that scatter the seeds, and others have fruits that are adapted to dispersal by water.

Wind dispersal

(a) **Censer mechanism.** Examples are the white campion, poppy (Fig. 7.22c) and antirrhinum. The flower stalk is usually long and the ovary becomes a dry, hollow capsule with one or more openings. The wind shakes the flower stalk and the seeds are scattered on all sides through the openings in the capsule.

(b) **"Parachute" fruits and seeds.** Clematis, thistles, willow herb and dandelion (Fig. 7.23) have seeds or fruits of this kind. Feathery hairs projecting from the fruit or seed increase its surface area so much that air resistance to its movements is very great. In consequence, it sinks to the ground very slowly and is likely to be carried great distances from the parent plant by slight air currents.

(c) **Winged fruits.** Fruits of the lime (Fig. 7.24), sycamore (Fig. 7.25), ash and elm trees have extensions from the ovary wall, or leaf-like bracts on the flower stalk which make wing-like structures. These cause the fruit to spin as it falls from the tree and so prolong its fall, increasing its chances of being carried away in air currents.

Animal dispersal

(a) **Mammals. Hooked fruits.** In herb bennet, hooks develop from the styles of the fruits; in agrimony (Fig. 7.26) they grow from the receptacle. The small hooks on goose-grass fruits grow out of the ovary wall and the larger ones on burdock are developed from bracts surrounding the inflorescence. In all these cases, the hooks catch in the fur of passing mammals or in the clothing of people, and later, at some distance from the parent plants, they fall off or are brushed or scratched off and the seeds may germinate where they fall.

(b) **Birds. Succulent fruits.** Fruits like the blackberry and elderberry are eaten by birds. The hard pips containing the seed inside are undigested and pass out with the faeces of the bird away from the parent plant.

Even if the seeds are not swallowed, the fruit is often carried away before the seeds are dropped, e.g. rose hip. Some seeds with a fleshy, sticky covering, e.g. yew and mistletoe, stick to the bird's beak and are discarded some distance from the parent plant, in the latter case often being wiped off on to a branch of the tree on which it will grow as a parasite.

The succulent texture and conspicuous colour of these fruits may be regarded as an adaptation to this method of dispersal.

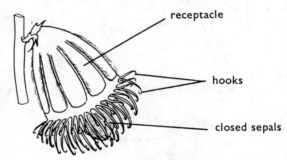

Fig. 7.26   Agrimony fruit

Explosive fruits

The pods of flowers in the pea family, e.g. gorse, broom, lupin and vetches, dry in the sun and shrivel. The tough, diagonal fibres in the pericarp shrink and set up a tension. When the carpel splits in half down two lines of weakness, the two halves curl back suddenly and flick out the seeds (Fig. 7.27).

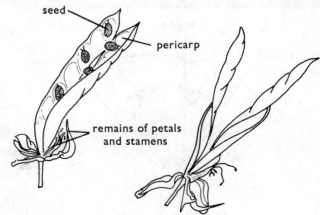

Carpel splits open and curls back, ejecting the seeds

Fig. 7.27   Lupin

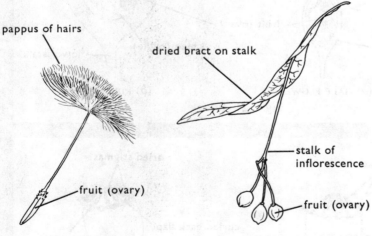

Fig. 7.23   Dandelion fruit   Fig. 7.24   Lime fruits

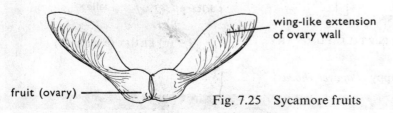

Fig. 7.25   Sycamore fruits

PRACTICAL WORK

Pollen on stigma. If the stigmas of a number of flowers are examined dry and by reflected light under the low power of the microscope, pollen grains may be seen adhering to them. If the stigmas are crushed with water between two slides, pollen tubes may be seen growing between the cells (see Fig. 7.14).

Pollen tubes. By placing pollen grains in a 10–20 per cent solution of cane sugar in a cavity slide and covering them with a glass cover slip the growth of pollen tubes may be seen after several hours. Sweet pea pollen has proved satisfactory.

Pollen. Pollen can be examined microscopically by dusting or squashing ripe anthers on to a slide.

QUESTIONS

1. What do you understand by the term "gamete"? What are the male and female gametes in a flowering plant?
2. What is "fertilization" and where does it occur in a flowering plant?
3. Pollination may occur without fertilization taking place but fertilization will not occur without pollination. Explain why this is so.
4. Why should a disease which causes the blossom to fall from apple trees in spring affect the yield of fruit in the autumn?
5. What part in reproduction is played by (a) petals, (b) stamens, (c) carpels?
6. How is self-pollination prevented in the white deadnettle?
7. Only large insects such as bees are likely to effect pollination in the lupin and deadnettle. Why are smaller insects unlikely to do so?
8. Which part of the fruit is considered edible in (a) runner beans, (b) tomato, (c) strawberry, (d) apple, (e) green peas?
9. Cucumber and marrow are "fruits" to the biologist, while rhubarb is not. Explain the basis for this distinction.
10. Most flowering plants produce many more seeds than are ever likely to grow to maturity. (a) What kind of adverse circumstances are likely to prevent successful germination and growth? (b) How does seed dispersal contribute to the survival of the species despite these hazards?
11. What kind of competition is likely to take place between seedlings growing closely together?
12. Structures which help to disperse fruits and seeds are modifications of structures present in all flowering plants. Name the structures from which the dispersive features are derived in the case of (a) lupin, (b) sycamore, (c) dandelion (see p. 32).
13. Distinguish between wind pollination and wind dispersal.

8 Seeds, Germination and Tropisms

SEEDS

Seed structure

A seed develops from an ovule after fertilization. It consists of a tough coat or *testa* enclosing an *embryo* which is made up of a *plumule*, a *radicle* and one or two *cotyledons*. In favourable conditions the seed can grow and become a fully independent plant, bearing flowers and seeds during its life cycle. In the embryo of the seed are all the potentialities of development and growth to a mature plant resembling other members of its species in almost every detail of leaf shape, cell distribution and flower colour and structure.

The **testa.** The integuments (p. 35) round the ovule form the testa, a tough, hard coat which protects the seed from fungi, bacteria and insects. It has to be split open by the radicle before germination can proceed.

The **hilum** is a scar left by the stalk which attached the ovule to the ovary wall.

The **micropyle** is the opening in the integuments through which the pollen tube entered at fertilization (p. 35). It remains as a tiny pore in the testa opposite the tip of the radicle and admits water to the embryo before germination.

The **radicle** is the embryonic root which grows and develops into the root system of the plant.

The **plumule** is the leafy part of the embryonic shoot. These leaves are attached to the embryonic stem, of which the part above the attachment of the cotyledons is called the *epicotyl* and the part below, the *hypocotyl* (Fig. 8.3b).

Cotyledons. The grasses, cereals and narrow-leaved plants such as iris and bluebell have seeds with only one cotyledon. Such plants are called monocotyledons. The other flowering plants, the dicotyledons, have two cotyledons in their seeds. These cotyledons are modified leaves attached to the epicotyl and hypocotyl by short stalks, and they often contain food reserves which are used in the early stages of germination. In

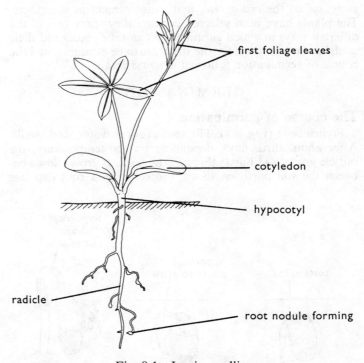

Fig. 8.1   Lupin seedling

most dicotyledonous plants the cotyledons are brought out of the testa and above the ground where they become green and make food by photosynthesis. The cotyledons eventually fall off, usually after the first foliage leaves have been formed. The cotyledon leaves bear no resemblance to the ordinary foliage leaf, the shape of which is first apparent when the plumule leaves open and grow (Fig. 8.1).

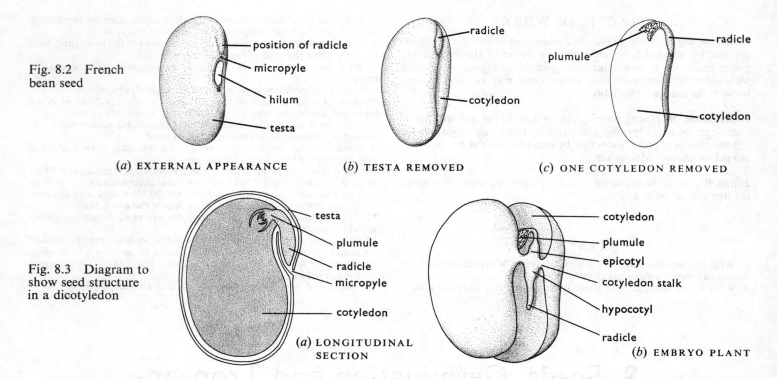

Fig. 8.2 French
bean seed

(a) EXTERNAL APPEARANCE   (b) TESTA REMOVED   (c) ONE COTYLEDON REMOVED

Fig. 8.3 Diagram to
show seed structure
in a dicotyledon

(a) LONGITUDINAL
SECTION

(b) EMBRYO PLANT

Structure of types of seed. This is best shown by the diagrams and drawings in Figs. 8.2–8.5. One important point of difference is that maize has only one cotyledon, and a separate food store, called the *endosperm*, that is not present in the others. The plants have been selected because they show two of the different ways in which germination can take place, and their seeds are large enough for the structure to be examined and the course of germination followed in some detail.

GERMINATION

The course of germination

French bean (Fig. 8.4). The seed absorbs water, and swells. After about three days, depending on the temperature, the radicle grows and bursts through the testa. It grows down between the soil particles, its tip protected by a root cap (*see*

p. 20). Root hairs appear in the region where elongation has ceased. Water and salts from the soil are absorbed by the root hairs on the radicle and pass to the rest of the seedling. Later, lateral roots develop from the radicle. Once the radicle is firmly anchored in the soil, the hypocotyl starts to grow. The rapid growth of the hypocotyl pulls the cotyledons out of the testa and through the soil. The plumule is still between the cotyledons and thus protected from damage during its passage through the soil. Sometimes the testa, still partly enclosing the cotyledons, is brought above the soil and pushed off later as the cotyledons separate. Once above the soil, the hypocotyl straightens and the cotyledons separate, exposing the plumule.

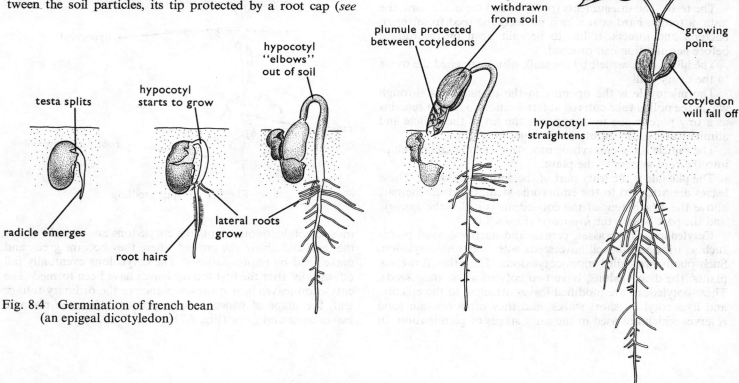

Fig. 8.4 Germination of french bean
(an epigeal dicotyledon)

The cotyledons become green and, presumably, photosynthesize for a day or two before shrivelling and falling off. Meanwhile, the epicotyl has extended and the plumule leaves have expanded and begun to photosynthesize, making the seedling independent. The germination of a seedling such as a french bean which brings its cotyledons above the ground is called *epigeal*.

In the early stages of germination, the food reserves in the cotyledons, mostly starch and protein, have been acted upon by enzymes and converted to soluble products which pass to, and are used by, the actively growing regions where new cells and new protoplasm are being made, and energy for these processes is being released. Glucose is formed from the stored starch, being utilized in various ways. Some is built up into cellulose and incorporated into new cell walls, and part is oxidized by respiration and releases energy which may be used in the many chemical activities taking place in the growing regions.

The conversion of starch to glucose (a form of sugar) also results in a fall of osmotic potential (*see* p. 63), which may assist the seedlings to take in water and their newly formed cells to extend during growth. When the first foliage leaves are above soil and their chlorophyll properly developed, the seedling can make its own food and is independent of the cotyledons.

Maize. (Fig. 8.6). The maize grain is really a fruit containing one seed, but the thin ovary wall does not interfere with germination. The fruit absorbs water, swells, and a radicle bursts through the *coleorhiza* (Fig. 8.5b) and fruit wall. Root hairs grow on the upper regions of the radicle. The plumule grows straight up and through the fruit wall, but the growing point and first leaves are protected by a sheath, the *coleoptile*, with a hard, pointed tip. From the base of the plumule grow adventitious roots. Once above the soil, the first leaves burst out of the coleoptile which remains as a sheath round the leaf bases. The cotyledon remains below the soil, absorbing food from the endosperm and transmitting it to the growing root and shoot. Eventually, both the cotyledon and the exhausted endosperm rot away.

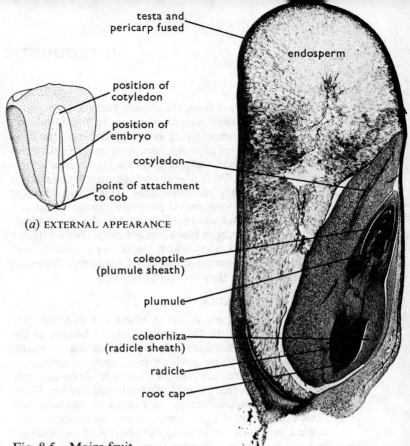

Fig. 8.5   Maize fruit

(*Brian Bracegirdle*)

(a) EXTERNAL APPEARANCE

(b) LONGITUDINAL SECTION (× 10)

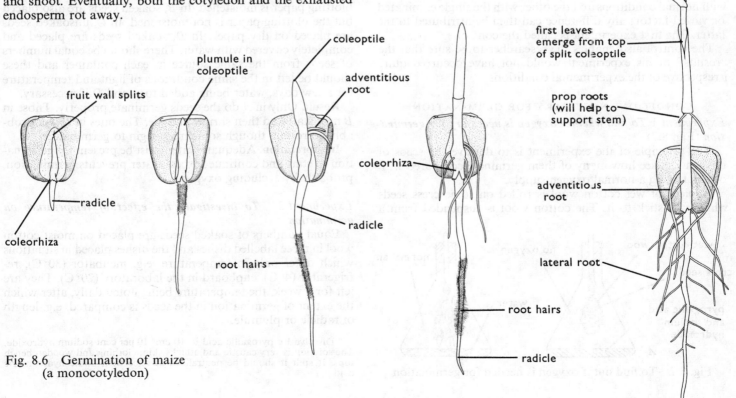

Fig. 8.6   Germination of maize
(a monocotyledon)

Dormancy

Most seeds when shed from the parent plant are very dry; about 10 per cent of their weight is water. In this condition all the chemical processes of living are very slow and little food is used. In this dry condition the seed may remain alive but dormant for long periods, without germinating but still retaining the power to do so. With a large enough number of seeds it could be shown that they are consuming oxygen and releasing carbon dioxide while dormant. If properly stored, wheat can still be germinated after about 15 years. On the other hand, the seeds of the Para-rubber plant can germinate only within a few days of being shed, after which the power is lost. Of any quantity of seeds the percentage which will germinate decreases with the length of time they are kept dormant.

Controlled experiments

A controlled experiment is one in which the experimenter controls the conditions. By this method he can be sure of the way in which these conditions influence the animal or plant. If, out of 50 seeds shed by a wild flower, only 20 germinate and only 10 of these produce mature flowers, it is possible only to guess at the factors that might have been responsible. These could include dead seeds, attack by fungus or bacteria, unfavourable conditions of light, temperature, moisture or air.

All these conditions are beyond the control of an observer because he is unable to influence them in the plant's natural environment and he does not know the variations in them that have already taken place during the development of the plants.

To find out the importance of a particular condition to the normal development and existence of an animal or plant, the usual experimental practice is to exclude or vary this particular condition, keeping all others constant, and observe the effect on the plant or animal. Most experiments involve placing the plant or animal in an unusual situation, in boxes, jars or cages, when it can be argued that the peculiar experimental conditions are responsible for the observed results. For this reason, it is necessary to set up two almost identical experiments, the one with normal conditions and the other with the single eliminated or varied factor; any difference can then be attributed to the latter. The first experiment is called the control.

The control also enables the researcher to be sure that the "results" of his experiment would not have occurred quite irrespective of the experimental conditions.

CONDITIONS NECESSARY FOR GERMINATION

Experiment 1. **To find whether oxygen is necessary for germination** (Fig. 8.7)

The principle of the experiment is to deprive the seeds of oxygen and see how many of them germinate compared with others having a normal oxygen supply.

A piece of wet cotton wool is rolled on some cress seeds which will stick to it. The cotton wool is suspended from a

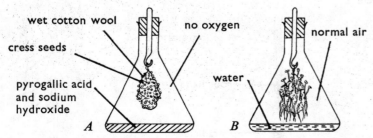

Fig. 8.7   To find out if oxygen is needed for germination

thread inside a tightly corked flask, *A*, which contains a solution of pyrogallic acid and sodium hydroxide.* This mixture absorbs oxygen from the air. The cotton wool must not touch the chemicals. It could be objected that seeds sown in such abnormal surroundings could hardly be expected to germinate anyway. To check on this a second apparatus is set up using the same size flask, *B*, seeds from the same source, but with water in the flask instead of the chemicals. This is the control, in which the experimental situation is the same except that the seeds in it are not deprived of oxygen. Both flasks are placed in the same conditions of light and temperature.

Result. After a few days most of the seeds in flask *B* will have germinated, while those in flask *A* will not, or if they have, they are fewer in number and much less advanced.

Interpretation. Without oxygen, germination cannot take place. To show that the chemicals have not killed the seeds the cotton wool from flask *A* can be transferred to *B*, when, after a few days the seeds will germinate successfully.

Note. Since sodium hydroxide absorbs carbon dioxide, flask *A* will lack both oxygen and carbon dioxide. Strictly speaking the control flask should contain sodium hydroxide solution and not water.

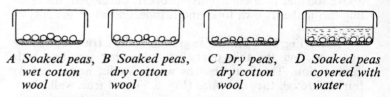

A Soaked peas,   B Soaked peas,   C Dry peas,   D Soaked peas
  wet cotton   dry cotton   dry cotton   covered with
  wool   wool   wool   water

Fig. 8.8.   The need for water in germination

Experiment 2. **To find whether water is necessary for germination** (Fig. 8.8)

Four dishes, e.g. margarine cartons, that can be covered to prevent evaporation, are labelled *A*, *B*, *C* and *D*. Cotton wool or blotting paper is placed in the bottom of each. In *A* are placed seeds that have been soaked overnight, and the blotting paper is moistened. In *B*, soaked seeds are also placed but the blotting paper is not moistened. In *C*, unsoaked seeds are placed on dry paper. In *D*, soaked seeds are placed and completely covered with water. There should be equal numbers of seeds from the same source in each container and these should be left in the same conditions of light and temperature for a few days, water being added to *A* and *D* if necessary.

Result. Only in *A* do the seeds germinate properly. Those in *B* may start and then shrivel and die. The ones in *D* will probably go rotten, though some may begin to germinate.

Interpretation. Adequate water must be present for germination to start and continue; excess water prevents germination, probably by excluding oxygen.

Experiment 3. **To investigate the effect of temperature on germination**

Equal numbers of soaked seeds are placed on moist cotton wool in three labelled dishes and the dishes placed in situations which differ only in temperature, e.g. incubator (30° C), refrigerator (4° C), cupboard in the laboratory (20° C). They are left for a week, the temperature being noted daily, after which the extent of germination in the seeds is compared, e.g. length of radicle or plumule.

* Dissolve 1 g pyrogallic acid in 10 cm³ 10 per cent sodium hydroxide. The solution is very caustic and attacks skin, clothing and wooden bench tops. If spilt it should be neutralized at once with dilute hydrochloric acid.

Result. It will be found that extremes of temperature do not favour germination. Low temperatures prevent it altogether. Higher temperatures accelerate it up to the point where either the protoplasm is killed, drying up is too rapid, or fungal growth is promoted.

Interpretation. Each species of seed probably has an optimum temperature for germination, but in this experiment the intervals between temperatures are too widely spaced to determine this optimum.

The seeds from the refrigerator should subsequently be allowed to germinate in a warm place to demonstrate that failure to germinate in the refrigerator was due to the retarding effect of a low temperature, and not to their being killed by the cold.

EXPERIMENTS ON THE SENSITIVITY OF PLANTS: TROPISMS

Seedlings are good material for experiments on sensitivity because their growing roots and shoots respond readily to the stimuli of light and gravity. Growth movements of this kind, in which the direction of growth is related to the direction of the stimulus, are called *tropisms*.

Experiment 4. *The effect of one-sided lighting on growing shoots*

Two potted seedlings, e.g. sunflower, at about equivalent stages of growth are selected. After watering, one is placed under a cardboard box with a window cut in one side so that the light reaches its shoot from one direction only (Fig. 8.9). The other is placed in an identical situation, but on a slowly rotating *clinostat*. This consists of an electric or clockwork motor which rotates a turntable about four times per hour, thus exposing all sides of the shoot equally to the source of light. This is a control.

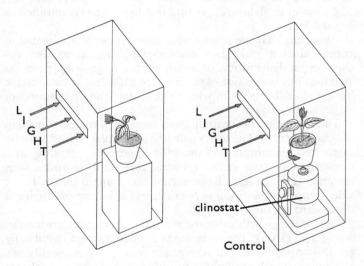

Fig. 8.9   Phototropism in shoots

Result. After a few days, the two plants are removed from the boxes and compared, when it will be found that the stem of the plant with one-sided illumination has changed its direction of growth and is growing towards the light.

Interpretation. The results suggest that the young shoot has responded to one-sided lighting by growing towards it. This tendency to grow in response to the direction of light is called *phototropism* and the shoot is *positively phototropic* because it grows towards the direction of the stimulus.

However, the results of an experiment with a single plant cannot be used to draw conclusions which apply to green plants as a whole. The experiment described is more of an illustration than a critical investigation. To investigate phototropisms thoroughly a large number of plants from a wide variety of species would have to be used.

Effect of light on growth of radicles.
Most roots are unaffected by one-sided illumination. Where they are influenced they are negatively phototropic, i.e. they grow away from the light.

Experiment 5. *The effect of gravity on shoots*

Two equivalent potted seedlings are selected as for Experiment 4. One is placed on its side so that the shoot is horizontal, while the other is placed in a clinostat so that, although the shoot is horizontal, all sides are exposed equally to the "pull" of gravity (Fig. 8.10). The lighting conditions should be the same for each shoot or else both experiments should be covered by cardboard boxes.

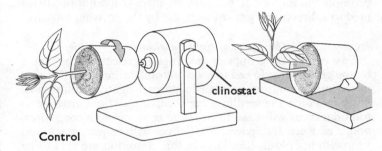

Fig. 8.10   Geotropism in shoots

Result. After about 24 hours, the shoot in the clinostat will still be growing horizontally while that of the stationary plant will have changed its direction of growth to vertically upwards.

Interpretation. The result illustrates that growing shoots tend to grow away from gravity. A response of this kind to gravity is called a *geotropism* and since the shoots grow away from the direction of the stimulus, the response is said to be negative, i.e. shoots are *negatively geotropic*. As with Experiment 4, the number of specimens used does not permit a general conclusion about plants as a whole.

Experiment 6. *The effect of gravity on roots*

Bean or pea seedlings with straight radicles are pinned to a large cork as shown in Fig. 8.11. The cork is then placed in the mouth of a jar which is left on its side for 48 hours. The radicles being horizontal are subjected to a gravitational force perpendicular to their length. A control with a clinostat is arranged as shown, so that gravity acts equally on all sides of the radicles in turn. In both cases, moist blotting paper lines the container to saturate the air with water vapour, and the apparatus is left in darkness to eliminate the possibility of a phototropic response.

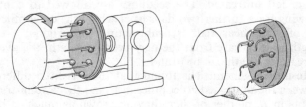

Fig. 8.11   Geotropism in roots

Result. The radicles of the seedlings on the clinostat continue to grow horizontally while those in the stationary jar have grown downwards.

Interpretation. If light, humidity and temperature are uniform in the containers and identical in each experiment, the radicles showing a growth curvature must have responded to the one-sided pull of gravity. This growth movement is *geotropism* and the roots are *positively geotropic*. Lateral roots, however, growing more or less horizontally, are clearly not positively geotropic.

Hydrotropism, the response to water. Experiments are sometimes quoted which purport to show that radicles respond positively to the "direction" of water by growing towards it. In practice it is difficult to design an experiment with a moisture gradient and the results are often susceptible to more than one interpretation. A root system in a soil which was not uniformly moist might well show a lop-sided distribution because the dry soil has inhibited the root growth while this would have been promoted in moist soil, but such an effect could not be attributed to a directional growth response by the growing root tips.

Experiment 7. *The effect of light on plants*

Two equivalent groups of seedlings are selected. The plants of one group are allowed to grow in total darkness while those in the other continue growth in normal lighting conditions. After two days the seedlings are compared. Those growing in total darkness will have taller, thinner stems with long internodes between the leaves. The leaves will be few, small and yellowish in colour. Seedlings in this condition are said to be *etiolated*. The plants growing in the light will have shorter, stouter, stems with short internodes between leaves which will be larger, more numerous and greener in colour. Light, therefore, seems to (*a*) reduce the rate of growth of stems, (*b*) promote the production of chlorophyll. The phototropic response of shoots is an outcome of this effect of light on growth rate; the side of the stem receiving more illumination will grow more slowly, so producing a curvature towards the light.

These growth responses to light can be seen to be advantageous to green plants. The shoots of seedlings which are partially obscured by other vegetation will grow rapidly until they reach the light, whereupon the leaves will expand, chlorophyll will be activated and photosynthesis will proceed.

Experiment 8. *Indoleacetic acid on wheat coleoptiles* (Fig. 8.12)

Ten soaked wheat grains are placed in each of four shallow dishes containing moist cotton wool. The fruits are allowed to germinate in darkness for six days after which the coleoptiles will be about 20 mm long. The dishes are labelled *A* to *D*. The seedlings in dishes *A*, *B* and *C* have 2 mm cut from the coleoptile tips and then the lengths of all the coleoptiles in each dish is measured and recorded. The average length in each dish is calculated. To the cut tips of the coleoptiles in *A* is added a small quantity of lanolin containing 0·1 per cent IAA. To the tips of the coleoptiles in *B* is added plain lanolin, while *C* and *D* are left untreated. The seedlings are allowed to continue growing for a further two days after which the coleoptiles are cut off and measured. By subtracting the original average length in each case from the new average length, the average increase in length can be found.

Result. In general it is found that there is little difference between the untreated coleoptiles in *C* and the lanolin-treated shoots in *B*, neither of them having grown as much as those in *A* and *D*. The decapitated coleoptiles treated with IAA in *A*

may well have grown more than those in *D*.

Interpretation. Removal of the coleoptile tip seems to cause retardation of growth in *C*. This might be attributed to the damage inflicted on the growing point. Since, however, growth is normal or above normal in *A* in which IAA is supplied, it looks as if removal of the tip in *C* deprived the coleoptile of IAA or a similar growth-promoting substance. That the active agent in the experiment is IAA and not the lanolin in which it is dissolved, is shown by the failure of the controls in *B* to grow significantly longer than those in *C*.

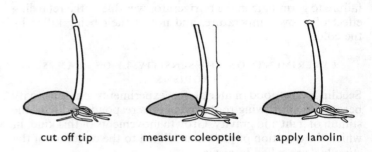

cut off tip   measure coleoptile   apply lanolin

Fig. 8.12   IAA on wheat coleoptiles

The auxin theory of tropistic response

There is evidence to suggest that the cells near the tip of some growing shoots produce a chemical, an auxin or growth substance, which in certain concentrations accelerates growth in length. The auxins achieve this effect probably by delaying the loss of plasticity in the walls of the cells in the region of extension so that at a time when the cells are osmotically active and taking in water, the increased pressure of the vacuole forces the cell wall to extend. (*See* Fig. 3.5, p. 11 and Plate 6, p. 21). The kind of evidence supporting this hypothesis is outlined in Fig. 8.13.

Assuming that the results with coleoptiles are applicable to other plants it looks as if one-sided lighting alters the production or distribution of auxin from the growing point so that the illuminated side of the shoot receives less auxin than the darker side. The way in which light alters the distribution of auxin is not known.

The same kind of reasoning can be applied to geotropism in shoots and roots. The lower side of a shoot placed horizontally might receive more auxin than the upper side, resulting in a curvature upwards in the growing region. With roots, it is assumed that the higher concentration of auxin on the lower side, retards rather than accelerates extension so producing a downward curvature.

One of the growth substances isolated from plants is indoleacetic acid (IAA). The term auxin is often applied specifically to this compound but there are many other regulatory substances which are known to influence not only growth but flowering, bud sprouting, leaf shedding and seed dormancy.

Indoleacetic acid promotes the growth of shoots in concentrations of about 10 parts per million. Roots respond to lower concentrations and different species of plants vary in their response to specific concentrations of auxin. When compounds related to IAA, e.g. 2–4 D, are sprayed on lawns as selective weed killers, the concentrations are chosen so that the broad-leaved plants are killed but the grasses are unaffected.

Plant "growth" substances are sometimes called "plant hormones", but they are not really comparable to animal hormones (p. 144) in their structure and mode of action.

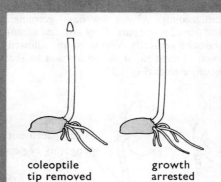

coleoptile tip removed   growth arrested

Interpretations: either
Coleoptile tip provides the cells for growth, or
Coleoptile tip produces growth-promoting chemicals, or
(c) *Desiccation of or damage to cut coleoptile stops growth*

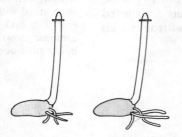

tip replaced but separated from shoot by foil   growth arrested

Interpretation. *Foil prevents cells and chemicals getting to coleoptile, but desiccation is ruled out now as a cause of arrested growth*

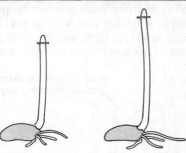

tip replaced but separated from shoot by agar jelly   growth resumed

Interpretation. *Cells cannot pass through the agar but chemicals can. A growth-promoting chemical from the tip has passed through the agar and caused the shoot to extend*

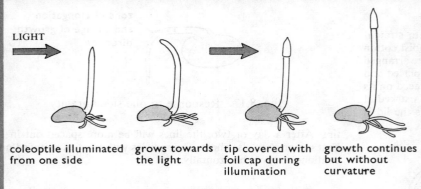

LIGHT

coleoptile illuminated from one side   grows towards the light   tip covered with foil cap during illumination   growth continues but without curvature

Interpretation. *The coleoptile tip is sensitive to one-sided light but the response occurs below the tip. There must be some form of communication between the tip and the rest of the coleoptile*

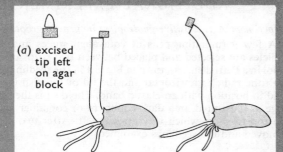

(a) excised tip left on agar block

(b) agar block placed asymmetrically on decapitated coleoptile

(c) coleoptile grows and bends as shown

Interpretation. *A growth-promoting chemical has diffused from the coleoptile tip into the agar In (b) the chemical reaches the right side of the coleoptile more than the left. The extra extension on the right side caused the curvature*

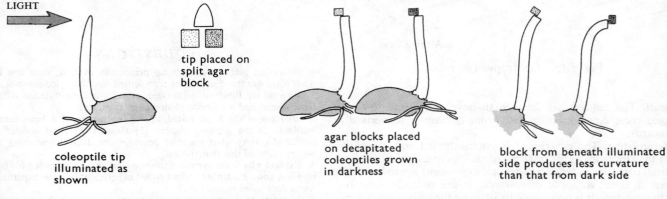

LIGHT

coleoptile tip illuminated as shown

tip placed on split agar block

agar blocks placed on decapitated coleoptiles grown in darkness

block from beneath illuminated side produces less curvature than that from dark side

Note. The coleoptile is a short-lived and specialized structure in grass and cereal seedlings. Its responses to growth substances may not be typical of plants as a whole. There is a good deal of evidence which does not support this simple auxin theory of tropisms.

Interpretation. *Light reduces the amount of hormone reaching shoot from coleoptile tip. Therefore illuminated side grows less than dark side so producing curvature*

(After Went & Thimann, Phytohormones, Macmillan, 1937)

Fig. 8.13   Some classical experiments to test the auxin theory

To obtain seedlings with straight radicles, soaked seeds are rolled in blotting paper or newspaper as shown in Fig. 8.14, placed in a beaker and kept moist. If several sets are started at daily intervals there will be abundant material from which to select for experiments on tropisms.

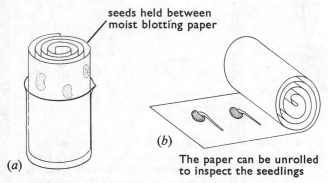

Fig. 8.14   To obtain seedlings with straight radicles

ADDITIONAL EXPERIMENTS

*Experiment A 1. **Positive geotropism in growing roots***

A few germinating peas at equivalent stages and with straight radicles are selected and placed between two strips of moist cotton wool in a Petri dish, as shown in Fig. 8.15. The seedlings are arranged with the radicles horizontal and the lid of the dish is replaced and held in position with an elastic band. The dish is then placed on its edge as illustrated, in a dark cupboard or container and marked at the top to show which way up it is left. After two days the lid is removed and the seedlings examined.

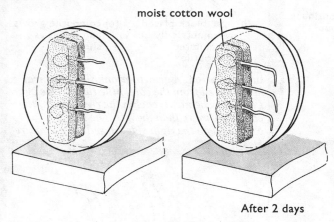

Fig. 8.15   Geotropism in roots

Result. The radicles will be seen to have grown and to have changed their direction of growth from horizontal to vertically downwards.

Interpretation. The experiment is not controlled and illustrates rather than investigates the response of radicles to unilateral gravity. A partial control is to set up an identical experiment but to leave the radicles directed vertically downwards. If the controls show no growth curvature, it is reasonable to attribute the response made by the experimental seedlings to the position in which the radicles were held.

*Experiment A 2. **To find the region of most rapid growth in radicles***

Straight radicles of bean seedlings after a few days' germination are marked with indian ink lines 2 mm apart. They are then arranged as in Fig. 8.15 but with radicles vertically downward and allowed to continue growth. The region of most rapid elongation will be shown by the subsequent spacing of the lines (Fig. 8.16).

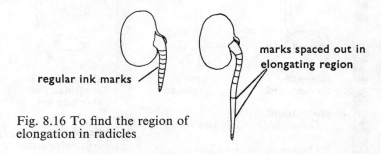

Fig. 8.16 To find the region of elongation in radicles

*Experiment A 3. **To find the region of response in the radicle*** (Fig. 8.17)

Straight radicles are marked as in *A* 2 and placed horizontally in Petri dishes as in *A* 1. Some of the radicles have 1 mm cut from their

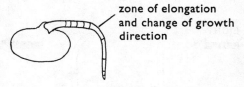

Fig. 8.17   Response to one-sided gravity

tips. After a day or two, the lines will be more spaced out in the bent region than elsewhere. The radicles with the tips removed may continue to grow horizontally.

*Experiment A 4. **Food tests on seeds***

To investigate the kind of food stored in the cotyledons or endosperm of seeds, the seeds should be crushed and heated with water. The food tests described on p. 85 can then be applied.

[*Note:* these experiments and several others are fully detailed in the laboratory manual *Germination and Tropisms—see* p. 1.]

QUESTIONS

1. Flowering plants are made up principally of root, stem and leaf. In what form are these structures represented in a dicotyledonous seed?
2. How do the functions of cotyledons differ in a dicotyledon such as the french bean and a monocotyledon such as maize?
3. (*a*) How is the food stored in the cotyledons of a bean seed made available to the growing region? (*b*) How is the food utilized by the seedling? (*c*) At what stage of development does the seedling become independent of this stored food?
4. Explain why the experiment to show that oxygen is needed for germination is a good example of a controlled experiment. Why is experiment *A* 1 not a controlled experiment?
5. In terms of the auxin hypothesis explain why (*a*) shoots deprived of light grow very tall and (*b*) shoots illuminated from one side grow towards the light source.

9 Respiration

RESPIRATION in living organisms is the series of chemical changes which release energy from food material. It involves a complicated chain of chemical breakdowns, accelerated by *enzymes* (*see* p. 85), but it can be regarded, for experimental purposes, as the breakdown of carbohydrates to form carbon dioxide and water, with a corresponding release of energy.

The energy so produced is used for such activities as muscular contraction, nervous conduction, secretion of enzymes and driving a great many chemical reactions in the living cell. Respiration is one of the most important aspects of the vital chemistry of living matter.

A distinction is usually made between two forms of, or stages in, respiration called *aerobic* and *anaerobic*. Aerobic respiration involves the use of oxygen in the breakdown of carbohydrates or fats which are eventually oxidized completely to carbon dioxide and water. Anaerobic respiration is the breakdown of carbohydrates to release energy without the use of oxygen. This is discussed more fully on p. 49.

The term "respiration" is also often used loosely in reference to breathing, as in "artificial respiration", "pulse and respiration rate" or in connexion with gaseous exchange, "the respiratory surface of a gill", "organs of respiration". For this reason, the "respiration" described in this chapter is sometimes called *tissue respiration* or *internal respiration* to distinguish it from either the breathing movements (ventilation) or the intake of oxygen and output of carbon dioxide (gaseous exchange).

From the equation above, it is apparent that for respiration to occur food and oxygen must be taken in and react together. Also, carbon dioxide and water, which are the end-products of the reaction, must constantly be removed.

Methods of demonstrating respiration. A demonstration of respiration in material is one indication that the material is living, and measurements of the rate of respiration in cells, tissues, organs or organisms gives some idea of the rate of chemical activity. Consequently, to the biologist, methods of measuring respiration rates are important.

The equation suggests (Fig. 9.1) that if an organism is respiring it will (*a*) use up carbohydrate, (*b*) take in oxygen, (*c*) give out carbon dioxide, (*d*) produce water or water vapour, and (*e*) release energy. With the exception of (*d*), if one or more of these changes are taking place the material is likely to be living and respiring.

(a) Decrease in dry weight (using up carbohydrate)

If living material is respiring and converting carbohydrate to carbon dioxide and water, its weight will decrease. However, it is the dry-weight which must be measured since any material, living or non-living, may lose weight by the evaporation of water into the atmosphere.

One hundred wheat seeds are soaked in water for 24 hours. Half of them are killed by boiling (controls). The 50 living seeds are placed in one dish with moist cotton wool and the 50 dead

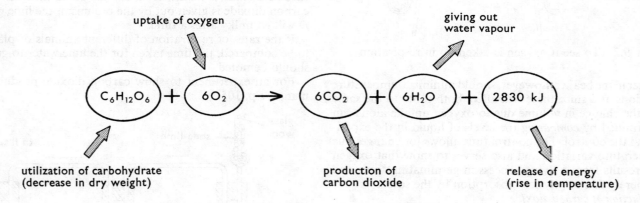

Fig. 9.1   Ways of detecting respiration

Aerobic respiration

The following equation summarizes the process of aerobic respiration, i.e. respiration which uses oxygen (the formulae represent the molecular weights of the substances in grams):

$$C_6H_{12}O_6 + 6O_2 \rightarrow 6CO_2 + 6H_2O + 2830 \text{ kJ}$$

   glucose   oxygen   carbon   water   energy
                     dioxide

seeds in identical conditions. Every other day for 10 days, 10 seeds or seedlings are selected from each dish and heated in an oven at 120° C for 12 hours to evaporate all the water. The two samples of 10 seeds are then weighed. In this way only the solid matter in the seeds is weighed. If the seeds are respiring, the solids in the food reserve of the endosperm should be decreasing as some of the food is oxidized to carbon dioxide which escapes to the atmosphere.

(b) *Uptake of oxygen* (Fig. 9.2)

The apparatus is arranged as shown in Fig. 9.2. After five minutes the tubes will have acquired the temperature of the water in the beaker and the screw clips are closed. If the seeds are respiring they will give out carbon dioxide and take in oxygen so there may be no effective change in the volume of gas in the tube. However, soda lime will absorb all the carbon dioxide produced so that any volume change may be attributed to the uptake of oxygen. If oxygen is absorbed by the seeds, the level of liquid in the capillary should be seen to rise within 20 minutes or so. Any change in the temperature of the tubes will cause the air in them to expand or contract and produce corresponding movements of the liquid in the capillary, which could be confused with the movements due to oxygen uptake.

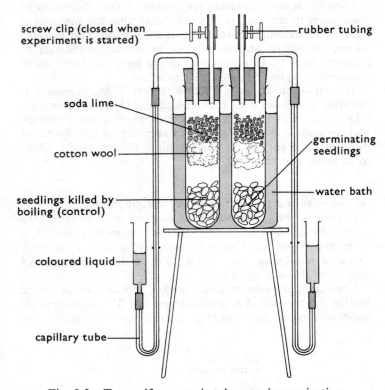

Fig. 9.2   To see if oxygen is taken up in respiration

The water in the beaker, however, should minimize temperature fluctuations and since these will affect both tubes to the same extent, the change in volume due to oxygen uptake alone can be determined by *comparing* the levels of liquid in the experiment and the control. The control thus allows for changes due to temperature variation and also serves to show that oxygen uptake results from a living process in germinating seeds and is not merely due to physical absorption by the seeds.

(c) *Production of carbon dioxide*

(i) *Germinating seeds* (Fig. 9.3)

Wet cotton wool is placed in two flasks A and B. Soaked seeds are added to A and an equal number of boiled seeds to B. Both groups of seeds are soaked for 15 minutes in sodium hypochlorite solution to prevent fungal or bacterial growth which might produce carbon dioxide. The flasks are securely corked and left in the same conditions of light and temperature until germination is clearly perceptible in A. The seeds in B should not germinate. The gases in each flask are then tested by removing the cork and tilting the flask over a test-tube of lime water and shaking up the test-tube.

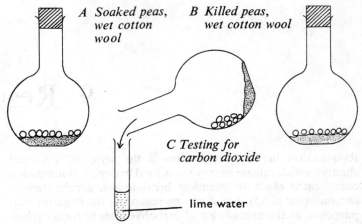

Fig. 9.3   Carbon dioxide production in germinating seeds

Result. The air from flask A should turn the lime water milky showing carbon dioxide is present. Air from B should have no effect.

Interpretation. The carbon dioxide must have been produced by the germinating seeds. B is a control and proves that it is not the cotton wool or anything other than the germinating seeds that give carbon dioxide.

(ii) *Animals and plants* (Fig. 9.4)

This experiment is suitable for giving fairly quick results with small animals but will also work, over a longer period, with plant material.

The animal or plant is placed in the vessel C. If it is a plant, the vessel must be "blacked out" to prevent photosynthesis occurring. If a potted plant is used, the pot must be enclosed in impermeable material so that the respiration of organisms in the soil does not affect the result. A stream of air is drawn slowly through the apparatus by means of a filter pump at E. In A, soda lime absorbs the carbon dioxide from the incoming air; the lime water in B should stay clear and so prove that carbon dioxide is absent from the air going into vessel C. If carbon dioxide is given out by the organism, the lime water in D will go milky after a time.

If the rates of respiration of different animals or plants are to be compared, the time taken for the lime water to go milky should be noted.

For an experiment to show carbon dioxide production in man, *see* p. 105.

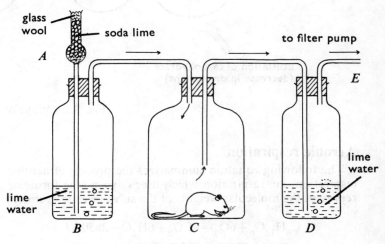

Fig. 9.4   Carbon dioxide production

(*d*) **Production of water vapour**

Since non-living matter may give off water vapour by evaporation, this is not a reliable test of respiration.

(*e*) **Release of energy in germinating seeds** (Fig. 9.5)

Heat production is a good indication of energy release. Sufficient wheat grains to fill two small vacuum flasks are soaked in water for 24 hours and half of them killed by boiling for 10 minutes. Both lots of wheat are soaked for 15 minutes in a solution of sodium hypochlorite (e.g. commercial hypochlorite diluted 1:4) to kill fungal spores on the fruit walls. The grains are rinsed with tap water; the living grains are placed in one flask, the dead grains in the other. Thermometers are inserted and the mouths of the flasks plugged with cotton wool.

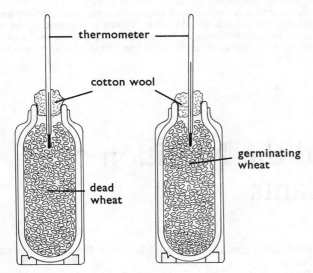

Fig. 9.5 Energy release in germinating wheat

Result. After a few days the temperature in the flask with living wheat will be considerably higher than in the control.

Interpretation. During the germination of wheat, heat energy is released. The results, however, do not justify the conclusion that the heat is the result of respiration rather than any other chemical process.

Anaerobic respiration

This is the release of energy from food material by a process of chemical breakdown which does *not* require oxygen. The food, e.g. carbohydrate, is not broken down completely to carbon dioxide and water but to intermediate compounds such as lactic acid or alcohol. The incomplete breakdown of the food means that less energy is made available during anaerobic respiration than is released during aerobic respiration.

Both processes may be taking place in cells at the same time. Indeed, the first steps in the breakdown of glucose in respiration are anaerobic.

glucose $\xrightarrow{anaerobic}$ lactic acid $\xrightarrow{aerobic}$ carbon dioxide and water

During vigorous activity, the oxygen supply to muscle cells may not be sufficient to oxidize food rapidly enough to meet their energy demands. Consequently the products of the initial, anaerobic, stages accumulate, e.g. lactic acid. These products are oxidized or converted back to carbohydrate so that even after vigorous activity has ceased, the uptake of oxygen continues at a high rate. The organism is said to have incurred an "*oxygen debt*" as a result of its excess of anaerobic respiration.

Certain bacteria and fungi derive all or most of their energy from anaerobic respiration and the end products are frequently alcohol and carbon dioxide; the process in this case is called fermentation.

Fermentation

The term fermentation is not applied exclusively to anaerobic respiration in which alcohol is produced; a variety of organic acids, e.g. citric, butyric, oxalic may be formed by the anaerobic respiration of micro-organisms and such fermentations are exploited commercially to produce these compounds.

The yeasts (unicellular fungi) and bacteria which bring about fermentation are able to employ their enzyme systems to release energy anaerobically from carbohydrates, particularly starch and sugar. Alcoholic fermentation on a commercial scale is usually brought about by yeasts (*see* p. 77) acting on sugar solutions such as the malt sugar prepared from germinating barley. The equation below summarizes the reactions:

$$C_6H_{12}O_6 \rightarrow 2CO_2 + 2C_2H_5OH + 118 \text{ kJ}$$
alcohol

If this equation is compared with the one on p. 47 it can be seen that far less energy is obtained from a gramme molecule of glucose during anaerobic respiration than is released when the sugar is completely oxidized. Unless the products of fermentation, in this case ethanol, are removed, they will reach a concentration which will eventually kill the organism producing them.

(*f*) **To show carbon dioxide production during anaerobic respiration (fermentation) in yeast**

Some water is boiled to expel all the dissolved oxygen and when cool is used to make up a 5 per cent solution of glucose and a 10 per cent suspension of dried yeast. 5 cm³ of the glucose solution and 1 cm³ of the yeast suspension are placed in a test-tube and covered with a thin layer of liquid paraffin to exclude atmospheric oxygen from the mixture. A delivery tube is fitted as shown in Fig. 9.6 and allowed to dip into clear lime water. After 10–15 minutes, with gentle warming if

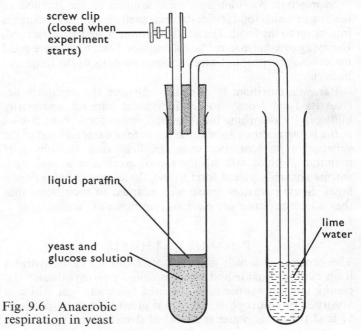

Fig. 9.6 Anaerobic respiration in yeast

necessary, there should be signs of fermentation in the yeast-glucose mixture and the bubbles of gas escaping through the lime-water turn it milky. The gas is therefore carbon dioxide.

A control can be set up in the same way but using a boiled yeast suspension which will not ferment. The fact that the living yeast produces carbon dioxide despite being deprived of oxygen is evidence to support the contention that anaerobic respiration is taking place.

If the experiment is repeated on a larger scale using 500 cm³ glucose solution and left for several days in a warm place, the alcohol can be distilled off, collecting the fraction that vaporizes between 70–80°C. The distillate can be identified as alcohol by its taste, odour and the fact that it can be ignited.

Metabolism

The thousands of enzyme-controlled chemical changes which take place in organisms and the cells of which they are composed are often referred to collectively as metabolism. The reactions concerned may be changing one compound into another more useful or more reactive, combining simple substances into more complex ones which can be built into the tissues, breaking down complex compounds to release their energy or to make them more easily transportable. Respiration is one manifestation of metabolism.

[See Experimental Work in Biology No. 7 for other experiments (p. 1)]

[See Experimental Work in Biology No. 7 for other experiments (p. 1)]

QUESTIONS

1. (a) Where does respiration occur? (b) What is the importance of respiration? (c) What materials does respiration (i) need, (ii) produce?
2. List the differences between aerobic and anaerobic respiration. Are these two forms of respiration mutually exclusive? Explain.
3. Which aspects of respiration can be measured or demonstrated?
4. In the mammal, which classes of food can be used to provide energy? Which ones provide the most energy? (See p. 82.)
5. An organism in the course of respiration takes in 50 cm³ oxygen. It is quite likely to give out 50 cm³ carbon dioxide in the same period so that there is no volume change in the gas surrounding it. In such a case how can one, in principle, design an experiment to show that oxygen is being taken up?

10 Photosynthesis and the Nutrition of Green Plants

TYPES OF NUTRITION

Holozoic nutrition is the method of feeding which is characteristic of animals. Complex food material is broken down, by digestion, to simple substances which are then absorbed and incorporated into body structures, or oxidized to obtain energy.

Holophytic nutrition means feeding like plants. Simple compounds, namely carbon dioxide, water and salts, are absorbed and built up into complex substances that can be oxidized for energy or synthesized into living protoplasm and cell walls.

Saprophytic nutrition is a term applied to the method of feeding in which food is digested externally by enzymes secreted into or on to the food. The soluble products are then absorbed. Decaying organic matter like rotting wood and humus are good materials for saprophytic feeders such as many of the fungi and bacteria.

Parasitic nutrition. In parasitic nutrition the organism derives its food from another individual without necessarily killing it. By absorbing blood, sap, digested food or the tissues of the living plant or animal the parasite obtains its food at the expense of its *host*. For example, fleas suck the blood of mammals, aphids take up the sap of green plants, and tapeworms absorb digested food in the alimentary canals of their hosts. Some parasites are so well adapted to their hosts that they do not produce any obvious symptoms of disease.

PHOTOSYNTHESIS

The process by which green plants build up carbohydrates from carbon dioxide and water is called photosynthesis. The energy for this synthesis is obtained from sunlight which is absorbed by chlorophyll. Oxygen is given off as a by-product. In land plants the water is absorbed from the soil by the root system and the carbon dioxide from the air through the stomata.

Photosynthesis goes on principally in the leaves, though any green part of the plant can photosynthesize.

The process may be represented by the equation:

$$6CO_2 + 6H_2O \xrightarrow[\text{absorbed by chlorophyll}]{\text{energy from sunlight}} C_6H_{12}O_6 + 6O_2$$

though it must be realized that this shows only the beginning and end of a very complicated chain of chemical reactions involving many intermediate compounds and numerous enzymes which promote the different chemical changes.

Photosynthesis in a leaf (see Figs. 4.9 and 4.10, pp. 18 and 19). The leaves of most green plants are well adapted to the process of photosynthesis taking place within them.

(a) Their broad, flat shape offers a large surface area for absorption of sunlight and carbon dioxide.

(b) Most leaves being thin, the distances across which carbon dioxide has to diffuse to reach the mesophyll cells from the stomata are very short.

(c) The large intercellular spaces in the mesophyll provide an easy passage through which carbon dioxide can diffuse.

(d) Numerous stomata on one or both surfaces allow the exchange of carbon dioxide and oxygen with the atmosphere.

(e) In the palisade cells the chloroplasts are more numerous than in the spongy mesophyll cells. The palisade cells being on the upper surface will receive most sunlight and this will be available to the chloroplasts without being absorbed by too many intervening cell walls. The elongated shape of many palisade cells may confer the same advantage.

(f) The branching network of veins provides a ready water supply to the photosynthesizing cells.

Photosynthesis in a palisade cell (Fig. 10.1). Water passes into the cell by osmosis (*see* Chap. 12) from the nearest vein; carbon dioxide from the adjacent air spaces diffuses through the cellulose wall and into the cytoplasm. In the chloroplast, molecules of carbon dioxide and water are combined by a series of chemical changes. For some of these changes the energy is provided in the chloroplast from sunlight absorbed by the chlorophyll. One of the final and easily recognizable products of photosynthesis is starch, and during daylight starch grains may be built up inside the chloroplast.

Oxygen is released during the process and diffuses out of the cytoplasm, through the cell wall into the air spaces and, finally, out of the stomata into the atmosphere.

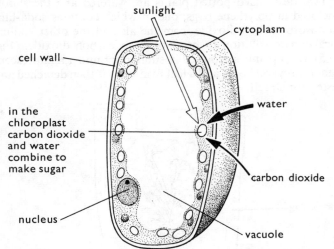

Fig. 10.1   Photosynthesis in a palisade cell

When photosynthesis is rapid, starch will accumulate in the cells, but enzymes are acting on the starch all the time, turning it into soluble carbohydrates like sucrose which pass out of the cells and are carried off in the sieve tubes of the phloem. The sucrose may travel to storage organs and be changed back into starch or pass to actively growing regions where it can be

(a) oxidized in respiration to provide energy for chemical reactions in the cell,

(b) concentrated in the cell sap, lowering the osmotic potential (*see* p. 63) of the vacuole which enables the cell to take in more water, increase its turgor and so extend,

(c) converted to cellulose to be built into new cell walls or to thicken existing ones,

(d) used as a basis for the synthesis of many other compounds, proteins, fats, pigments etc.

In darkness all the starch in the leaf is converted to sugar and removed.

During rapid photosynthesis the production of sugars faster than they could be removed by the sieve tubes could result in the osmotic potential of the cells being lowered to a harmful or disruptive level. Starch, being insoluble, can accumulate without causing osmotic disturbances.

From the explanations given above, it is apparent that if starch is accumulating in a leaf, photosynthesis is probably going on and in the experiments which follow, a positive result to the iodine test on a leaf is regarded as evidence of photosynthesis. Some plants, e.g. those in the iris and lily family, do not form starch in their leaves and in these cases the starch test would prove nothing.

Gaseous exchange and compensation point. It can be seen that the simplified equation given above is the reverse of that on p. 47 representing respiration. During respiration, oxygen is used up, carbohydrates are broken down and water and carbon dioxide are excreted. In photosynthesis carbon dioxide and water are taken in, carbohydrates are built up and oxygen is excreted. In green plants, as in all living organisms, respiration goes on all the time.

In dim light, e.g. at dawn and dusk, the rate of photosynthesis may become equal to the rate of respiration so that all the carbon dioxide produced by respiration is used in photosynthesis and all the oxygen produced by photosynthesis is used in respiration. The rate of carbohydrate breakdown in respiration is equalled by the rate of carbohydrate build-up by photosynthesis and there is no net gaseous exchange with the atmosphere. In this state, the plant is said to have reached its *compensation point.*

In bright light, respiration continues but will be exceeded by the rate of photosynthesis so that more oxygen will be produced than is used and more carbon dioxide used than is produced. The net gaseous exchange with the atmosphere will thus be carbon dioxide taken in and oxygen given out. In darkness there will be no photosynthesis and the plant will take in oxygen and give out carbon dioxide as a result of continuing respiration.

EXPERIMENTS ON PHOTOSYNTHESIS

Destarching. Since the presence of starch is regarded as evidence of photosynthesis, the experimental plants must have no starch in their leaves at the start of the experiment. If they are potted plants, they are destarched by leaving them in a dark cupboard for two or three days. Experiments conducted on plants in the open should be set up the day before, since during the night most of the starch will be removed from the leaves. Preferably, the selected leaves should be destarched by wrapping in aluminium foil for two days and one such leaf tested to ensure that no starch is present.

Testing a leaf for starch.

1. The leaf is detached and dipped in boiling water for half a minute. This kills the protoplasm by destroying the enzymes in it, and so prevents any further chemical changes. It also makes the cell more permeable to iodine solution.

2. The leaf is boiled in methylated spirit, using a water-bath (Fig. 10.2), until all the chlorophyll is dissolved out. This leaves a white leaf and makes colour changes caused by interaction of starch and iodine much easier to see.

3. Alcohol makes the leaf brittle and hard, but it can be softened by dipping it once more into boiling water, then spreading it flat on a white surface such as a glazed tile.

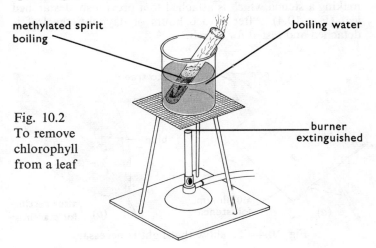

Fig. 10.2
To remove
chlorophyll
from a leaf

4. Iodine solution is placed on the leaf. Any parts which turn blue have starch in them. If no starch is present the leaf is merely stained brown by iodine.

The following controlled experiments can be used to investigate or verify the conditions necessary for photosynthesis.

Experiment 1. **Is chlorophyll necessary for photosynthesis?**

As with the experiments on germination, the basis of these experiments is to eliminate the condition under investigation to see if photosynthesis can take place without it.

It is not possible to remove chlorophyll from a leaf without killing it, and so a leaf, or part of a leaf, which has chlorophyll only in patches is used. Such variegated leaves are found in Tradescantias, some Pelargoniums ("geraniums") and the Cornus illustrated in Fig. 10.3a. After a period of destarching, the leaf on the plant is exposed to daylight for a few hours. It is then detached, drawn carefully to show the distribution of chlorophyll, and tested for starch as described above.

Result. Only the parts that were previously green turn blue with iodine. The parts that were white stain brown (Fig. 10.3b).

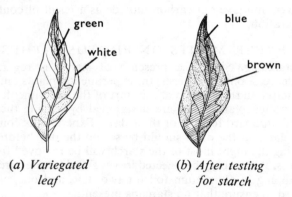

(a) Variegated leaf   (b) After testing for starch

Fig. 10.3   To show that chlorophyll is necessary

Interpretation. Since starch is present only in the parts which were green, it seems reasonable to suppose that photosynthesis goes on only in the presence of chlorophyll. It must be remembered, however, that there are other possible interpretations which the experiment has not eliminated, e.g. starch is made in the green parts and sugar in the white parts. Such alternative explanations can be tested by further experiments.

Experiment 2. **Is light necessary for photosynthesis?**

A simple shape is cut out from a piece of aluminium foil making a stencil which is attached to a previously destarched leaf (Fig. 10.4). After 4 to 6 hours of daylight, the leaf is detached and tested for starch.

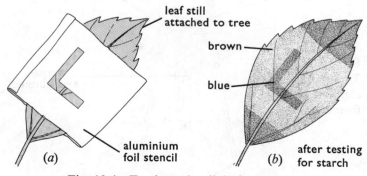

(a)   (b)

Fig. 10.4   To show that light is necessary

Result. Only areas which received light go blue with iodine.

Interpretation. As starch has not accumulated in the areas without light, it may be assumed that light plays an essential part in starch formation and hence in photosynthesis. It may be objected, however, that the aluminium foil prevented carbon dioxide from reaching the leaf and that it was shortage of this gas rather than absence of light which prevented photosynthesis. Against this it can be argued that a leaf produces carbon dioxide by its own respiration but a control may be designed using a transparent material, e.g. polythene, instead of the aluminium foil stencil.

Experiment 3. **Is carbon dioxide needed for photosynthesis?**

Two destarched potted plants are watered and the shoots enclosed in polythene bags, one of which contains soda-lime to absorb carbon dioxide from the air and the other sodium bicarbonate solution to produce extra carbon dioxide. (Fig. 10.5). Both plants are placed in sunlight or under a fluorescent light for several hours and a leaf from each is then detached and tested for starch.

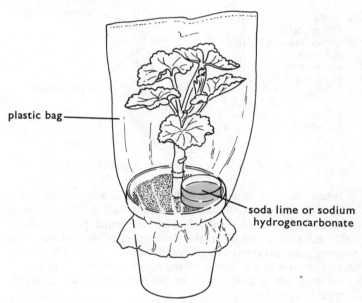

Fig. 10.5   To show that CO_2 is necessary

Result. The leaf deprived of carbon dioxide will not turn blue, while that from the carbon dioxide enriched atmosphere will turn blue.

Interpretation. The fact that no starch is made in the leaf deprived of carbon dioxide suggests that the latter must be necessary for photosynthesis. The control rules out the possibility that high humidity or temperature in the plastic bag prevents normal photosynthesis.

Experiment 4. **Is oxygen produced during photosynthesis?**

A short-stemmed funnel is placed over some Canadian pond weed in a beaker of water, preferably pond water, and a test-tube filled with water is inverted over the funnel-stem (Fig. 10.6). The funnel is raised above the bottom of the beaker to allow free circulation of water. The apparatus is placed in sunlight, and bubbles of gas soon appear from the cut stems, rise and collect in the test-tube. When sufficient gas has collected, the test-tube is removed and a glowing splint is inserted. A control experiment should be set up in a similar way but placed in a dark cupboard. Little or no gas should collect.

Result. The glowing splint bursts into flames.

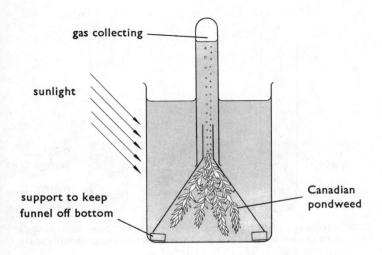

Fig. 10.6   To show that oxygen is set free

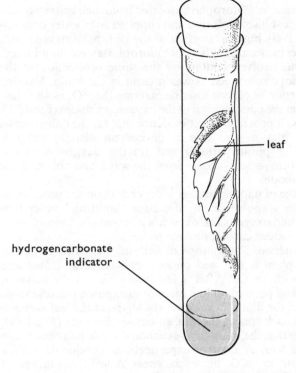

Fig. 10.7   Gaseous exchange during photosynthesis

Interpretation. The relighting of a glowing splint does not prove that the gas collected is *pure* oxygen but it does show that in the light, this particular plant has given off a gas which is considerably richer in oxygen than is atmospheric air.

Experiment 5. *Gaseous exchange during photosynthesis*

Three test-tubes are washed with tap water, distilled water and hydrogencarbonate indicator before placing 2 cm³ hydrogencarbonate indicator* in each. A green leaf is placed in tubes 1 and 2 so that it is held against the walls of the tube and does not touch the indicator (Fig. 10.7). The three tubes are closed with bungs, tube 1 is covered with aluminium foil and all three are placed in a rack in direct sunlight or a few centimetres from a bench lamp for about 40 minutes.

Result. The indicator (which was originally orange) should not change colour in tube 3, the control; that in tube 1, with the leaf in darkness, should turn yellow; and in tube 2 with the illuminated leaf, the indicator should be scarlet or purple.

Interpretation. Hydrogencarbonate indicator is a mixture of dilute sodium hydrogencarbonate solution with the dyes cresol red and thymol blue. It is a pH indicator in equilibrium with the atmospheric carbon dioxide, i.e. its original colour represents the acidity produced by the carbon dioxide in the air. Increase in atmospheric carbon dioxide makes it more acid and it changes colour from orange to yellow. Decrease in atmospheric carbon dioxide makes it less acid and causes a colour change to red or purple.

Thus, the results provide evidence that in darkness (tube 1) leaves produce carbon dioxide (from respiration), while in light (tube 2) they use up more carbon dioxide in photosynthesis than they produce in respiration. Tube 3 is the control, showing that it is the presence of the leaf which causes a change in the atmosphere in the test-tube.

The experiment can be criticized on the grounds that the hydrogencarbonate indicator is not a specific test for carbon dioxide but will respond to any change in acidity or alkalinity. In tube 2 there would be the same change in colour if the leaf produced an alkaline gas such as ammonia, and in tube 1, any acid gas produced by the leaf would turn the indicator yellow. However, a knowledge of the metabolism of the leaf suggests that these are less likely events than changes in the carbon dioxide concentration.

* *See* p. 211.

Isotopic labelling

The experiments described above are easy to perform in a school laboratory but they offer only indirect evidence about the chemistry of photosynthesis. Direct evidence can be obtained by isotopic labelling.

Most elements exist in more than one form, each form differing from the others only in the weight of its atoms and not in chemical properties. These alternative forms are called *isotopes*. Naturally occurring magnesium, for example, is a mixture of three isotopes having atomic weights of 24 (79 per cent), 25 (10 per cent) and 26 (11 per cent) respectively. Some isotopes of certain elements, e.g. uranium, are unstable and break down spontaneously, emitting radiation and sub-atomic particles. These are *radio-active isotopes* and their presence can be detected by electronic equipment (geiger counter) or because they blacken photographic plates. Radio-active isotopes of common elements such as carbon and phosphorus can be made artificially by bombarding the element with neutrons.

By supplying radio-active isotopes to organisms, biologists are able to trace many of the chemical steps in metabolism. For example, if rats are fed with glucose containing radio-active carbon, ^{14}C, it is found that the carbon dioxide they exhale is radio-active.

$$^{14}C_6H_{12}O_6 + 6O_2 \rightarrow 6^{14}CO_2 + 6H_2O$$

This is direct evidence that the carbon in glucose is eventually converted to carbon dioxide in respiration.

Although radio-isotopes are perhaps the easiest isotopes to detect, it is not necessary for the isotope to be radio-active for its presence and quantity to be discovered, e.g. the isotopes of oxygen $^{18}O_2$ and nitrogen ^{15}N are stable but can be estimated by a piece of equipment called the mass spectrometer.

Direct evidence for photosynthesis

Light reaction. Chloroplasts isolated from their cells were suspended in a suitable solution and supplied with water containing a relatively high proportion of the isotope of oxygen $^{18}O_2$ in the water molecules. When the chloroplasts were illuminated, oxygen was evolved containing the same proportion of the heavy isotope of oxygen as was present in the water. Varying the proportion of carbon dioxide carrying the $^{18}O_2$ isotope had no effect on the composition of the oxygen produced (Fig. 10.8).

This result provides direct evidence that (*a*) the chloroplasts and light play a vital role in photosynthesis, (*b*) water is essential for photosynthesis and (*c*) the oxygen produced during photosynthesis comes from the water and not from the carbon dioxide.

This stage of photosynthesis is dependent on the presence of light and is responsible, in effect, for "splitting" water into hydrogen and oxygen. The oxygen is released and the hydrogen is used to reduce carbon dioxide to sugars.

Dark reaction. The isotope of carbon, ^{14}C, is radio-active. A potted plant is enclosed under a bell jar, illuminated and supplied with $^{14}CO_2$. After a few hours, one of the leaves is detached and pressed on to a piece of unexposed photographic film. When the film is developed, the shape of the leaf is represented as a dark area on the film, an *auto-radiograph* (Fig. 10.9). This shows that the radio-active carbon dioxide had been taken up by the leaf. A control experiment is conducted with a similar plant in $^{14}CO_2$ but in darkness. A leaf from this plant does not affect the photographic film.

More detailed information is obtained by supplying a culture of unicellular green plants, algae, with $^{14}CO_2$ and illuminating them for 5 seconds, 30 seconds, 90 seconds, 5 minutes etc. and then analysing the contents of the cells to see which of their carbohydrates is radio-active. In this way, the intermediate steps in the building up of sucrose from carbon dioxide can be traced.

Provided enough hydrogen atoms are made available from the light process described above, the synthesis of carbohydrates from carbon dioxide can take place in darkness.

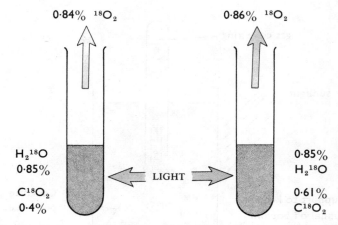

The percentage of $^{18}O_2$ isotope given off by the algae corresponds closely to the percentage of $H_2^{18}O$ and does not change significantly when the percentage of $C^{18}O_2$ is altered

Fig. 10.8   The source of oxygen from photosynthesis

PROTEIN SYNTHESIS IN PLANTS

Green plants use the sun's energy for only certain stages in the build-up of carbohydrates, but from these all the other compounds necessary for protoplasm and cell walls can be made. The sugar molecules are linked up into longer molecules to make cellulose. Additional elements such as nitrogen and sulphur are combined with carbohydrates to make proteins. The necessary energy for these processes comes from respiration, and they go on continuously in the plant, day and night.

Essential elements. Besides carbon, hydrogen and oxygen assimilated during photosynthesis, the plant takes up other elements from the soil. The chlorophyll molecule needs magnesium; many enzyme systems need phosphorus; calcium is used in the material between cell walls and potassium plays some part in controlling the rates of photosynthesis and respiration. These elements and the nitrogen and sulphur used in making proteins are obtained in the form of soluble salts from the soil.

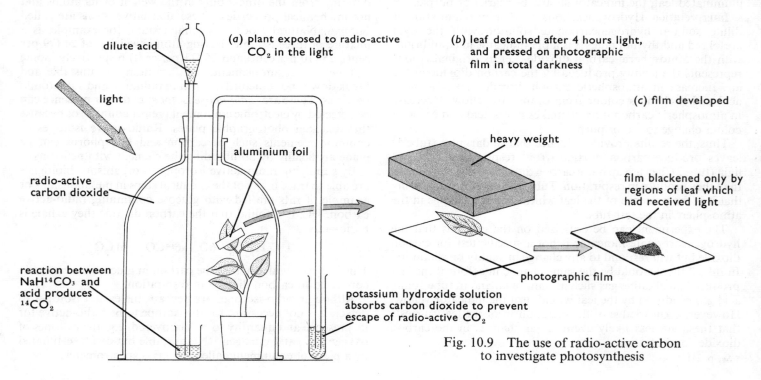

Fig. 10.9   The use of radio-active carbon
to investigate photosynthesis

Complete   no Na   no S   no Mg   no P   no Fe   no Ca   no K   no N

Plate 13. BEAN PLANTS GROWING IN CULTURE SOLUTIONS

They are absorbed in very dilute solution by the root system. The plant must be able to collect these salts from the soil and concentrate them. The following are examples of soluble salts containing the essential elements mentioned above, but these are not necessarily the ones that occur in the soil:

Potassium nitrate contains potassium and nitrogen.
Magnesium sulphate contains magnesium and sulphur.
Potassium phosphate contains potassium and phosphorus.
Calcium nitrate contains calcium and nitrogen.

Many other elements, e.g. copper, manganese, boron, in very small quantities are also needed for healthy growth. These are often referred to as *trace elements*.

Water cultures. Since a plant can make all its vital substances from carbon dioxide, water and salts, it is possible to grow plants in water containing the necessary salts. These solutions are called water cultures, and by carefully controlling the salts present it is possible to find out the relative importance of particular elements in the growth of the plant.

Experiment 6. To demonstrate the importance of certain elements in normal plant growth (Fig. 10.10)

A suitable culture solution can be made from the salts listed above by dissolving 2 g of calcium nitrate, 0·5 g of the others and a trace of ferric chloride in 2000 cm³ of distilled water. To make a solution lacking sulphur, magnesium chloride replaces sulphate; for one deficient in calcium, potassium nitrate is used instead of calcium nitrate; to eliminate phosphorus, potassium sulphate replaces the phosphate; ferric chloride is omitted to exclude iron; the nitrates are replaced by chlorides for a nitrogen-free solution; calcium salts replace potassium salts to eliminate potassium; potassium sulphate instead of magnesium sulphate excludes magnesium.

Five test-tubes are labelled and filled with experimental solutions as follows: (*a*) complete culture solution, (*b*) distilled water, (*c*) culture solution lacking nitrate, (*d*) a solution lacking

phosphate, and (*e*) one lacking calcium. Other solutions, deficient in different elements can be used instead of, or in addition to, those listed.

Seedlings are germinated as suggested on p. 46 and five of these at about equivalent stages of development are selected and placed, one in each of the culture solutions, supported by cotton wool as shown in Fig. 10.10. It is best to select seeds with only a little food reserve in their cotyledons or endosperm, e.g. wheat, so that the reserves are quickly exhausted and the plants come to depend on the culture solution for

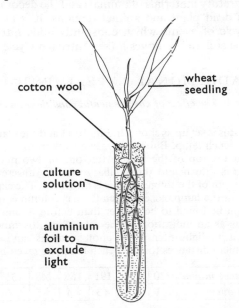

cotton wool   wheat seedling

culture solution

aluminium foil to exclude light

Fig. 10.10   To set up a water culture

their minerals. Light is excluded from the solutions to prevent growth of algae which might affect the mineral content. The tubes are placed where the shoots receive sunlight equally, and the solutions kept topped-up with distilled water.

Result. After a few weeks the seedlings are compared by counting and measuring their leaves and roots as far as possible. The one in distilled water will have grown hardly at all and is stunted and weak; the one in complete culture solution will be large and sturdy with dark green leaves. The others are all likely to be smaller than that in complete culture, with fewer and smaller leaves, frequently with a much paler green colour (Plate 13, p. 55).

For various reasons, e.g. impurities in the chemicals used to prepare the culture solutions, the results may not be so clear-cut as predicted above. However, all the seedlings from the class experiments can be collected together in batches, i.e. all seedlings from complete culture together, all seedlings from distilled water together, etc. The sorted batches of seedlings are then placed in labelled beakers and heated in an oven at 120 °C for twelve hours to dry them. The dried seedlings are then weighed and the dry weights compared. The dry weights represent the amount of new matter synthesized by the seedling and it is very likely that the seedlings provided with the full range of minerals will show the greatest increase in dry weight.

Interpretation. A supply of mineral salts containing the elements listed above is necessary for the normal growth and existence of green plants.

Note. Although nitrogen is abundant in the atmosphere it cannot be utilized directly by green plants.

Source of salts, particularly nitrates

Since plants are continually removing salts from the soil, and rain-water washes out those that are soluble, it follows that the salts must be replaced or all vegetation would die.

Rock consists of minerals, some of which are dissolved out by the action of slightly acid rain-water containing dissolved carbon dioxide. The salts from these weathered rocks soak into the soil and are eventually washed into streams and rivers and, finally, into the sea

A great proportion of the essential salts in the soil comes from the excretory materials of animals and the decomposition products of dead plant and animal remains. It is possible to trace the cycle of events which constantly adds nitrogenous matter to the soil and removes it (*see* Nitrogen Cycle p. 59).

ADDITIONAL EXPERIMENT

*Experiment A 1. **The effect of environmental conditions on the rate of photosynthesis***

The apparatus is set up as shown in Fig. 10.11 and the plant illuminated with a bench lamp. Bubbles of gas escape from the cut stem and collect at the top of the bulb. After one or two minutes the bubble of gas is drawn into the capillary stem by unscrewing the clip and the length of the column of gas is measured. The experiment is repeated with the lamp closer. When the gas column is measured this time it will be found to be greater than before. Using the production of gas as an indication of the rate of photosynthesis* the effect of increasing light intensity on photosynthesis can be investigated. Convenient distances for the bench lamp are given below:

distance in mm	330	233	191	165	148	135
light intensity	×1	×2	×3	×4	×5	×6

Similar experiments can be carried out with changes in temperature or, with more difficulty, changes in carbon dioxide concentration.

* The scientific basis for this experiment is rather flimsy since the oxygen content of the gas given off by the pond weed may vary from 20 to 80 per cent or more but it usually gives results clear enough to form the basis for a reasonable interpretation.

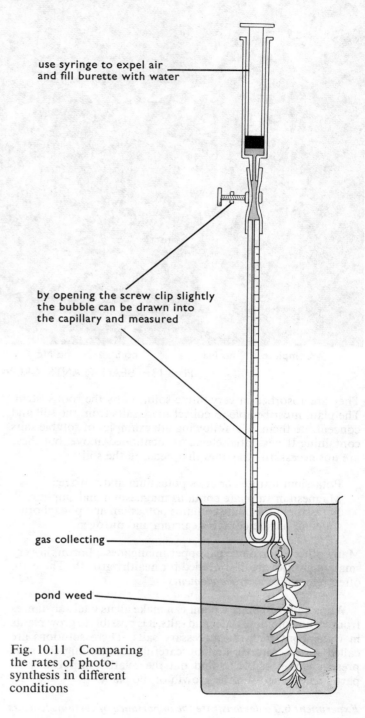

use syringe to expel air and fill burette with water

by opening the screw clip slightly the bubble can be drawn into the capillary and measured

gas collecting

pond weed

Fig. 10.11 Comparing the rates of photosynthesis in different conditions

Limiting factors. If a graph is plotted of rate of photosynthesis against light intensity, it usually appears something like Fig. 10.12a. If the experiment is repeated with different concentrations of carbon dioxide, the graphs appear as in Fig. 10.12b.

Clearly, in each case, increasing the light intensity raises the rate of photosynthesis only up to the point where the plant can absorb carbon dioxide no faster. Carbon dioxide is thus limiting the increase in photosynthesis. Similarly, if the carbon dioxide concentration is increased while holding the light intensity at a steady level, the light intensity will be limiting.

Temperature can be another limiting factor, as well as the physical properties of the leaf itself, e.g. restriction of the flow of carbon dioxide through the stomata, the rate of diffusion of carbon dioxide through the cell walls, the available chloroplast surface for the absorption of sunlight.

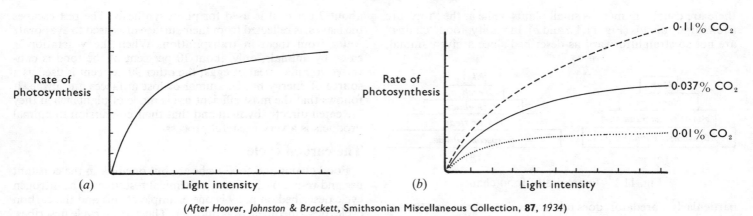

(*After Hoover, Johnston & Brackett*, Smithsonian Miscellaneous Collection, **87**, *1934*)

Fig. 10.12   Limiting factors in photosynthesis

[*Note:* the experiments in this chapter and many others are described in greater detail in the laboratory manual *Photosynthesis—see* p. 1.]

Photosynthesis—see p. 1.

QUESTIONS

1. What are the requirements for photosynthesis? How are these requirements met in (*a*) a land plant, (*b*) an aquatic plant?
2. What is meant by "destarching" a plant? In what circumstances can the presence of starch in a leaf be regarded as evidence that photosynthesis has occurred?
3. If a plant was producing carbon dioxide and taking in oxygen why would you *not* be justified in assuming that no photosynthesis was taking place?
4. It can be claimed that the sun's energy is indirectly used to produce a muscle contraction in your arm. Trace the steps in the conversion of energy which would justify this claim.
5. Proteins contain carbon, hydrogen, oxygen, nitrogen and often sulphur. Name the source, for green plants, of each of these elements.

11 Food Chains and the "Balance of Nature"

ALL animals derive their food either directly or indirectly from plants. Carnivorous animals feed on other animals which themselves feed on smaller animals but sooner or later in such a series we come to an animal which feeds on vegetation. For example, pike eat perch, perch eat stickleback which feed on water fleas; the water fleas feed on microscopic plants in the pond. This kind of relationship is called a *food chain*. The basis of food chains on land is vegetation in general, but particularly grass and other leaves. In water, the basis is the *phytoplankton* (Plate 14)—the millions of microscopic plants living near the surface of the sea, ponds and lakes. These need only the water round them, the dissolved carbon dioxide and salts and sunlight to make all their vital substances. Feeding on these microscopic plants are tiny animals, *zooplankton* (Plate 15), such as water fleas and other crustacea and the larvae of many kinds of animal. The small animals of the zooplankton are eaten by surface-feeding fish such as herring; the herring, in turn, forms part of the diet of man (Fig. 11.1*a*). At the base of a food chain

Plate 14.   PHYTOPLANKTON (×100)
These microscopic plants are diatoms

(*Dr. D. P. Wilson*)

Plate 15.   ZOOPLANKTON (×15)
Mostly adult and larval crustacea from the sea

(*Dr. D. P. Wilson*)

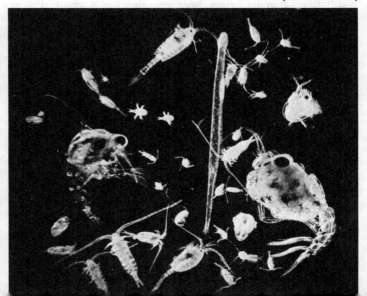

there are usually numerous small plants while at the "top" are a few large animals (Fig. 11.1 a and b). In reality, food "chains" are not so straightforward as described since a given animal,

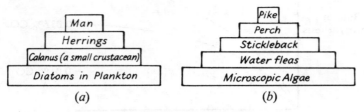

Fig. 11.1   Examples of food chains

particularly a predator, does not live exclusively on one type of food, e.g. stoats may eat rats, rabbits or birds. This more complex relationship can be shown as a food "web" (Fig. 11.2) but even this is greatly simplified and generalized.

The green plants at the base of a food chain are sometimes referred to as the *producers* while the animals which eat them

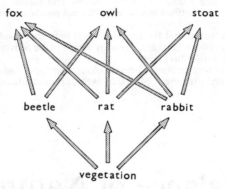

Fig. 11.2   A food web

are called *consumers*. A *first order* consumer eats vegetation; a *second order* consumer eats animals which feed on vegetation. Consumers such as fungi which are involved in the processes of decay are classed as *decomposers*.

If the population of one of the animals or plants in a food chain is altered, all the others are affected. If pike and perch are removed from a pond, the stickleback might increase for a time at the expense of the water fleas. If diatoms in marine plankton flourish, herrings may become plentiful as a result of the increase in the crustaceans that eat the diatoms. When, in 1954/5, the rabbits in Britain were almost exterminated by myxomatosis, the vegetation in what had been rabbit-infested areas changed, and sheep could graze where rabbits had previously eaten all the available grass; trees that were hitherto nipped off as seedlings began to grow to maturity with the result that what had once been grassland, e.g. chalk downs, started to become scrub and eventually woodland. Foxes ate more voles, beetles and blackberries than before and attacked more lambs and poultry.

Ultimate sources of energy. All the energy released on the Earth, apart from atomic energy and tidal power, comes from the sun, via food oxidized in animals and plants, coal from fossilized forests, petroleum from marine deposits, or hydroelectric power from rain-water in lakes. There are ways of using the sun's energy directly and making it heat water to produce steam, but, at the moment, one of the best ways of trapping sunlight is to grow trees and other plants from which food and other energy-rich products can be collected.

Of the sunlight which reaches an area of grassland, only about 2 per cent is used for photosynthesis. The rest escapes the leaves, is reflected from their surface or is used to evaporate water from them in transpiration. When the vegetation is eaten by animals, only about 10 per cent of the food is converted to milk, meat or eggs; the other 90 per cent is used as a source of energy by the animal or lost in faeces and urine. It follows that the most efficient use is made of plants when they are eaten directly by man and that their conversion to animal products is a very wasteful process.

The carbon cycle

Food chains and food webs are but one link in the constant use and re-use of the Earth's chemical resources. The nitrogen cycle described on p. 59 is one example of this and the carbon cycle, described below, is another. The carbon cycle describes, in essence, the processes which increase or decrease the carbon dioxide in the environment (Fig. 11.3).

Removal of carbon dioxide from the atmosphere. Green plants, by their photosynthesis (p. 50) remove carbon dioxide from the atmosphere or from the water in which they grow. The carbon of the carbon dioxide is incorporated at first into carbohydrates such as sugar or starch and eventually into the cellulose of cell walls, and the proteins, pigments and other organic compounds which comprise living organisms. When the plants are eaten by animals the organic plant matter is digested, absorbed and built into compounds making the animals' tissues. Thus the carbon atoms from the plant become an integral part of the animal.

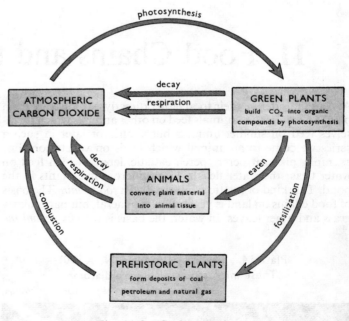

Fig. 11.3   Carbon cycle

Addition of carbon dioxide to the atmosphere.

(a) *Respiration.* Plants and animals obtain energy by oxidizing carbohydrates in their cells to carbon dioxide and water (*see* p. 47). These products are excreted and the carbon dioxide returns once again to the environment.

(b) *Decay.* The organic matter of dead animals and plants is used by bacteria and fungi as a source of energy. The microorganisms decompose the plant and animal material, converting the carbon compounds to carbon dioxide.

(c) *Combustion.* In the process of burning carbon-containing fuels such as wood, coal, petroleum and natural gas, the carbon is oxidized to carbon dioxide. The hydrocarbon fuels originate from communities of plants such as prehistoric forests or deposits of marine algae which have only partly decomposed over the millions of years since they were buried.

Thus, an atom of carbon which today is in a molecule of carbon dioxide in the air, may tomorrow be in a molecule of cellulose in the cell wall of a blade of grass. When the grass is eaten by a cow, the carbon atom may become one of many in a protein molecule in the cow's muscle. When the protein molecule is used for respiration the carbon atom will enter the air once again as carbon dioxide. The same kind of cycling applies to nearly all the elements of the Earth. No new matter is created but it is repeatedly rearranged. A great proportion of the atoms of which you are composed will, at one time, have been an integral part of many other organisms.

Today, man's activities affect these cycles. For example, the nitrogen present in his excretory products is not usually re-cycled to the land producing his food; the carbon fuels are being burned in ever-increasing quantities, depleting their sources and adding more carbon dioxide to the atmosphere.

The nitrogen cycle (Fig. 11.4)

When a plant or animal dies its tissues decompose, largely as a result of the action of enzymes and bacteria. One of the important products of this decomposition is ammonia, which is washed into the soil where it forms ammonium compounds.

Nitrifying bacteria. In the soil are many other bacteria and certain of these, e.g. *Nitrosomonas*, oxidize the ammonium compounds to nitrates(III). Others, e.g. *Nitrobacter*, further oxidize nitrates(III) to nitrates(V), and these can be taken up in solution by plants. The faeces of animals contain organic matter that is similarly broken down, while their urine is rich in nitrogenous waste products such as ammonia that can be oxidized to nitrates by soil bacteria.

The bacteria derive energy from these oxidative processes

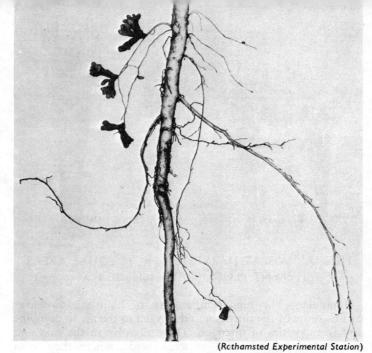

Plate 16.   ROOT NODULES ON LUCERNE

in much the same way as plants and animals derive energy from respiration by oxidizing carbohydrates to form carbon dioxide and water.

Nitrogen-fixing organisms. Although green plants cannot utilize the nitrogen in the atmosphere, there are bacteria and blue-green algae in the soil which absorb and combine it with other elements, so making nitrogen compounds. This is called the *fixation of nitrogen.* Some nitrogen-fixing bacteria, as well as existing free in the soil, are also found in special root swellings, or *nodules* (*see* Fig. 8.1 and Plate 16), in plants of the pea family such as clover, beans and lucerne, and since these plants increase the nitrogen content of the soil they are included in the three- or four-year crop rotation used in agricultural practice.

Denitrifying bacteria. There are also bacteria in the soil that obtain energy by breaking down compounds of nitrogen to gaseous nitrogen which consequently escapes to the atmosphere.

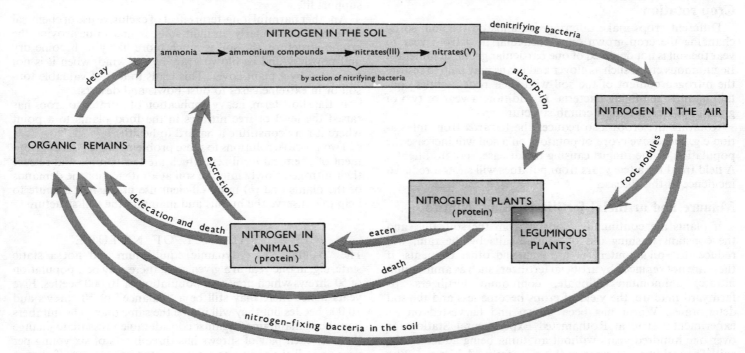

Fig. 11.4   Nitrogen cycle

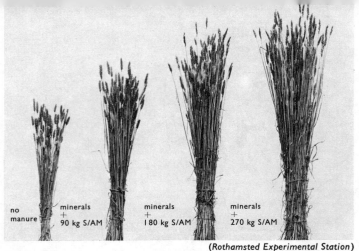

(*Rothamsted Experimental Station*)

Plate 17. WHEAT HARVESTED FROM EQUAL AREAS OF DIFFERENT PLOTS (S/AM=sulphate of ammonia)

Lightning. The high temperature of lightning discharge causes some of the nitrogen and oxygen in the air to combine and form oxides of nitrogen. These dissolve in the rain and are washed into the soil as weak acids, where they form nitrates. Although several million tonnes of nitrate may reach the earth's surface in this way each year, this forms only a tiny fraction of the total nitrogen being recycled.

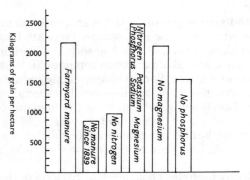

Fig. 11.5 Rothamsted experimental station, Broadbalk Field: average yearly wheat yields from 1852 to 1925

Crop rotation

Different crops make differing demands on the soil, so by changing the crop grown on a particular field from year to year the soil is not depleted of one particular group of minerals. Leguminous crops such as clover and beans may help to restore the nitrogen content of the soil with their root nodules containing nitrogen-fixing bacteria. In addition, a year or two of grass improves the soil's crumb structure.

Rotating the crops also reduces the hazards from infestation e.g. successive crops of potatoes on a soil will increase the population of the fungus causing the disease "potato blight". A field freed for a few years from potatoes will show a reduced incidence of this disease.

Manure and artificial fertilizers in agriculture

If plants are continually cropped from the soil, this, and the constant washing out of soluble salts by the rain, will reduce the soil's content of nitrogen and other elements. If these are not replaced by artificial fertilizers such as ammonium nitrate, ammonium sulphate, compound fertilizers, or farmyard manure, the yield of crops become less and the soil deteriorates. Wheat has been grown and harvested on an experimental strip at Rothamsted Experimental Station for over one hundred years without anything being added to the soil. The yield has dropped in this time from 14·7 kg to 6·9 kg per 100 m². The soil nitrogen, however, has remained at a

steady concentration over the last eighty years, and this is probably attributable to the nitrogen-fixing bacteria and blue-green algae present in the soil. (*See* Plate 17.)

Hazards of using chemical fertilizers

Long-term experiments show that by applying to the soil a programme of crop rotation and using farmyard manure, the yield of crops can be increased over a fifty-year period. The long-term effects of using chemical fertilizers, in particular soluble salts of nitrogen, are far less satisfactory. Whereas organic manure contributes to the humus content of the soil (*see* p. 78), helping to maintain its crumb structure and hence its porosity and permeability to air, the continued exclusive use of chemicals leads to a loss of the organic humus, a deterioration of crumb structure and a decrease in porosity. As a consequence the plant roots are deprived of oxygen in the poorly aerated soils and cannot absorb the salts effectively. The unabsorbed nitrates are washed by rain from the soil and eventually drain into rivers and lakes. Here they stimulate excessive growth of microscopic plants, algae. The algae grow quickly, die and decompose. The bacterial decomposition of their cells requires a supply of oxygen from the dissolved gases in the water. Eventually the oxygen supply is so depleted that fish and other aquatic animals are suffocated and die, and the semi-decomposed organic remains of the algae form a foul mud. This over-fertilization or *eutrophication* of lakes and rivers results not only from the excess of nitrates added to the soil but from the sewage effluents of cities, intensive animal rearing units, and the phosphates present in certain detergents.

Lake Erie in America receives water draining from 30,000 square miles of farmland and the effluents of large cities such as Detroit and Cleveland. In its waters eutrophication has produced an organic mud with a large oxygen deficit. A layer of iron oxide separates the organic mud from the water above but some scientists think that further removal of oxygen could dissolve this layer, allowing the organic matter to mix with the water and so produce anaerobic conditions which could not support life.

Another harmful long-term effect of exclusive use of chemical fertilizers, particularly on light soils, is that in destroying the crumb structure, the soil is much more likely to become dry and powdery and be blown away by the wind, when it is not protected by a plant cover. This leads to loss of valuable topsoil or in extreme cases to dust-bowls and deserts.

In the short term, heavy application of nitrates to crops has raised the level of free nitrates in the food plants to a point where it may constitute a hazard to health.

Two possible solutions to these problems are (*a*) the development of chemical fertilizers which are less soluble and release their nitrogen slowly into the soil at a rate to suit the demands of the plants and (*b*) more efficient use of organic manure to help to conserve the humus and maintain the soil structure.

THE "BALANCE OF NATURE"

The "balance" is a dynamic equilibrium and not a static state, e.g. in one year in a given area, there may be a population of 50 shrews which prey on a population of 10,000 beetles. Five years hence there may still be a "balance" of 50 shrews and 10,000 beetles but they will not be the same ones. The numbers have been maintained against considerable pressure to change them. If each pair of shrews has three litters of six young per year and all the youngsters survive, the population of shrews after only one year would be 500. The beetles' reproductive capacity is greater still. There must be a high mortality of both animals for the numbers to remain in balance from year to year.

In a natural environment the vegetation will support the herbivorous creatures which, in turn, will feed the carnivores; the natural cycles replenish the soil and nourish the plants. If the predators become too numerous, the herbivores will be eaten in greater numbers and their population will diminish; the vegetation, temporarily relieved from grazing will grow more densely. Eventually, the population of predators, having depleted the herbivores which constitute their food, will decline; the reduction of predators and the more luxuriant vegetation will permit the herbivores to flourish once more. Thus, the populations remain basically constant with relatively minor fluctuations.

Civilized man, however, with his technology and agriculture, interrupts the natural cycles and disturbs the balance of nature to an extent which gives cause for concern.

(a) **Deforestation.** Trees may be cut down to make way for agriculture or to use the timber. If carried out in the right areas to the correct extent and with provision for reafforestation, a balanced community can be maintained.

The soil on sloping ground, however, is often thin and once the tree cover is removed, it is no longer protected by the leaf canopy from the forces of wind and rain. The topsoil is washed away, silting up rivers and lakes and causing floods (*see* Fig. 16.5, p. 81).

(b) **Erosion.** Repeated ploughing of soils, exclusive use of artificial fertilizers, overgrazing of pastures can, in certain areas, lead to soil erosion as described on p. 80, ultimately making the land incapable of supporting life of any kind.

(c) **Eutrophication.** This is the overgrowth of aquatic plants resulting from an excess of nitrogenous salts reaching rivers. The nitrates may come from farmland where heavy application of soluble nitrogenous fertilizers is taking place or from the effluent of treated sewage. Its effects are discussed more fully on p. 60.

(d) **Monoculture.** A natural environment usually has a wide variety of vegetation at different levels, flowering and fruiting at different times. This vegetation is exploited in different ways by the animals living there, e.g. deer browse on the leafy branches, rabbits crop the turf, squirrels take berries and nuts from the trees, and worms consume the leaves which fall from them. Agricultural practice involves removal of the natural plant and animal community and its replacement with large populations of a single species of plant or animal: arable fields given over to wheat, pastures supporting sheep exclusively. This practice obviously makes the environment unsuitable for the majority of its original inhabitants, indeed it is meant to; a mixed population of cereal and "weeds" is commercially undesirable. The practice of monoculture, however, has its disadvantages.

Parasites and pests, which in a mixed community find their hosts well spaced, in a monoculture can spread rapidly since suitable hosts are growing closely together. In many parts of Africa there is evidence to suggest that more protein could be obtained by harvesting the mixed populations of wild animals living in a natural environment than is derived from the herds of sickly cattle which replace them and destroy their habitat.

(e) **Pesticides.** To protect the plants in a monoculture from the depredations of insects the crops are often sprayed with insecticides such as the chlorinated hydrocarbon DDT. This prevents the loss of hundreds of tonnes of food but in some cases it upsets the dynamic balance of life in unexpected ways. For example, the chemicals affect harmful and beneficial insects alike, so that after spraying orchards to eliminate the codling moth whose larva burrows into apples, enormous numbers of red mites appeared because the spray had killed the spiders which normally preyed upon them.

For a few years after it had been discovered and developed, DDT seemed to be the perfect insecticide. Used to kill body lice and mosquitoes it must have saved thousands of lives by eradicating typhus and malaria, spread respectively by these insects in certain areas and conditions. The concentrations used seemed harmless to man and other animals though, in higher concentration, it was known to be poisonous, particularly to fish. Unfortunately, however, when DDT is taken in with food and water, it is not all eliminated from the body, a proportion being retained and accumulated in the fat deposits of the body. When, in some animals, the fat is mobilized for respiration, harmful quantities of DDT may be released into the blood.

The animals at the end of food chains are particularly vulnerable when DDT is used to kill insect pests on a large scale. In America, DDT was used to kill the beetle which transmits dutch elm disease. The spray and the sprayed leaves reached the soil where the DDT was taken up by worms. When birds, notably the American robin, ate the worms, they accumulated lethal doses of DDT and whole populations of birds were wiped out.

A similar event occurred when an insecticide was used to kill gnat larvae in Clear Lake, California. At a concentration of 0·015 part per million of insecticide in the lake water the fish were unharmed. After five years, however, the western grebes on the lake were dying in large numbers. Although the water contained only 0·015 ppm, the plankton living in the waters had accumulated the compound to a level of 5 ppm. The small fish which fed on the plankton contained 10 ppm and the predatory fish even higher concentrations. The grebes which fed on the larger fish had as much as 1600 ppm in their body fat.

DDT is a stable compound and its effects last for a long time; a good property for an insecticide but potentially disastrous for the balance of nature. When applied to the crops, it reaches the soil and destroys the insect life there. Eventually it reaches the rivers, lakes and oceans where, if sufficient accumulated, it could begin to poison the fish and other marine life.

There is no end to the examples of man's wilful and unwitting depredation of the Earth: the excessive killing of animals for food or profit, to a point where they are exterminated; irrigation schemes which make dry areas more productive but spread the water snails which carry the disease bilharziasis; clearing tropical forests for agriculture and so providing conditions in which the tsetse fly can breed and spread sleeping sickness.

One is forced to the conclusion that the demands made on the biological systems of the world to support an ever increasing human population and a rising standard of life will, unless controlled, eventually destroy all natural resources and make the Earth's surface uninhabitable.

QUESTIONS

1. Trace the food chains involved in the production of the following articles of man's diet: eggs, cheese, bread, meat, wine. In each case show how the energy in the food originates from sunlight.
2. Discuss the advantages and disadvantages of man's attempting to exploit a food chain nearer to its source, e.g. the diatoms of Fig. 11.1a.
3. Construct a diagram, on the lines of the carbon cycle (Fig. 11.3) to show the cycling process for hydrogen.
4. How do you think evidence is acquired in order to assign animals such as a badger and a wood pigeon to their position in a food web?
5. What would be the desirable qualities of an insecticide to control, e.g., greenfly on sugar beet leaves? What might be the disadvantages of total eradication of an insect pest such as the greenfly?

12 Diffusion and Osmosis

Diffusion

All substances are made up of minute particles called *molecules*, e.g. the smallest particle of carbon dioxide is a molecule consisting of one atom of carbon joined to two atoms of oxygen. In a solid the molecules are packed relatively closely together with little or no freedom to move; in liquids the molecules are close together but are free to move; the molecules of a gas are very much further apart and are moving about at random colliding with each other and the walls of whatever contains the gas. Because of this constant random movement the molecules of a gas tend to distribute themselves evenly throughout any space in which they are confined. The same principle holds true for substances which dissolve in a liquid, e.g. if a crystal of copper sulphate is placed at the bottom of a beaker of water the blue colour of the solid will eventually spread throughout the water as the copper sulphate dissolves. This molecular movement in gases or liquids which tends to result in their uniform distribution is called *diffusion*.

The following experiments illustrate the process of diffusion in air and water.

Experiment 1. *Diffusion of a gas*

Squares of wetted red litmus paper are pushed with a glass rod or wire into a wide glass tube, corked at one end, so that they stick to the side and are evenly spaced out (Fig. 12.1). The open end of the tube is closed with a cork carrying a plug of cotton wool saturated with a strong solution of ammonia. The alkaline ammonia vapour diffuses along inside the tube at a rate which can be determined by observing the time when each square of litmus paper turns completely blue. If the experiment is repeated using a more dilute solution of ammonia the rate of diffusion is seen to be slower.

Experiment 2. *Diffusion in a liquid*

Diffusion in a liquid is very slow and liable to be affected by convection currents or other physical disturbances in the liquid. In this experiment the water is "kept still" so to speak, by dissolving gelatin in it. 10 g gelatin is dissolved in 100 g hot water and the solution is poured into test-tubes to half fill them. Some of the liquid gelatin remaining is coloured with methylene blue and when the first layer of gelatin in the test-tube has set firmly, a narrow layer of blue gelatin is poured into it. When the blue layer of gelatin is cold and firm, the test-tube is filled with cool but liquid gelatin and cooled quickly so that the blue gelatin is sandwiched between two layers of clear gelatin (Fig. 12.2). After a week, the blue dye is seen to have diffused into the clear gelatin, upwards and downwards to equal extents.

The rate of diffusion of a substance depends to a large extent on the size of the molecule, the temperature of the substance and its concentration. The larger the molecule, the more slowly it diffuses and the warmer the substance, the more rapidly it diffuses. Experiment 1 shows that the more concentrated the source of a substance, the more rapidly it diffuses and that the direction of diffusion is from the region of high concentration to the region of low concentration. The difference in concentration which results in diffusion is called a *diffusion gradient* and the "steeper" the diffusion gradient the more rapid is the resulting diffusion. For example, when rapid photosynthesis is going on in a palisade cell of a leaf (p. 51) carbon dioxide is being removed from the cytoplasm and incorporated into carbohydrate. The carbon dioxide concentration in the cell falls and so sets up a diffusion gradient between the air in the intercellular spaces and the cell. As a result, carbon dioxide diffuses into the cell from the intercellular space. The same cell will be producing oxygen in the course of photosynthesis, creating an oxygen gradient in the opposite direction to that for carbon dioxide. Oxygen consequently diffuses out of the cell into the intercellular space.

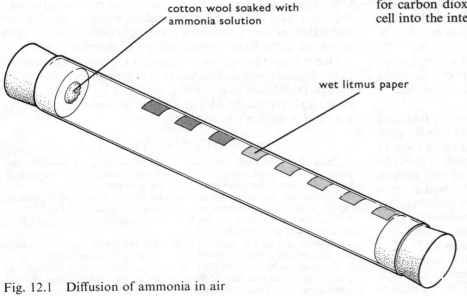

cotton wool soaked with ammonia solution

wet litmus paper

Fig. 12.1 Diffusion of ammonia in air

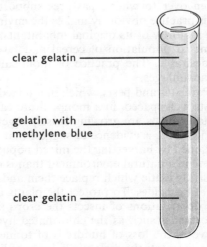

clear gelatin

gelatin with methylene blue

clear gelatin

Fig. 12.2 Diffusion in a liquid

Diffusion in living organisms. Diffusion plays a part in most cases of uptake or expulsion of substances within an organism or between the organism and the environment. Sometimes, as in microscopic single-celled organisms, diffusion may be rapid enough to account entirely for the uptake of oxygen and the removal of carbon dioxide and other excretory products (*see* p. 178). In many cases, however, diffusion is too slow to meet the demands of the living tissues and is superseded by, on a large scale, such processes as the circulation of blood and, on a cellular scale, "active transport" whereby substances are moved by some form of chemical activity into, across or out of cells.

Osmosis

Osmosis can be regarded as a special case of diffusion. When a solute dissolves in water, each ion or molecule attracts a small number of water molecules around itself. The solute molecules are then said to be *hydrated*.

The hydration process means that some of the water molecules in a solution are no longer free to move independently. This has the effect of reducing the concentration of free water molecules in a solution. A concentrated solution of sugar will contain fewer free water molecules than the same volume of a dilute solution. If two such solutions were in contact, the hydrated sugar molecules would diffuse from the concentrated to the dilute solution, but the free water molecules would diffuse from the dilute to the concentrated solution.

Fig. 12.4 shows the two solutions separated by a thin membrane that stops them from mixing but allows individual molecules to diffuse through pores in the membrane. More water molecules will pass through the membrane from left to right because there are more free water molecules in the dilute solution. More hydrated sugar molecules will pass from right to left because there are more of them in the concentrated solution.

The small water molecules diffuse faster than the large sugar molecules and so the most obvious effect is that water diffuses from the dilute to the concentrated solution.

The pores in the membrane which permit the passage of molecules are very small, but quite big enough to allow sugar and water molecules through. Nevertheless, because water molecules diffuse through the membrane more rapidly than sugar molecules do, the membrane is called "*selectively permeable*" or, sometimes "semi-permeable".

A definition of osmosis, for our purposes, is the *diffusion of water through a "selectively permeable" membrane from a dilute to a concentrated solution* (Experiment 8).

If you find the explanation of osmosis difficult or confusing, you need only remember this definition in order to understand the effects of osmosis in living organisms.

*Experiment 3. **Demonstration of osmosis** (Fig. 12.3)*

A length of dialysis tubing (cellophane) is filled with a concentrated solution of syrup or sugar and fitted over the end of a capillary tube with the aid of an elastic band. The dialysis tube is lowered into a beaker of water and the tube clamped vertically. In a few minutes, the level of liquid is seen to rise up the capillary tube and may continue to do so for a metre or more according to the length of the tube.

Interpretation. The most plausible interpretation is that water molecules have passed through the cellophane tubing into the sugar solution, increasing its volume and forcing it up the capillary tube. This movement should theoretically continue until the pressure of the syrup in the capillary is equal to the diffusion pressure of water entering the dialysis tube.

Osmotic potential. The result of Experiment 3 shows that pressure builds up in the syrup solution and forces liquid up the capillary. If the capillary were blocked, the pressure would burst the dialysis tubing. The pressure results from the rapid diffusion of water molecules from the dilute to the concentrated solution. The *osmotic potential* of a solution is a measure of the tendency for water molecules to diffuse out of it. A concentrated solution, having relatively few free water molecules, has a low osmotic potential, while a dilute solution with a larger proportion of free water molecules has a high osmotic potential. Pure water has the highest possible osmotic potential.

capillary tube

first level

elastic band

dialysis tube containing syrup

water

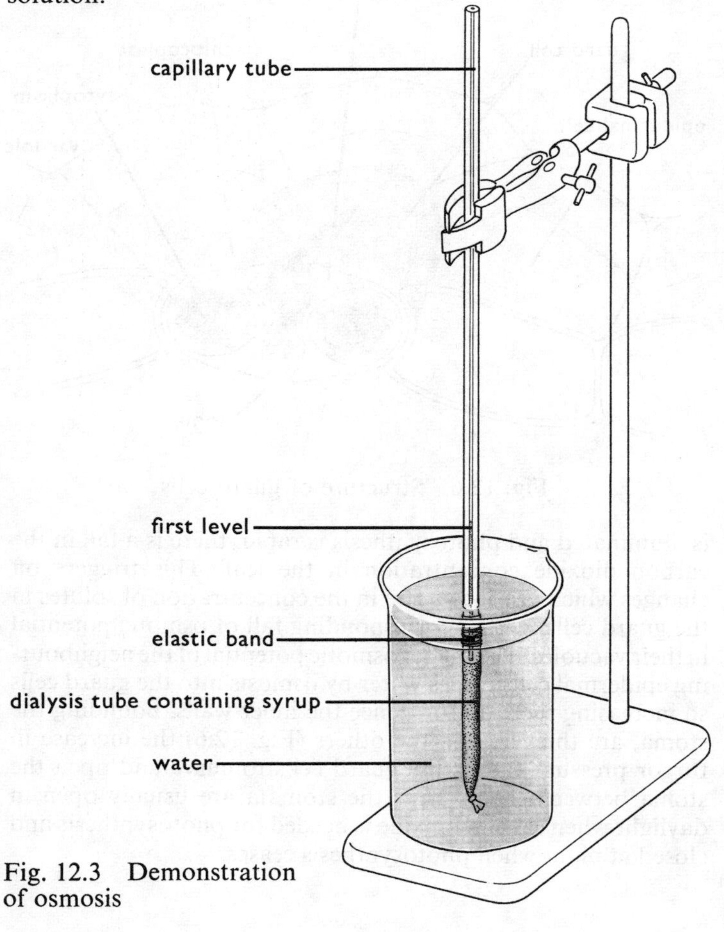

Fig. 12.3 Demonstration of osmosis

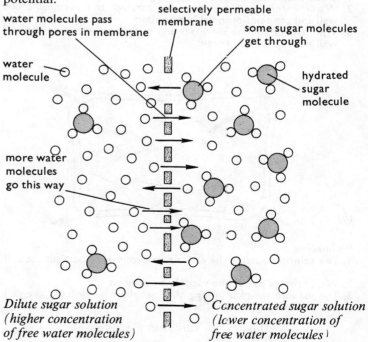

water molecules pass through pores in membrane

selectively permeable membrane

some sugar molecules get through

water molecule

hydrated sugar molecule

more water molecules go this way

Dilute sugar solution (higher concentration of free water molecules)

Concentrated sugar solution (lower concentration of free water molecules)

Fig. 12.4 The diffusion theory of osmosis

OSMOSIS IN PLANTS

The surfaces of plants and animals and the membranes in their cells frequently have selectively permeable properties. When these organisms or their individual cells are surrounded by fluids weaker or stronger than their own, osmotic forces are set up.

Turgor

The cellulose wall of plant cells is freely permeable to water and dissolved substances. The cytoplasm, however, behaves as a selectively permeable membrane, while the cell sap in the vacuole, since it contains salts and sugars, has an osmotic potential less than pure water. If an isolated plant cell is surrounded by water, osmosis would cause water to enter the cell sap. The vacuole would expand, pushing the cytoplasm against the cell wall (Fig. 12.5a). Eventually the outward pressure of the vacuole would be equalled by the resistance of the inelastic cell wall and the cell could take in no more water. Such a cell is *turgid* and the vacuole is exerting *turgor pressure*. The normal source of water for producing turgidity in cells is the xylem vessels or neighbouring cells.

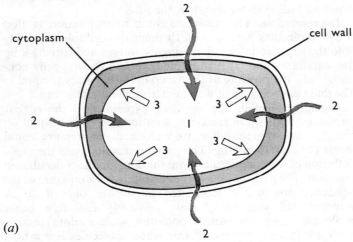

(a)

Turgor
1. Dissolved salts and sugars in cell sap give it a low osmotic potential
2. Water enters by osmosis passing through the permeable cell wall and the semi-permeable cytoplasm
3. The cell sap volume increases and pushes outwards on the cell wall making the cell turgid

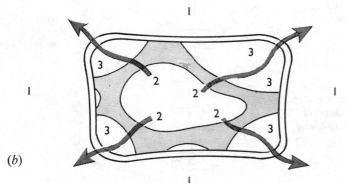

(b)

Plasmolysis
1. The solution outside the cell is more concentrated than the cell sap
2. Water passes out of the vacuole by osmosis
3. The vacuole shrinks, pulling the cytoplasm away from the cell wall and leaving the cell flaccid

Fig. 12.5   Turgor in plant cells

A plant structure made of turgid cells is resilient and strong; the plant stem stands upright and the leaves are held out firmly. Many young plants depend entirely upon this turgidity for their support but in older plants, woody and fibrous tissues take over this function.

Growth. In the growing regions of plants, the cell walls are still fairly plastic. After new cells have been produced at the growing point, vacuoles begin to form in the new cell's cytoplasm (*see* Fig. 3.5, p. 11). Water enters the cell by osmosis and the vacuoles join up and increase in volume. Since the cell wall is plastic, the cell is extended as the vacuole pushes outwards on the walls. Hundreds of cells extending in this way produces expansion or extension growth (*see* pp. 20 and 44).

Wilting. When plants are exposed to conditions in which they lose water to the atmosphere faster than it can be obtained from the soil, water is lost from the vacuoles. The turgor pressure of the vacuoles decreases and they no longer push out against the cell wall. The cell becomes limp or *flaccid* (like a deflated football). A plant structure made of such cells is weak and flabby, the stem droops and the leaves are limp; in other words, the plant is *wilting*.

Plasmolysis. If a plant cell is surrounded by a solution more concentrated than the cell sap, water passes out of the vacuole to the outside solution. Loss of water causes the vacuole to shrink and pull the cytoplasmic lining away from the cell wall (Fig. 12.5b). There is now no pressure outwards on the cell wall and the cell is flaccid. This condition, called *plasmolysis*, can be induced experimentally in living cells without necessarily harming them, but it is an extreme condition and rarely occurs in nature.

Stomata (*see* p. 19). Although the details of the stomatal mechanism are not fully worked out, it seems that when a leaf

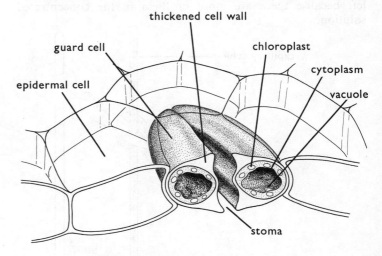

Fig. 12.6   Structure of guard cells

is illuminated and photosynthesis is rapid, there is a fall in the carbon dioxide concentration in the leaf. This triggers off changes which produce a rise in the concentration of solutes in the guard cells, with a corresponding fall of osmotic potential in their vacuoles. The higher osmotic potential of the neighbouring epidermal cells forces water by osmosis into the guard cells so increasing their turgor. Since the inner walls, bounding the stoma, are thicker than the others (Fig. 12.6) the increase in turgor pressure causes the guard cells to curve and open the stoma between them. Thus the stomata are usually open in daylight when carbon dioxide is needed for photosynthesis and closed at night when photosynthesis ceases.

Movement of water in a plant

Cells. Osmosis plays a part in the passage of water from one cell to another in a plant. Imagine two adjacent cells A and B (Fig. 12.7). A has a lower concentration of sugar in its cell sap and hence a higher osmotic potential than B. The more dilute cell sap in A will thus force water by osmosis into cell B. The water entering cell B will dilute its cell sap and raise both its osmotic potential and its turgor pressure so tending to force water out into the next cell in line. Thus water passes from cell to cell down an osmotic gradient.

However, if cell B is fully turgid it is unable to expand and so can take in no more water even though it has a lower osmotic potential than its neighbour A. In fact, if cell A is not fully turgid and still capable of expansion, the high wall pressure of cell B will force water out into cell A against the osmotic gradient. The movement of water between plant cells depends, therefore, not only on the osmotic potential of their vacuoles but also on how turgid they are.

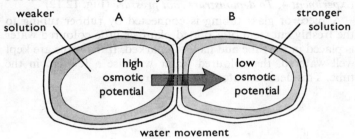

Fig. 12.7   Movement of water through cells

Roots. In this way, water may be absorbed from the soil by the roots, though this is not established fact, but rather one of the more widely held theories. Fig. 12.8 shows the general direction of water-flow, and Fig. 12.9 represents a few cells along the radius of the root.

The root hairs are thin-walled extensions from the cells of the outer layer of a root (*see* p. 20). They grow out, pushing between the soil particles to which they adhere closely. The film of water which surrounds the soil particles also surrounds the root hairs. Although soil water has mineral salts dissolved in it, they make only a very weak solution, and the cell sap of the root hair is more concentrated than this. Water passes from the soil through the cell wall and its thin cytoplasmic lining into the vacuole of the root hair. This extra water raises the turgor pressure of the vacuole and so forces water out into the cell walls towards the inside of the root. If the cell next to the root hair, on the inside, has a lower turgor pressure than the root hair, water may pass into it by osmosis. Thus, some of the water absorbed by the root hairs will pass from cell to cell across the root to the xylem vessels in the centre.

In fact, it is now thought that only a small amount of water passes through the cells in this way and that the bulk of it travels *along* or *between* the cell walls, from cell to cell, without entering the cytoplasm or vacuole of each cell (Fig. 12.9).

Once it reaches the xylem vessels, the water is drawn up the root to the stem by the "suction" effect of the transpiration stream, described in the next chapter.

The water passing along the cell walls will also carry dissolved salts from the soil. The process of osmosis by which root hairs take up water from the soil and by which some of the water passes through the root, cannot account for salt uptake by the cells since, by definition, osmosis is the movement of water. Salts may enter cells by diffusion or, more likely, by active transport.

In some plants at certain times of the year, it is possible to demonstrate a positive pressure, *root pressure*, in the xylem. For example, sap oozes from the stump of a tree or branch if it is cut down in the spring. If a glass tube containing a little

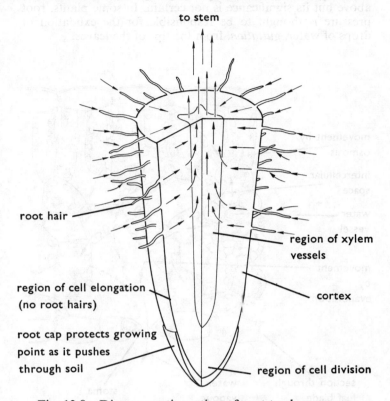

Fig. 12.8   Diagrammatic section of root to show passage of water from the soil

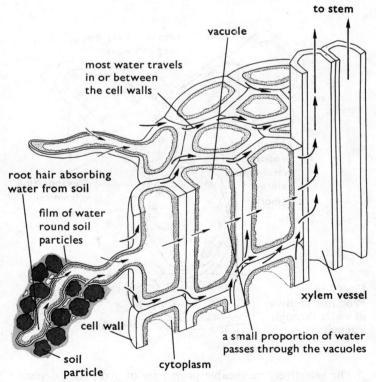

Fig. 12.9   Diagram to show the probable pathway of water from soil to xylem vessel in a root

water is fitted to the cut stem of a well-watered potted plant, as shown in Fig. 12.12, the water can be seen to rise several centimetres in the tube. There is some evidence that this root pressure derives from the sort of osmotic activity described above but its significance is not certain. In some plants, root pressure is thought to be responsible for the exudation of drops of water, *guttation*, from the tips of the leaves.

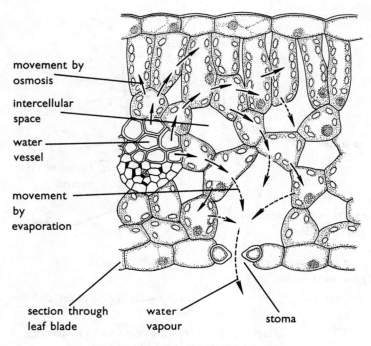

movement by osmosis

intercellular space

water vessel

movement by evaporation

section through leaf blade

water vapour

stoma

Fig. 12.10   Movement of water through a leaf

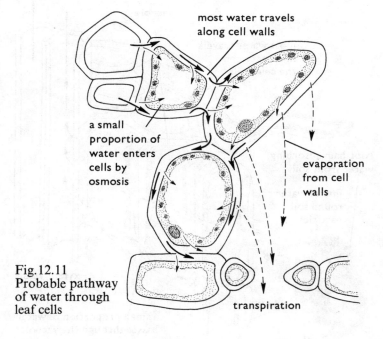

most water travels along cell walls

a small proportion of water enters cells by osmosis

evaporation from cell walls

Fig.12.11 Probable pathway of water through leaf cells

transpiration

The selectively permeable properties of cytoplasm depend on its being alive. Anything which kills the cytoplasm also destroys its selective permeability. At the same time, of course,

it will destroy all transport systems which depend on living processes such as active transport or movements of solutes in the phloem.

Leaves. Fig. 12.10 shows the water movement in a small part of a leaf blade. The mesophyll cells are losing water by evaporation through their cell walls. A small amount of water is also being used up by photosynthesis. This loss of water makes the cell sap more concentrated and its osmotic potential falls. Water may then enter from neighbouring cells with a higher osmotic potential and these cells absorb water from the nearest xylem vessels.

By far the greatest flow of water, however, is thought to take place along the cell walls without entering the cytoplasm or vacuole (Fig. 12.11). About 98 per cent of this water is lost by evaporation (*see* "Transpiration", p. 69), some 1·5 per cent is taken into the vacuoles to maintain the cells' turgor and only about 0·2 per cent is actually used for photosynthesis.

Experiment 4. ***To demonstrate root pressure*** (Fig. 12.12)

A piece of glass tubing is connected by rubber tubing to the freshly cut stem of a potted plant. A little coloured water is placed in the tube and its level marked. If the roots are kept well watered, the coloured water will rise a few cm in the tube. This demonstrates root pressure.

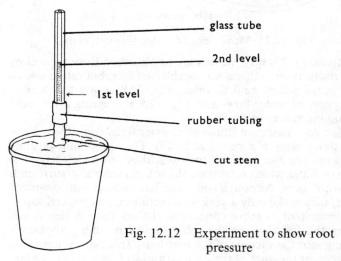

glass tube

2nd level

1st level

rubber tubing

cut stem

Fig. 12.12   Experiment to show root pressure

Experiment 5. ***Turgor in potato tissue***

Two cylinders of potato tissue are prepared as follows: a no. 4 or no. 5 cork borer is pushed into a large potato, the core of tissue extracted by pushing with the flat end of a pencil and the ends trimmed square with a scalpel or razor blade. Both cylinders should be the same length and as long as possible (50 mm or more). The length of the potato cylinders are measured and recorded. One cylinder is placed in a test-tube of water and the other in a test-tube of strong sugar solution. After 24 hours the cylinders are removed and measured again.

The cells from the potato were probably not fully turgid, so that the tissue which has been in water will have taken up water by osmosis, increased its length by 2 mm or more and will feel firm. The tissue in sugar solution will lose water by osmosis and shrink slightly; it will also feel limp and flabby.

OSMOSIS IN ANIMALS

The skins of many aquatic animals are more or less selectively permeable and this results in osmotic effects between them and their surroundings.

Fresh-water animals. The blood of fishes and amphibians and the body fluids of invertebrates that live in fresh water are more concentrated than the pond or stream water that surrounds them. Water tends to enter their bodies by osmosis, through their skins or such patches of skin as are exposed. If this water were not removed continually, the animal's blood would be diluted and the whole creature would swell and become water-logged. Vital processes in the body would be interfered with and the animal would die. In a great many fresh-water animals it has been possible to show that certain organs are able to eliminate the excess water, or *osmoregulate* (*see* p. 177) and keep the concentration of the body fluids constant. In frogs and fish the kidneys extract excess water from the blood as it passes through them, and it is passed out of the body as a dilute urine.

Impermeable coverings, like the cuticle on the exoskeleton of beetles and other aquatic insects and crayfish, greatly reduce the surface over which water might be absorbed. In some fresh-water animals it is not clear how osmosis affects them or how their osmo-regulation is carried out.

Salt-water animals. Sea water is a more concentrated solution than the blood of many marine fish, in consequence there is a tendency for water to pass out of their bodies and into the sea, by osmosis. Sea-water fish swallow water and in some way absorb it through their alimentary canals. There is also evidence that they can eliminate the excess salts taken in.

Animals in estuaries have to withstand extreme osmotic changes, being alternately covered with fresh and salt water.

Land animals. Osmotic effects occur within the bodies of all animals (*see* p. 98), but in general, land animals lose water from their body surfaces by evaporation and gain it from their food and drink. The concentration of the blood is nevertheless kept very constant by the regulatory action of the kidneys and other organs (*see* pp. 106 and 108). The impermeable cuticle of insects, the fur of mammals and the feathers of birds reduce water-loss by evaporation and contribute to the success of these groups on land.

PRACTICAL WORK

1. *Plasmolysis in cells.* Some of the red epidermis from a rhubarb petiole is mounted on a slide with some water. A few drops of salt solution are added. The red vacuoles will be seen to shrink away from the cell walls as the cells lose water by osmosis. If the solution is replaced with water, the cells will take up water once again and the vacuoles will swell to their normal size.

2. *To show osmotic pressure in cells.* A dandelion flower stalk is cut into six strips a few cm long. Since the epidermis tends to shrink, while the cortex tends to expand, the strips will curl outwards. One strip is placed in water, another in sugar solution and a third left as a control. The first strip will absorb water by osmosis. The cells of the cortex will expand while the inelastic epidermal cells will not, and the strip will curl up more. In the sugar solution, water diffuses from the cells, which lose their turgor; the unequal stresses in the strip disappear and it straightens.

By placing the straight strips in a range of solutions from 5 to 15 per cent in strength, an estimate can be made of the osmotic potential in the cell. The solution which only just produces some curling must approximate in strength to the osmotic potential of the cell sap.

[*Note:* more experiments relevant to this chapter are to be found in the laboratory manual *Diffusion and Osmosis—see* p. 1.]

QUESTIONS

1. A solution of salt is separated by a selectively permeable membrane from a solution of sugar of equal concentration (equimolecular).* Discuss whether osmosis would take place and justify your conclusions.
2. Pieces of fresh beetroot are washed and left in water. Similar pieces of cooked beetroot are washed and left in water. After an hour or two, the water round the cooked beetroot is coloured red while that round the fresh beetroot is not. Explain this result.
3. What features, for example, a fish gill, would help to maintain a steep diffusion gradient of oxygen between it and its immediate environment?
4. In an actively photosynthesizing cell, the sugars being formed are quickly converted to starch. What is the biological advantage of this in terms of osmosis?
5. What activities in man are likely to increase and decrease the osmotic potential of his blood and body fluids?
6. What osmotic problems confront the salmon, which is hatched in a river, grows to maturity in the sea and returns to the river to lay its eggs?

* The same number of gram molecules per litre.

[*Note.* Since *Introduction to Biology* was first published, the accepted terminology relating to osmosis in cells has changed several times so that the terms in previous impressions, though meaningful in the context of the book, are now at variance with more advanced usage. *Osmotic potential* is a measure of a solution's tendency to *give out* water and therefore a very *concentrated* solution has a *low* osmotic potential. The term, *water potential*, which has not been used in this impression, relates to a plant cell's tendency to *expel* water. The concentration of its cell sap and the inward pressure of its cell wall both contribute to its water potential. If a cell has a dilute cell sap and a high wall pressure it will tend to expel water to other cells and its water potential is therefore high. The use, in previous impressions, of "water potential" as a measure of a cell's tendency to *take in* water is not satisfactory.]

13 Translocation and Transpiration

TRANSLOCATION is the movement of dissolved substances through a plant. Transpiration is the evaporation of water from the leaves and the subsequent movement of water through the xylem.

Translocation

In very general terms, water and dissolved salts from the soil travel upwards through the xylem vessels while food made in the leaves passes downwards or upwards in the sieve tubes of the phloem. (*See* Figs. 3.8*c*, p. 13 and 4.3*b*, p. 15.)

The sucrose produced by photosynthesis in the leaves is carried in the phloem out of the leaf and into the stem. It may then travel up the stem to actively growing regions or maturing fruits and seeds or downwards to the roots and underground storage organs. It is quite possible for substances to be travelling both upwards and downwards at the same time in the phloem.

Mechanism of translocation in the phloem. The mechanism, in fact, is not known though it does depend on the fact that the sieve tubes, unlike the xylem vessels, contain living cells. Anything which kills the phloem cells, severely interferes with the movement of food. This is illustrated by the experiment in Fig. 13.1. The leaf makes sucrose from the radio-active carbon dioxide (*see* p. 53). When the phloem of the stem below the leaf is killed by a jet of steam the substances containing radio-active carbon are found to move up the stem. When the phloem above the leaf is killed, conduction is down the stem. If the phloem above and below the leaf is killed, the radio-active substances do not appear anywhere in the stem. Similarly if the oxygen supply to the phloem is cut off, translocation of sugars ceases.

Xylem vessels consist of dead cells which are unaffected by heat or oxygen shortage. If transport of sucrose took place in the vessels, heat treatment of the stem would not be expected to interfere with the movement of sugar.

One of the most widely held theories to account for the movement of solutes in the phloem is the *mass flow hypothesis* depicted diagrammatically in Fig. 13.2. The "root cell" could,

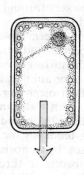

(1) LEAF CELL

(a) Accumulation of sugar
(b) Fall in osmotic potential of cell sap
(c) Intake of water by osmosis
(d) High turgor pressure

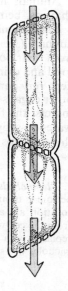

(3) SIEVE TUBES

Liquid forced from region of high turgor pressure to region of low turgor pressure

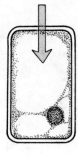

(2) ROOT CELL

(a) Sugar used up in respiration or stored as starch
(b) Rise in osmotic potential of cell sap
(c) Low turgor pressure

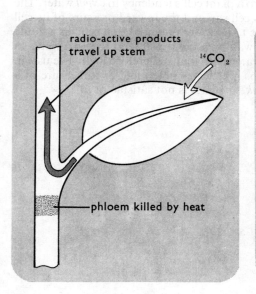

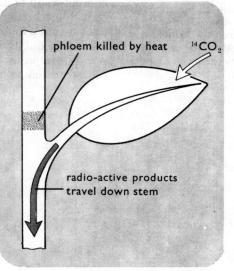

Fig. 13.1 Experimental evidence for transport in phloem

(After Rabideau & Burr, Amer. J. Bot. **32** 1945)

Fig. 13.2

in practice, be any cell in which sucrose is being used up in respiration or converted to starch and removed from solution.

There are several objections to the mass flow hypothesis. For example, how can it account for the simultaneous movement of substances both up and down the phloem and why must the sieve tube cells be alive for the process to go on?

Movement of salts in the xylem. It can be shown that if a ring of bark including phloem (*see* Fig. 4.7, p. 17) is removed from a stem, the upward movement of salts is little affected. If a core of xylem is removed, however, the upward movement of salts is arrested. Similarly, killing the phloem by heat treatment does not significantly affect the upward movement of salts. More detailed information is provided by the experiment shown in Fig. 13.3.

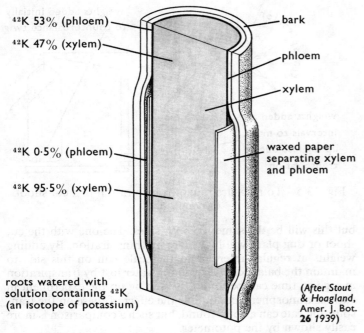

^{42}K 53% (phloem) — bark

^{42}K 47% (xylem) — phloem

— xylem

^{42}K 0.5% (phloem) —

^{42}K 95.5% (xylem) — waxed paper separating xylem and phloem

roots watered with solution containing ^{42}K (an isotope of potassium)

(*After Stout & Hoagland,* Amer. J. Bot. **26** 1939)

Fig. 13.3 Evidence for transport of salts by xylem

The bark and phloem are carefully raised from a willow twig, separated from the xylem by a layer of waxed paper and carefully replaced. When radio-active potassium (^{42}K) is supplied to the roots of the plant the salt is subsequently found in the xylem and hardly at all in the phloem. Controls are set up to ensure that the separation of xylem and phloem does not have a harmful effect in the shoot.

When the contents of the xylem and phloem are analysed above the region of separation by waxed paper, it is found that the radio-active potassium is almost equally abundant in both. This evidence suggests that substances can easily move sideways across the stem from the xylem to the phloem.

The forces moving the salts through the xylem in the transpiration stream are described below under "Transpiration".

Uptake of salts by the roots. There is, as yet, no wholly convincing explanation of the uptake of mineral salts from the soil by roots. It may be that diffusion from a relatively high concentration in the soil to a lower concentration in the root cells accounts for some uptake of salts, but it has been shown (*a*) that salts can be taken from the soil even when their concentration is below that in the roots and (*b*) that anything which interferes with respiration impairs the uptake of salts. It looks, therefore, as if "active transport" plays an important part in the uptake of salts.

"Active transport" is itself only a hypothetical process. By expenditure of energy in respiration it is thought that enzyme-like substances, *carriers*, might combine with the salts, carry them across the cytoplasm and release them into the vacuole.

Transpiration

Transpiration is the process by which plants lose water as water vapour into the atmosphere. Most of this loss takes place through the leaves but evaporation also occurs from the stem and flowers.

Turgor pressure in the mesophyll cells (*see* p. 19) forces water outwards through the cell walls. From the outer surface of the cell walls, the water evaporates into the intercellular spaces and diffuses out of the stomata into the atmosphere. Closure of the stomata greatly reduces, but does not entirely prevent, evaporation from the leaf.

Significance of transpiration. Transpiration is probably an inevitable consequence of photosynthesis. For adequate photosynthesis to take place, a large surface area must be exposed to the atmosphere to absorb sunlight and carbon dioxide. A leaf which is permeable to carbon dioxide will also be permeable to water vapour. It seems, therefore, that evaporation of water must inevitably accompany photosynthesis. Nevertheless, transpiration produces effects which may be regarded as beneficial to the plant.

(*a*) *Transpiration stream.* Evaporation of water from the leaf cells causes their turgor to fall and the concentration of their cell sap to rise and consequently produces an increase in water potential (*see* p. 65). Cells in this condition will absorb water from their neighbours and eventually from the xylem vessels in the leaf. Withdrawal of water by osmosis from the xylem vessels produces a tension, i.e. the water is submitted to pressures below atmospheric. This tension draws water up the vessels of the stem from the roots. This flow of water is called the transpiration stream and is dependent on the rate of evaporation from the leaves (Fig. 13.4).

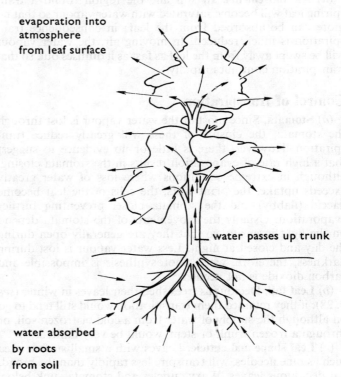

evaporation into atmosphere from leaf surface

water passes up trunk

water absorbed by roots from soil

Fig. 13.4 The transpiration stream

It is easy to envisage a wire or a string being subjected to tension along its length without breaking but one would expect a column of water under tension to break up, leaving gaps filled with water vapour. The *cohesion theory* outlined above, however, supposes that the cohesive forces between water molecules in very thin columns of water are not so easily overcome. This theory, then, offers an explanation of the movement of water up the stems of plants, including trees nearly 100 metres high.

A tree on a hot day may evaporate hundreds of litres of water from its leaves. Of all the water passing through the plant, only a tiny fraction is retained for photosynthesis and to maintain the turgor of the cells.

(b) *Transport of salts.* The transpiration stream undoubtedly carries mineral salts from the roots to the leaves but the rate of uptake of salts from the soil is not directly dependent on the rate of transpiration.

(c) *Cooling.* The rapid evaporation of water from the leaf surface and the consequent absorption of latent heat from the leaf tissues is almost certainly of value in keeping the temperature of a leaf below harmful levels in the direct rays of the sun.

Conditions affecting transpiration rate

(a) **Light intensity.** When light intensity increases, the stomata open and allow more rapid evaporation.

(b) **Humidity.** When the atmosphere is saturated with water vapour, little more can be absorbed from the plants, and transpiration will be reduced. In a dry atmosphere, transpiration will be rapid.

(c) **Temperature.** A high temperature increases the capacity of the air for water vapour; hence transpiration increases. When the leaf itself becomes warm, evaporation from it occurs more rapidly. Direct sunlight without a warm atmosphere will have this effect, since the leaf absorbs radiant energy and its temperature rises.

(d) **Air movements.** In still air, the region round a transpiring leaf will become saturated with water vapour so that no more can be absorbed from the leaf; in consequence transpiration is much reduced. In moving air, the water vapour will be swept away from the leaf as fast as it diffuses out, so that transpiration continues rapidly.

Control of transpiration

(a) **Stomata.** Since most of the water vapour is lost through the stomata, the closure of these will greatly reduce transpiration. However, there is little or no evidence to suggest that a high rate of evaporation results in the stomata closing, although in extreme conditions where loss of water greatly exceeds uptake, the plant wilts, the cells of the leaf become flaccid (flabby) and the stomata close, preventing further evaporation. Usually the movements of the stomata depend on the light intensity, so that they are generally open during the day and closed at night. Less water vapour is lost during darkness, therefore, when photosynthesis is impossible and carbon dioxide is not needed.

(b) **Leaf fall.** Deciduous trees shed their leaves in winter (*see* p. 28): if they retained them transpiration would still tend to go on although the supply of water from a cold or frozen soil, or through a frozen trunk or stem, would be very limited.

(c) **Leaf shape and cuticle.** Leaves with a small surface area such as pine needles, will transpire less rapidly than the broad, flat deciduous leaves. Waxy cuticles and stomata sunk below the epidermis level are also modifications thought to be associated with reduced transpiration. They are often found in plants which grow in dry or cold conditions or in situations where water is difficult to obtain. Most evergreen plants have one or more of these leaf characteristics and this probably plays a part in their retention of leaves during the winter months in temperate climates.

EXPERIMENTAL WORK

Experiment 1. *To measure rate of transpiration by loss of weight*

(a) **Cut shoot or uprooted small plant** (Fig. 13.5). The only direct method of measuring transpiration is to determine what weight of water the plant loses in a given time.

Both test-tubes will lose water vapour from their open ends

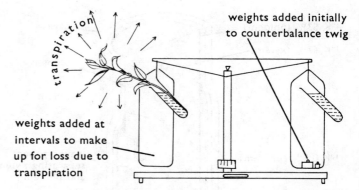

Fig. 13.5   To measure rate of transpiration in a cut shoot or small plant

but this will be the same for both sides. The one with the cut shoot or dug plant will lose more by transpiration. By adding weights at regular intervals to the scale pan on this side to maintain the balance, the weight of water lost by transpiration in a given time can be found. By setting up the experiment in various atmospheric conditions the effect of these on the transpiration rate can also be found, but such a comparison is more easily shown by the potometer.

(b) **Potted plant.** After the soil has been watered, the plant and pot are weighed at intervals to find the loss of weight due to transpiration. To prevent evaporation directly from the soil or pot, some impermeable material, polythene or rubber sheet, is wrapped round the pot and tied firmly round the plant's stem. In both experiments, it is assumed that increases of weight resulting from photosynthesis or decreases from respiration are small compared with losses from transpiration.

Experiment 2. **To show uptake of water by a cut leafy shoot** (Fig. 13.6)

This is done by the potometer. The type of potometer shown has the advantage of using only very small volumes of water, so that changes in temperature are likely to produce in the water only negligible contractions or expansions which otherwise would confuse or falsify the results.

To set up the potometer a shoot is cut from a tree and at once placed in water* to prevent air being taken into the water vessels of the stem. The potometer is filled with water, and the cut end of the shoot fitted into the rubber tubing, care being taken to avoid the inclusion of any air bubbles.

As water evaporates from the leaves, more is drawn from the stem which, in turn, draws it from the potometer tubes. The

* This precaution may not be necessary but is fairly easy to carry out. (*See The School Science Review*, 1957, 135, **38**, 269.)

tap below the funnel is closed so that this water will be with-drawn from the capillary tube. Here, the meniscus at the air/water boundary can be seen moving quite rapidly as water is withdrawn and air is drawn in behind the retreating water column. By timing this water-column movement over a fixed distance on the scale the rate of water uptake can be deter-mined. When the water column reaches the end of the capillary it can be sent back by momentarily opening the tap below the funnel. Any air that is drawn into the apparatus by allowing it to work too long will collect in the stem of the funnel and will not reach the cut shoot.

It must be emphasized that the potometer does not measure water lost by transpiration but only water taken up as result of it. Not all the water taken up will escape into the atmosphere. Some of it will be combined with carbon dioxide to make starch during photosynthesis. Since, in conditions of constant light intensity, this will be a fixed weight, the potometer can be used most effectively for comparing rates of transpiration of the same shoot in different situations.

Experiment 3. *Comparing rates of transpiration*

In various atmospheric conditions increase or decrease in the rate of water uptake will relate directly to changes in the rate of evaporation. By placing the potometer in different situations, e.g. in a warm room with still air, in front of a fan, in a cold room, etc., the difference in rate of water uptake can be measured, and this will indicate the difference in transpira-tion rate. The intensity of the light should not vary greatly.

Before readings are taken in any situation, the apparatus should be left for several minutes in the new conditions to allow the new rate of transpiration to become established.

Further experiments can be carried out to find more about transpiration. Different shoots of approximately the same leaf area can be compared, say laurel and beech. Leaves and flowers can be removed one by one, or their surfaces smeared with petroleum jelly.

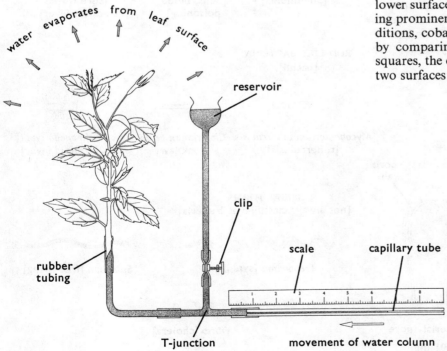

Fig. 13.6   The potometer

Experiment 4. *To show that water vapour is given off during transpiration* (Fig. 13.7)

The shoot of a recently watered potted plant, or a plant in the garden, is completely enclosed in a transparent, polythene bag which is tied round the base of the stem. The plant is allowed to remain for an hour or two in direct sunlight. The water vapour transpired by the plant will soon saturate the atmosphere inside the bag and drops of water will condense on the inside. The bag is removed and all the condensed water shaken into a corner so that it can be tested with anhydrous copper sulphate. A control experiment is set up using a shoot in a similar situation but from which all the leaves and flowers have been detached.

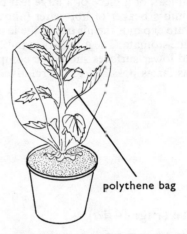

Fig. 13.7   To show that water is given off during transpiration

Experiment 5. *To find which surface of a leaf loses more water vapour* (Fig. 13.8)

Small squares of cobalt chloride paper taken straight from a desiccator are stuck as quickly as possible to the upper and lower surfaces of a leaf blade by means of "Sellotape", avoid-ing prominent veins to ensure an air-tight seal.* In damp con-ditions, cobalt chloride paper changes from blue to pink, and by comparing the time needed for such a change in both squares, the difference in the loss of water vapour between the two surfaces can be estimated.

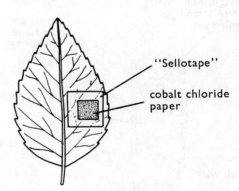

Fig. 13.8   To find which surface of a leaf loses more water vapour

* See J. F. Eggleston, *The School Science Review*, 1962 151, **43**, 723.

Experiment 6. *Transport in the vascular bundles*

Several small leafy plants are picked and the cut ends of their stems placed in a coloured solution such as neutral red or methylene blue. One of the plants is removed after five minutes, one after ten minutes and so on. By slicing across the base of the stem with a razor blade the water vessels will be seen to have been stained by the coloured solution. By cutting sections higher and higher up the stem until the colour can no longer be seen in the vessels, the height to which the solution has risen in the stem can be determined.

Experiment 7. *To investigate the distribution of stomata in a leaf*

A small leaf, or a piece of a large leaf is held in forceps and plunged into a beaker of hot water (about 80° C). The rise in temperature expands the air inside the leaf and this will escape through the stomata. The numbers of bubbles appearing on the upper and lower surfaces should be compared and the results related, as far as possible, to the relative numbers of stomata

seen under the microscope in comparable pieces of epidermis stripped from the upper and lower surfaces.

A control should be devised to try and eliminate the possible claim that the air bubbles come not from inside the leaf but from the air dissolved in the water merely collecting on the leaf surface.

QUESTIONS

1. Why should cutting a deep ring of bark from a tree cause its death in view of the facts that (*a*) water and salts can travel up to the leaves in the xylem and (*b*) the leaves can still manufacture food by photosynthesis?
2. Explain how it could come about that a plant could (*a*) take in more water than it is losing by transpiration and (*b*) lose more water by transpiration than it is taking in from the roots, at least for a time.
3. Outline the path taken by water from the soil through the roots, stem and leaves of a plant and into the atmosphere as vapour. Explain briefly the forces causing its movement at each stage.
4. Outline the path taken and changes undergone by a nitrogen atom in a solution of nitrate in soil-water as it passes into a plant, is built into an amino acid in a leaf and is finally stored in the cotyledon of a developing seed.

14 Bacteria

Structure (Fig. 14.1*a*)

Bacteria are very small organisms consisting of single cells, rarely more than 0·01 mm in length and visible only under the higher powers of the microscope. Unlike plant cells they do not have cellulose walls or chloroplasts. Their cell walls are of proteinaceous and fatty substances, and the single DNA strand (p. 186) is not enclosed in a nuclear membrane. Granules of glycogen, fat and other food reserves may be present. Bacteria reproduce rapidly by cell division, as often as once in 20 minutes, and may form chains of individuals, clumps or films over the surface of static water. The individuals may be spherical, rod-shaped, or spiral (Fig. 14.2) and some have filaments protruding from them called *flagella*, the

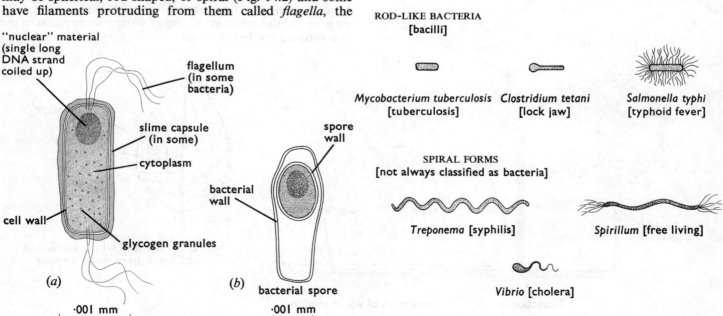

Fig. 14.1   Generalized diagram of a bacterium

Fig. 14.2   Bacterial forms

lashing action of which moves the bacterium about. Some bacteria need oxygen to respire while others can respire anaerobically, that is, obtain their energy by breaking down compounds without using oxygen (p. 49).

Spores (Fig. 14.1*b*). Some bacteria can form spores, a resting stage in which the protoplasm is concentrated in one part of the cell and surrounded by a thick wall. Bacterial spores are very widespread, being easily distributed by air currents, and occur on the surface of most objects, in soil, in dust and in the air. These spores are very resistant to extremes of temperature, and some can withstand the temperature of boiling water for long periods. Most normal bacteria can, however, be killed by temperatures above 50°C although they can live at very low temperatures. In favourable conditions, the spores break open and the bacteria reproduce and grow in the normal way.

The part played in nature by bacteria

Most bacteria live freely in the soil, in water and in decaying organic matter where they obtain their food saprophytically (p. 50), bringing about the decay of dead material. Enzymes made in the bacterial cells are secreted into the food, dissolving it to form soluble organic compounds such as amino acids which can be absorbed through the cell wall. Some bacteria obtain their energy by converting these organic compounds to inorganic substances, such as ammonia, which can then be absorbed by different bacteria and oxidized further to nitrates, releasing energy for their metabolism.

Thus the dead remains of plants and animals are converted once more to simpler substances which can be taken in by plants and used to synthesize food. In this way the resources of the Earth are re-cycled and made available for re-use by living organisms. (*See also* Nitrogen Cycle, p. 59 and Carbon Cycle, p. 58.)

The activities of decay bacteria are harnessed in the sewage works. Anaerobic bacteria act on the organic solids in the settling tanks and release soluble organic compounds from them. By spraying this liquid from the tanks over filter beds or by agitating it with paddles, it is oxygenated sufficiently for aerobic bacteria to flourish and convert the soluble organic compounds to inorganic salts such as nitrates and phosphates which are released with the effluent into rivers. Any industrial wastes which harm the bacteria in the sewage plant severely interfere with the effective treatment of sewage. Similarly, substances which cannot be used as a source of food by the bacteria will pass unchanged through the sewage works and contaminate the effluent. Hence the urgent need to find, for example, *bio-degradable* detergents which can be metabolized by the bacteria.

Even the "harmless" mineral salts in the sewage effluent can create hazards (*see* Eutrophication, p. 61).

Some types of bacteria live in the alimentary canals of animals. A few of these cause diseases such as cholera and typhoid but the majority are harmless or even beneficial. For example, in the intestine of man there are bacteria which appear to play a part in the synthesis of vitamin K and the vitamins of the B$_2$ complex (p. 84). In the alimentary canals of many herbivorous animals there is a dense population of bacteria which play a part in digestion. Few animals can produce an enzyme which digests cellulose, yet many herbivores feed exclusively on vegetation, the most nourishing part of which is enclosed in cellulose cell walls. These walls are broken open partly by efficient chewing and partly by the enzymes of the bacteria and single-celled animals in the gut; in the *caecum* and *appendix* of the rabbit, for example, and the *rumen* (or paunch) of cows and sheep.

Certain bacteria are able to produce food by photosynthesis and some of these can use H$_2$S (hydrogen sulphide) instead of H$_2$O as a source of hydrogen atoms.

Parasitic bacteria live in or on living animals or plants and are often harmful to the organism.

Harmful bacteria. Some bacteria which live parasitically in plants and animals do a great deal of harm on account of the damage to tissues and highly poisonous proteins, *toxins*, that they produce, e.g. diphtheria and tetanus. For example, five millionths of a gram of dried *tetanus* toxin will kill a mouse and 0.00023 g is fatal for a man. Certain bacteria cause diseases such as typhoid, cholera, plague, and tuberculosis. The diseases are "caught" by eating food on which are living bacteria or their spores; by "droplet infection" (inhaling bacteria in droplets of moisture in the air breathed, coughed or sneezed out by infected persons), particularly in a crowded, humid place such as a cinema; by touching the skin of an affected individual, or by the bacterial invasion of wounds. Since bacteria can reproduce so rapidly a single bacterium may give rise to a million offspring in a few hours.

Resistance to disease

The invasion of harmful bacteria is counteracted by the blood of healthy mammals in two ways.
(*a*) The skin forms a defence against the invasion of the blood and body cavities by bacteria. This is effected partly by the physical barrier of the dead, cornified, outer layer of the skin and partly by production of chemicals which destroy bacteria (*see* Vitamins, p. 84). The eyes, for example, are protected by an enzyme present in normal tears. The lining of the alimentary canal and respiratory passages also resist the entry of bacteria.
(*b*) The white cells, *leucocytes*, engulf the bacteria and secrete chemicals that kill them (p. 93). Other chemicals, *antitoxins*, made in the blood, effectively neutralize the poisonous proteins that the bacteria give out. Once an animal has recovered from a bacterial disease, its blood is much better able to combat the bacteria and neutralize its toxins; it is said to have acquired *immunity* (p. 94).

Prevention of infection

(*a*) **Food.** Most uncooked food contains bacteria, but not usually in sufficient numbers to cause disease, and the lining of the gut and the hydrochloric acid in the stomach will often prevent their development. The techniques of food preservation, involving the destruction of bacteria whose activities would cause the food to decay, also serve to destroy disease bacteria, e.g. typhoid, which could be transmitted by food. *Refrigeration* does not kill bacteria but prevents their multiplication. The high temperatures of *cooking* kill most bacteria and their spores, if continued long enough. *Bottling* and *canning* kill bacteria by the high temperature used and then prevent their access to the food by sealing it in a vacuum. Poisons such as sulphur dioxide may be added to canned or bottled fruit in a concentration sufficient to kill bacteria but too weak to harm the consumer.

Today, experiments are being conducted on food exposed to radiation from radio-active cobalt. This seems to destroy bacteria effectively without harming the food. Drying, curing, smoking and salting are all methods of food preservation.

In the case of salting and preservation with sugar, the preservation depends on the low osmotic potential of the food which plasmolyses (p. 64) any bacteria which arrive in it.

Food should be eaten soon after it has been cooked or removed from a container and so give invading bacteria, or any not destroyed during cooking, little opportunity to multiply. Flies, notorious carriers of bacteria, should never be allowed to settle on food, and people who handle food that is to be eaten without subsequent cooking should take great care that their hands do not carry harmful bacteria.

(*b*) **Water supplies.** Water for human consumption is filtered through beds of sand in which films of protozoa form and remove bacteria. Any bacteria which escape the filters are destroyed by chlorinating the water.

(*c*) **Sewage disposal.** Faeces and urine from people whose intestines or kidneys contain disease bacteria are highly infectious on account of the bacteria and spores that they contain. They are also breeding grounds for the small numbers of harmful bacteria that may be present even in the excreta from healthy people or which reach them later. Sewage must therefore be disposed of in such a way that the bacteria present cannot be transmitted by flies or other means to food or people. Modern sewage disposal methods do this effectively. Less modern methods must employ strong disinfectants or fairly deep burial.

(*d*) **Antiseptics and antibiotics.** The skin is a barrier to invasion by bacteria, but when it is broken by a cut, graze or burn, special precautions must be taken to remove some of the bacteria by washing and inactivate the remainder by applying mild antiseptic such as proflavine. Further bacterial invasion is prevented by bandages or surgical tape.

Antibiotics such as penicillin are chemicals, extracted from bacteria or fungi, which destroy many forms of harmful bacteria. They can be used internally against infections of the tissues where they kill the bacteria without harming the cells of the body. In such a way, an antibiotic like penicillin can be used to control an infection of the middle ear, either by the patient's eating the antibiotic or by its injection into his circulatory system.

(*e*) **Cleanliness.** Bacteria and bacterial spores are present all over the surface of the skin. Regular washing and bathing remove many of these and so reduce the chances of infection should the skin be damaged. The hands, particularly, should be kept clean since they handle food and manipulate articles that will be used or have been used by other people. Doctors and nurses who move from sick to healthy patients have to be scrupulously clean, and their instruments repeatedly sterilized, to avoid transmitting infectious bacteria.

(*f*) **Sterilization.** The removal of bacteria from an object by killing them, using high temperatures such as steam under pressure in an *autoclave*, or by chemicals, is termed sterilization. Many chemicals, such as strong acids, will kill bacteria, but where bacteria are to be destroyed in food, in wounds or in the blood stream, chemicals must be used of such a nature or in such concentrations that they are lethal to bacteria but harmless to man. No such "perfect" chemical has yet been found but there are many that approximate to these conditions, at least for certain groups of bacteria.

(*g*) **Isolation.** People carrying a disease which can be transmitted by touch or droplet infection should not mix with others if the disease is serious, and as little as possible if the disease is mild. In particular they should not mingle with crowds where they could pass on the bacteria to a great many people. Quarantine regulations are designed to prevent the spread of diseases by infected persons travelling from one country to another.

Wearing gauze masks across the nose and mouth, or merely placing a handkerchief across when coughing or sneezing, helps to reduce the chances of inhaling or exhaling bacteria.

(*h*) **Immunization.** By deliberately infecting a person with a mild form of disease, antibodies are formed in his blood and these will protect him against later infections (p. 94).

Bacteria culture

In order to identify and investigate the bacteria it is necessary to culture them. This is done by mixing fruit juice, meat extract or other nutrients with agar jelly to form a medium in or on which the bacteria can grow and reproduce. The bacteria multiply and form visible colonies whose colour and chemical reactions make them identifiable. By including or excluding certain substances from the culture medium, the essential conditions for the growth and reproduction of a particular bacterium can be found, and the effects of chemicals and antiseptics in various concentrations can be investigated.

Viruses

Viruses are disease-producing agents far smaller than bacteria; they seem to consist of nuclear chemicals enclosed in a protein coat. They can pass through filters which trap bacteria and some can be crystallized. They can live and reproduce only in living animals or plants and cannot be cultured except in living tissues. In the right environment they can multiply rapidly, and in man cause diseases such as poliomyelitis, influenza, smallpox and the common cold. They can be transmitted in ways similar to those described for bacteria.

PRACTICAL WORK

Bacteria can be obtained by allowing cooked potato or other vegetables to rot in water. After a few days, drops of the water examined under the high power of the microscope should show bacteria. Gentian violet will stain bacteria.

Cultures. The apparatus must be sterilized by superheated steam in an autoclave or pressure cooker for 15 minutes at 1 kg/cm^2, so that unwanted bacteria on the glassware are destroyed. $1 \cdot 5$ g agar is stirred into 100 cm^3 of hot distilled water and 1 g beef extract, $0 \cdot 2$ g yeast extract, 1 g peptone and $0 \cdot 5$ g sodium chloride added. The mixture is sterilized in an autoclave and poured into sterilized Petri dishes which are covered at once and allowed to cool.

Each dish is then exposed to bacteria in one of various ways; for example, *A* is left open outside for 5 minutes, *B* is placed in a classroom for the same length of time, *C* is coughed into, in *D* a fly is allowed to walk once over the medium, *E* is left open in a dust-bin for 5 minutes. One, the control, is left covered to prove that sterilization has been effective. After they have all been covered they are incubated at $36°C$ for two days, or simply left in a warm place, when colonies of bacteria and fungi will appear. No growth should appear on the control.

Before washing up glassware in which bacteria have been growing, it should be sterilized in an autoclave.

QUESTIONS

1. How does a knowledge of the methods of transmission of bacteria lead to the variety of aseptic precautions taken in a modern operating theatre?
2. Substances such as wood, paper and cloth will rot away in time. Materials like polystyrene and polythene will not. Why should there be this difference and why is it a cause for concern in industrialized countries?

3. Exposure to mild infectious diseases such as measles is considered to be a normal or even desirable hazard of communal living. Cases of serious infectious diseases such as typhoid are kept in isolation so that they cannot infect other people and yet an injection of dead typhoid bacteria can produce resistance to the disease.

Discuss the biological principles underlying these social attitudes and activities.

4. To sterilize substances which would be decomposed by the temperatures in an autoclave, the substances are heated to 100°C for one hour on three successive days and incubated at 25–30°C between each steaming.

What is the biological basis for this method of sterilization?
5. What kind of health checks are desirable on people who are preparing and serving food to a large number of others, or working in food processing factories?

15 Fungi

Structure. The fungi are sometimes included in the plant kingdom but in many ways they are quite different from green plants. The basic unit of fungus is a *hypha* (Figs. 15.1 and 15.2), not a cell, although in some species the hollow tubes of the hyphae may be divided by cross-walls. Some fungal hyphae have walls containing cellulose but in most, the wall consists mainly of an organic nitrogenous compound, *chitin*. The composition may vary with age and environmental conditions. Cytoplasm fills the tips of the growing hyphae but in older regions there may be a central vacuole. Many very small nuclei are present in the cytoplasm.

There are no chloroplasts or chlorophyll in the fungi and the food particles in the cytoplasm may be oil droplets or glycogen, but not starch. The hyphal threads spread out over and into the food material making a visible mesh or *mycelium* (Fig. 15.2). In some fungi they are massed together to make the familiar "fruiting bodies" of mushrooms and toadstools, though the organization and division of labour in the hyphae is never so complex as in the flowering plants.

Feeding. The absence of chlorophyll means that fungi cannot synthesize food from simple substances such as carbon dioxide and water, as is done in photosynthesis, but must take in organic matter derived from other living organisms. Fungi are either saprophytic or parasitic. The parasitic ones live on or in the tissues of another living organism, the host, absorbing nourishment from its body. Some of the most devastating diseases of crops are due to parasitic fungi, e.g. potato blight.

The saprophytes derive their food from dead and decaying materials. Examples are the moulds which develop on stale, damp food and the many fungi which live in the soil and feed on the humus there. Both parasitic and saprophytic fungi can

produce enzymes at the growing tips of the hyphae. These enzymes enable the parasitic hyphae to penetrate the cell walls of the host and, in both parasites and saprophytes, to break down the organic matter externally by digestion. The simpler, soluble substances so formed can be absorbed into the protoplasm through the cell walls.

Reproduction is typically of two kinds, asexual and sexual. In asexual reproduction, a great many tiny spores are produced and scattered into the air. If they land in a suitable situation they grow out into new hyphae and mycelia. Sexual reproduction involves the fusion of nuclei in special hyphal branches. The product sometimes represents a resting rather than a dispersive stage.

Mucor

Mucor is the name given to a genus of moulds which grow on the surface of decaying fruit, bread, horse manure and other organic matter. The fungus grows rapidly and in a few days covers the surface of its food with a dense, white or grey mass of hyphae. The hyphae have no cross-walls but branch repeatedly and give rise to the mycelium.

Reproduction. Asexual reproduction takes place rapidly after the establishment of a mycelium. Vertical hyphae, a little thicker than the rest, grow up and become swollen at their ends. These swellings contain dense protoplasm and many nuclei (Fig. 15.2). The swelling becomes the *sporangium* and the

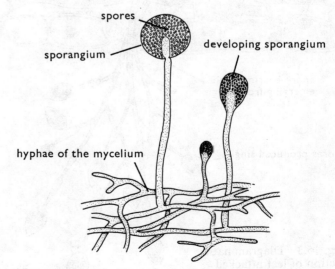

Fig. 15.2   Mucor, asexual reproduction

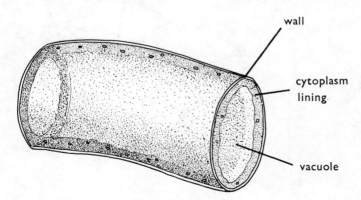

Fig. 15.1   Part of fungal hypha

protoplasm inside breaks up into elliptical *spores*, each with several nuclei and forming its own wall. In some species of Mucor and Rhizopus (a related genus) the sporangium wall breaks away and the powdery spore mass is dispersed by air currents. In other species of Mucor, the sporangium wall lique-fies, the method of spore dispersal is uncertain, though they may be carried by insects. The wall of the spore breaks open when a suitable situation is reached, and the protoplasm inside grows out into a new hypha and eventually a new mycelium. The germination of spores is greatly favoured by warm, damp conditions. The spores are said to be very resistant to adverse conditions and can remain dormant for years.

Antibiotics

Certain moulds, notably species of *Penicillium*, have become of economic importance owing to the anti-bacterial substances, *antibiotics*, they produce. These moulds are grown on nutrient broths, and the antibiotic chemicals such as penicillin are extracted from the fluid and purified.

Many antibiotics, e.g. *streptomycin*, come from soil-dwelling micro-organisms, *actinomycetes*, which exhibit characteristics of both fungi and bacteria. It may be that by producing anti-biotic chemicals in their natural environment these actino-mycetes restrict the growth of bacterial colonies in the soil and so reduce the competition for food. There is little direct evidence, however, to support this assumption.

Penicillin seems to attack the bacterial cell wall but the mode of action of antibiotics in general against bacteria has not been elucidated.

Parasitic fungi

Parasitic fungi are the principal disease-causing organisms in plants and fungal attacks can result in devastating agri-cultural losses.

The hyphae of the parasitic fungus *Phytophthora infestans*, which causes the disease potato blight, spread internally through the leaves. Short branches from the hyphae penetrate the cell walls, presumably with the aid of enzymes, and absorb nutriment from the cell contents (Fig. 15.3). The cells are eventually killed and then the leaves and finally the whole shoot dies. Before this happens, branching hyphae grow out of the stomata and the tips of the hyphae constrict to cut off individual spores which are blown away in air currents. If a spore lands on a leaf of a healthy potato plant, in warm moist conditions a new hypha grows out from it and penetrates the leaf via a stoma (Fig. 15.4). When spores fall on the ground, they may be washed into the soil by rain, so reaching and infecting the potato tubers, causing them to rot. The close proximity of the plants in the potato field allows very rapid spread of the fungus from one individual to the next.

Agricultural research is constantly trying to find varieties of food plants which are resistant to the many types of fungus disease and to develop sprays which will destroy the fungus without causing harmful side effects on the crop or on the other organisms in the area.

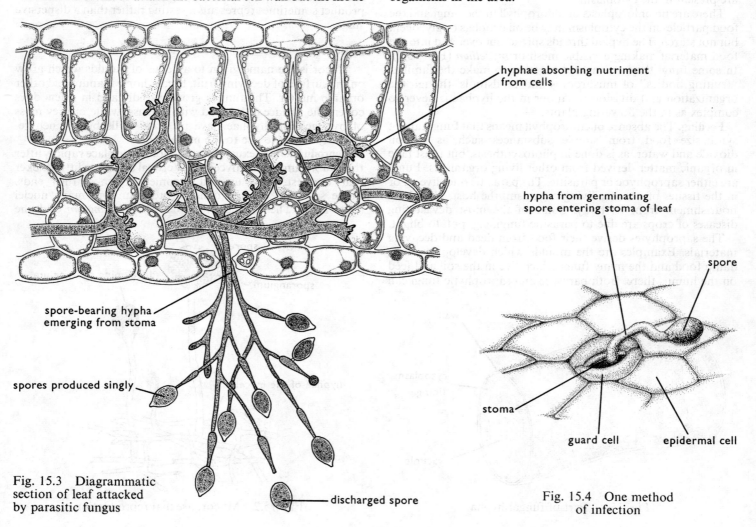

hyphae absorbing nutriment from cells

hypha from germinating spore entering stoma of leaf

spore-bearing hypha emerging from stoma

spores produced singly

discharged spore

spore

stoma

guard cell   epidermal cell

Fig. 15.3 Diagrammatic section of leaf attacked by parasitic fungus

Fig. 15.4 One method of infection

Yeast

The yeasts are a rather unusual family of fungi. Only a few of the several species can form true hyphae; the majority of them consist of separate, spherical cells, which can be seen only under the microscope. They live in situations where sugar is likely to be available, e.g. the nectar of flowers or the surface of fruits.

Structure. The thin cell wall encloses the cytoplasm, which contains a vacuole and a nucleus. In the cytoplasm are granules of glycogen and other food reserves.

Reproduction. The cells reproduce by budding, in which an outgrowth from a cell enlarges and is finally cut off from the parent as an independent cell. When budding occurs rapidly the individuals do not separate at once, and as a result, small groups of attached cells may sometimes be seen (Fig. 15.5).

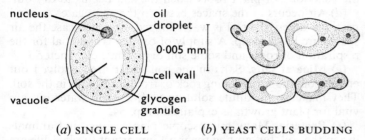

(a) SINGLE CELL   (b) YEAST CELLS BUDDING

Fig. 15.5   Yeast

In certain conditions, two cells may *conjugate*, that is, they join together and their cell contents fuse. Later, the cell contents divide into four individuals, each developing a thick wall (Fig. 15.6). These are spores and may constitute a resting stage. When the old cell wall enclosing them breaks open, the

Fig. 15.6   Resting stage with division into spores

spores are set free and can germinate to form normal, budding cells. Such spores often arise without any previous conjugation.

Fermentation. Yeasts, in nature, live on the surface of fruits and in other similar situations. They are of economic importance, however, in promoting alcoholic fermentation. Yeast cells contain many enzymes, one group of them being called, collectively, *zymase*. By means of these enzymes they can break down sugar into carbon dioxide and alcohol. This chemical change makes available energy which the yeast cells can use for their vital processes in a way similar to respiration, only far less energy is set free in this case:

$$C_6H_{12}O_6 \rightarrow 2CO_2 + 2C_2H_5OH + 118 \text{ kJ}$$
$$\text{alcohol}$$

Alcoholic fermentation is in fact similar to anaerobic respiration (p. 49), but unless the yeast is supplied with sugar, oxygen is necessary for the preliminary conversion of other carbohydrates to a suitable form for releasing energy.

In brewing, barley is allowed to germinate and in so doing converts its starch reserves into maltose. The germinating barley is killed and the sugars extracted with water. Yeast is added to this solution and brings about its fermentation to alcohol and carbon dioxide. In making beer, hops are added to give it a bitter flavour and the liquid is bottled under pressure so that the carbon dioxide is still very much in evidence. For making spirits like whisky the fermentation is allowed to go further and the alcoholic product is distilled off.

Many fruit juices will ferment of their own accord if the crushed fruit is left in suitable conditions and the yeasts present on the skin allowed to develop. If too much oxygen is admitted or the wrong species of yeasts, together with fungi and bacteria, allowed to enter, the oxidation of the alcohol may be continued until acetic (ethanoic) acid is produced, as in vinegar.

In baking, yeast is added to uncooked dough to make the dough "rise", as a result of the carbon dioxide bubbles given off, before the bread is baked.

Yeast as food. Yeasts are becoming important as a source of food for man and his farm animals. Given only sugar and inorganic salts, these micro-organisms will grow and reproduce very rapidly, converting the sugar and salts to make their protein. As one eminent biologist has expressed it, "In 24 hours, half a tonne of bullock will make a pound of protein; half a tonne of yeast will make 50 tonnes and needs only a few square metres to do it on." Yeasts contain most of the essential amino acids and vitamins (p. 83) needed by man, and at present the yeasts produced on a commercial scale are used to supplement diets inadequate in these respects. Yeast is also used specifically to remedy vitamin B deficiency diseases.

Economic importance of fungi. From the information in this chapter it can be seen that fungi are of considerable importance in a variety of ways. Their harmful effects include the crop losses caused by parasitic fungi and, to a much smaller extent, the harm done by fungal diseases in man and his domestic animals. Mould fungi cause damage to stored food, and those species which can digest wood bring about decay and rot in buildings.

On the positive side, the digestive activities of fungi on organic matter break down dead remains and help in recycling nutrients (p. 59). Fungi are important sources of antibiotic drugs and are used increasingly in industrial processes, apart from baking and brewing, to bring about chemical changes, e.g. the large scale production of citric acid and enzymes.

PRACTICAL WORK

Mucor and related species of fungi can be grown in the laboratory by placing moistened pieces of bread, apple, etc., on a Petri dish and covering them with a beaker. In the humid atmosphere colonies of mould grow in a few days. It is best to dip the apple in boiling water for a few seconds and then allow it to cool in the air before covering it with the beaker.

Pure cultures of any particular species can be obtained by transferring spores to a sterilized culture dish such as is described on page 74.

See also experiment (f), p. 49.

QUESTIONS

1. When something "goes mouldy" what is actually happening to it?
2. Suggest why bread, wood and leather are to be seen going mouldy while metal, glass and plastic are not.
3. Suggest why toadstools may be found growing in very dark areas of woodland where green plants cannot flourish.
4. The fungi are not easy to classify as plants but they bear little resemblance to animals. Discuss the ways in which they (a) resemble, (b) differ from green plants and animals.

16 Soil

Components

Soil consists of a mixture of: (a) particles of sand or clay, (b) humus, (c) water, (d) air, (e) dissolved salts, (f) bacteria.

(a) **Inorganic particles** are formed from rocks which have been weathered and broken down. Particles from 2 to 0·02 mm diameter constitute sand, 0·02 to 0·002 mm form silt and less than 0·002 mm clay. Chemically, sand is silicon oxide while clay may be various complexes of aluminium and silicon oxides. Iron oxide may give a red or brown coating to the particles.

Aggregates of these inorganic particles together with humus produce *crumbs* up to about 3 mm in diameter, which form the "skeleton" of the soil. The crumb structure of a soil depends on the proportion of clay, sand and humus and the activities of micro-organisms and plant roots; a good crumb structure is one of the most important attributes of a soil.

(b) **Humus** is the finely divided organic matter incorporated into the crumbs; it originates mainly from decaying plant remains. The presence of humus in the crumbs affects the colour and physical properties of the soil. Humus is black, structureless, often forms a coating round sand particles and may be important in "glueing" particles together to form soil crumbs. A sandy soil deficient in humus tends to have a poor crumb structure and is easily blown away if exposed by ploughing. The exclusive use of chemical fertilizers on certain soils in dry climates may lead to the formation of dust bowls or the advance of desert margins. (*See* p. 60.)

The bacterial decay of humus and the organic matter from which it originates produces the nitrates and other mineral salts needed for plant growth.

(c) **Water** is spread over the sand particles or clay aggregates as a thin film which adheres by capillary attraction. It may also penetrate the aggregates and be held to the clay particles by chemical forces. When a soil contains as much water as it can hold by capillary and chemical attraction (i.e. any more would drain away by the force of gravity), it is said to be at *field capacity*. Capillary attraction will tend to distribute water from regions above field capacity to drier regions. The forces holding water in the soil also set up considerable opposition to the "suction" of plant roots when the soil begins to dry out.

(d) **Air** occurs in the spaces between the aggregates or sand particles unless the soil is water-logged, in which case the air spaces are blocked up. A supply of oxygen is essential for the respiration of roots and some soil organisms, e.g. bacteria.

(e) **Mineral salts.** Salts in the soil water are dissolved out either from the surrounding rock or from the humus in the soil. They make a very dilute solution with the soil water but are vital for plant growth as explained on p. 54.

(f) **Bacteria.** Many microscopic plants, fungi and animals live in the soil, but among the most important to plant life are the bacteria which break down the organic matter and humus to form soluble salts which can be taken up in solution by roots. Other bacteria convert atmospheric nitrogen to organic compounds of nitrogen. For further details, *see* p. 59.

Heavy and light soils

Heavy soils. A soil in which clay particles predominate and which has a poor crumb structure will be sticky and difficult to dig or plough. This results partly from chemical and capillary forces acting on the very large surface area of the minute clay particles, making them difficult to separate. When dry, the soil forms hard clods which do not break up readily during cultivation.

The small distances between particles tend to produce poor aeration and drainage, but the large surface presented by the particles retains a high proportion of water in conditions of drought. There is also less tendency for soluble minerals to be washed away as they are held chemically to the clay particles.

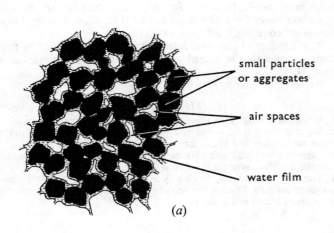

small particles or aggregates

air spaces

water film

(a)

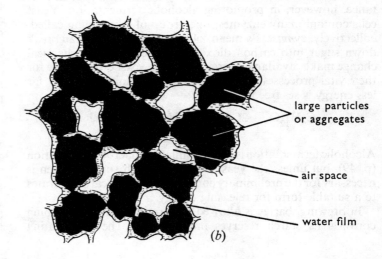

large particles or aggregates

air space

water film

(b)

Fig. 16.1   Comparison of structure of light and heavy soils

A heavy soil can be made lighter, more workable and permeable to water and air by adding organic matter or lime. The lime makes the particles clump together or *flocculate*, the clumps of particles behaving as the larger particles of a light soil. The crumb structure of a clay soil can be improved by growing grass on it for a year or two.

Light soils. The large inorganic particles of a light soil give it its sandy texture (Fig. 16.1*b*). The wider separation of the particles leads to better aeration and drainage but there is a smaller surface for the water film. Such a reduced surface lessens the surface forces of the water and makes it easier to separate the particles in ploughing and digging, and the clumps break up easily when dry.

The mineral salts are more liable to be washed out from a lighter soil and it loses water rapidly in dry conditions. Its water-holding properties and nitrogen content can be improved by adding farmyard manure or compost.

Loam. A soil with a balanced mixture of particle sizes, a good humus content and stable crumb structure is called a loam. Loams are the most productive soils in agriculture.

Acid soil. In some water-logged soils, the lack of air prevents the activities of aerobic bacteria that break down the plant remains to form humus. As a result the dead, partly decomposed vegetation accumulates in deep layers and becomes acid as a consequence of the incomplete decomposition of the plant remains. This is not the only way a soil becomes acid, but peat bogs are an example of this type of soil. Repeated applications of ammonium sulphate will make a normal soil acid unless accompanied by lime, which neutralizes the acidity. Only a restricted range of plants can grow in acid conditions and very acid soils are of little value for crop growing.

EXPERIMENTAL WORK

Experiment 1. *The weight of water in a soil sample*

A sample of soil is placed in a weighed evaporating basin which is then reweighed. The basin is heated over a water-bath for some hours or days, according to the weight and nature of the sample, to drive off the soil water. The basin is then allowed to cool in a desiccator and weighed again. Heating and reweighing are continued until two weighings give identical results, showing that all the soil water has evaporated.

The difference between the first and final weighings gives the weight of water that was originally present. Higher temperatures than the water-bath cannot be used as they will cause the burning of the humus in the soil and so give an additional loss of weight.

Experiment 2. *Weight of organic matter in the soil*

The dry soil from the previous experiment is heated in the same evaporating basin, but this time on a sand-tray or gauze, until it loses no further weight. All the organic matter has now been burned and oxidized to carbon dioxide and water so that the loss of weight represents the weight of humus and other organic matter originally present.

Experiment 3. *Volume of air in the soil* (Fig. 16.2).

Two jam or fruit tins of equal size and of measured volume are needed, both with cleanly cut tops. One is perforated at the bottom, driven down, open-end first, into the soil, dug out without disturbing the soil in it, and the soil cut level at the top. The other tin is placed empty in a large vessel of water and the level *A* marked. The tin is now removed full of water and the level drops to *B*. The tin, full of soil, is now placed in the water and the level at first returns to *A*, but air bubbles soon escape from the soil and the level drops to *C*. All the air should be forced out by breaking up the soil in the tin with a stick under the water.

Water is now added from a measuring cylinder to bring the level up to *A* once more. The volume of water added will be equal to the volume of air that has escaped from the soil.

Note. The quantitative results from these experiments will be much more instructive if widely different soil samples under the same conditions are compared, or alternatively, the same soil under different conditions.

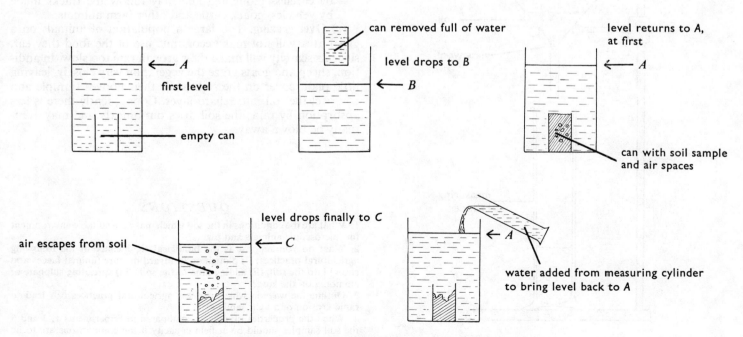

Fig. 16.2   To measure the volume of air in a soil sample

*Experiment 4. **A rough estimate of the solid components of a soil*** (Fig. 16.3).

About 20 cm³ of soil is thoroughly shaken with 200 cm³ of water in a measuring cylinder and then allowed to settle. The particles will settle according to their surface area and density, small stones first, then sand and clay so that more or less distinct layers appear. Some of the fine particles of clay will stay suspended in the water and large organic particles will float to the top. The depth of the various layers should be measured and compared with other soils similarly treated.

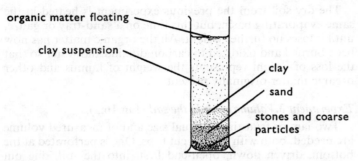

Fig. 16.3   To show, roughly, the composition of a soil

*Experiment 5. **Permeability of soil to water*** (Fig. 16.4)

Two glass funnels are plugged with glass wool and half-filled with equal volumes of sandy and clay soils respectively. Both are then covered with water, the level of which is kept constant by topping up throughout the experiment so that there is no difference in pressure between the two. The water that runs through in a given time is collected in a measuring cylinder.

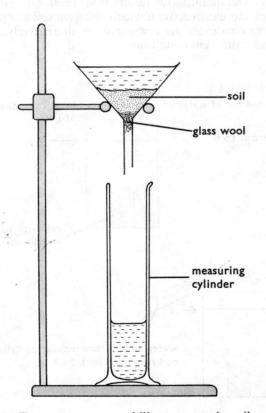

Fig. 16.4   To compare permeability to water in soil sample

*Experiment 6. **To show the presence of micro-organisms in the soil***

Prepare two Petri dishes of sterile, nutrient agar as described on p. 74. In one of them sprinkle some particles of soil and in the other, as a control, some particles of clean sand that have been sterilized by heating. If bacteria or fungi are present in the soil, they will appear as colonies in the surface of the agar within two days. The absence of such colonies from the control will prove that the bacteria were in the soil and did not come from the air, the dish, the medium or the instrument used to introduce the soil.

[*Note*: these experiments and several others are fully detailed in the laboratory manual *Soil—see* p. 1.]

Soil erosion (Fig. 16.5)

Soil erosion means the removal of top-soil, usually by the action of wind and rain.

(*a*) **Deforestation.** The soil cover on steep slopes is usually fairly thin but can support the growth of trees. If the forests are cut down to make way for agriculture, the soil is no longer protected by a leafy canopy from the driving rain. Consequently, some of the soil is washed away into the rivers which tend to become choked with silt and overflow their banks.

(*b*) **Poor farming methods.** Ploughing loosens the soil and destroys its natural structure. Failure to replace humus after successive crops, and burning the stubble or weeds, reduces the water-holding properties, so that the soil dries easily and may be blown away as dust. On sloping ground, such soil may be eroded by water.

(i) *Sheet* erosion is the imperceptible removal of thin layers of soil but usually leads to:

(ii) *Rill* erosion, in which the water cuts channels. The channels deepen as the volume of run-off increases and so become gulleys.

(iii) *Gulley* erosion. The gulleys so formed reach enormous proportions so that thousands of hectares of top-soil are carried off. Gulley erosion is often accentuated by careless ploughing and may follow the tracks made by vehicles, goats, cattle and other farm animals.

(*c*) **Over-grazing.** Too large a population of animals on a given area will not make economic use of the food they eat, since its scarcity will make their growth rate too slow. In addition, sheep and goats graze the vegetation very closely, leaving little plant cover on the soil, while their hooves trample and compact the soil into a hard layer. Consequently there is less absorption by rain; the soil dries out quickly and may eventually be blown away.

QUESTIONS

1. What are the conditions in the soil which make it a suitable environment for microscopic animals and plants?

2. What do you suppose is the biological significance of the following agricultural practices: (i) ploughing farmyard manure (animal faeces and straw) into the soil, (ii) adding lime to the soil, (iii) spreading sulphate of ammonia on the land?

3. Outline the ways in which careless agricultural practices can lead to rapid erosion of a light sandy soil.

4. When the properties of soils are compared in Experiments 1, 3 and 5 the soil samples should be at field capacity if the comparisons are to be valid. Explain why this is necessary.

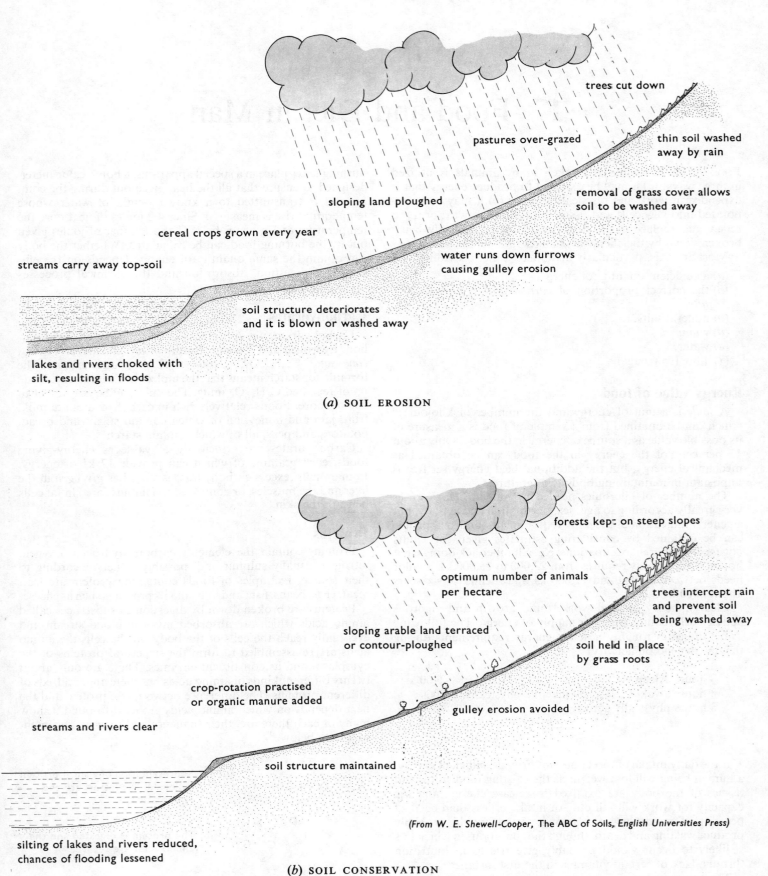

(a) SOIL EROSION

trees cut down

thin soil washed away by rain

pastures over-grazed

removal of grass cover allows soil to be washed away

sloping land ploughed

cereal crops grown every year

water runs down furrows causing gulley erosion

streams carry away top-soil

soil structure deteriorates and it is blown or washed away

lakes and rivers choked with silt, resulting in floods

forests kept on steep slopes

optimum number of animals per hectare

trees intercept rain and prevent soil being washed away

sloping arable land terraced or contour-ploughed

soil held in place by grass roots

crop-rotation practised or organic manure added

gulley erosion avoided

streams and rivers clear

soil structure maintained

silting of lakes and rivers reduced, chances of flooding lessened

(From W. E. Shewell-Cooper, The ABC of Soils, English Universities Press)

(b) SOIL CONSERVATION

Fig. 16.5

17 Food and Diet in Man

FOOD made by plants and taken in by animals is utilized in three ways: (a) it may be oxidized to produce energy that is expended in work and physical exercise; (b) it may be incorporated into new cells and tissue to produce growth; or (c) to renew and replace parts of tissue that are constantly being broken down by the chemical changes that occur during life.

Most animals, particularly man, need a diet providing:

(a) a sufficient quantity of energy,
(b) the correct proportion of *carbohydrates*, *proteins* and *fats*,
(c) mineral salts,
(d) *vitamins*,
(e) water,
(f) fibre (roughage).

Energy value of food

A *joule* is a unit of energy and the number of kilojoules* which can be obtained from a sample of food is a measure of its possible value as a source of energy in the body. Only about 15 per cent of the energy in the food can be obtained as mechanical energy, but the additional heat energy set free is important in maintaining body temperature.

The number of kilojoules needed by human beings varies very greatly according to age, sex, occupation and activity, but a general indication of the daily requirement of normal people can be obtained by considering estimates given for adults engaged in different occupations, e.g. a lumberjack doing eight hours' work per day requires from 23,000 to 25,000 kJ, a tailor needs between 10,000 and 11,000 kJ and a child of six years about 8000 kJ.

Clearly, heavy manual work, rapid growth and vigorous activity will require a greater supply of energy. The following Table indicates roughly how the energy requirements during the day depend on activity:

	kJ
8 hours asleep	2400
8 hours awake; relatively inactive physically	3000
8 hours physically active	6600
Total	12,000

If the daily intake of food does not provide sufficient energy, a human being will lose weight as the existing food stores and tissues of the body are oxidized to release energy, and the capacity for work will fall off. An intake of less than 6300 kJ per day, if maintained for a long period, would probably produce wasting and death, though the conditions in which this is likely to occur would probably give rise to malnutrition through lack of certain vitamins in the first instance.

The energy value of food is calculated by burning a known weight of it completely to carbon dioxide and water. The burning takes place in a special apparatus, a bomb calorimeter, designed to ensure that all the heat given out during the combustion is transmitted to a known weight of water whose temperature rise is measured. Since 4·2 joules of heat raise the temperature of 1 g water by 1°C, the number of joules given out by the burning food can be calculated. Whether the body can obtain the same quantity of energy depends on the efficiency of digestion, absorption and the chemical processes undergone by the food.

Carbohydrates

Carbohydrates are substances containing the elements carbon, hydrogen and oxygen. Examples are *glucose*, $C_6H_{12}O_6$; *cane sugar*, $C_{12}H_{22}O_{11}$; *starch*, $(C_6H_{10}O_5)_n$ and *cellulose*. The formula for starch means that the molecule is a large one, made up of repeated $C_6H_{10}O_5$ units. The "n" may be equivalent to 300 or more. Foods relatively rich in carbohydrates are milk, fruit, jam and honey, all of which contain sugar, and bread, potatoes and peas, all of which contain starch.

Carbohydrates are principally of value as energy-giving foods, each gramme of which can provide 17 kJ of energy. In mammals, excess carbohydrate is stored as *glycogen* in the liver and the muscles, or converted to fat and stored in fat cells beneath the skin.

Proteins

Proteins contain the elements carbon, hydrogen, oxygen, nitrogen, usually sulphur and, possibly, others according to their source. Examples of foods containing protein are lean meat, eggs, beans, fish, and milk and its products such as cheese.

Proteins are broken down by digestion to substances called amino acids which are absorbed into the blood stream and eventually reach the cells of the body. In the cells the amino acids are re-assembled to form the structural proteins of the cytoplasm and its constituent enzymes. There are only about twenty different kinds of amino acids but there are hundreds of different proteins. The difference between one protein and the next depends on which amino acids are used to build it, how many of each there are, their sequence and their arrangement as shown below.

A-B-C-D-E-F-G-H-I-J-K

(a) Representation of a small protein molecule. The letters are the amino acids

F B J D A
I H E C G K

(b) The protein is digested and the amino acids are set free

B-J-K-D-A-F
|
E-C-G-H-I

(c) The same amino acids are built up into a different protein

Plants can build all the amino acids they need from carbo-hydrates, nitrates and sulphates but animals cannot. They must therefore obtain their amino acids from proteins already made by plants or present in other animals and the diet must therefore include a minimum quantity of protein of one sort or another. A diet with a sufficient energy content of fats and carbo-hydrates and rich in vitamins and salts will lead to illness and death because of its lack of proteins. Proteins are particularly important during periods of pregnancy and growth when new cytoplasm, cells and tissues are being made.

Although animals cannot make amino acids they can, in some cases, convert one amino acid into another. There are, however, ten or more amino acids which animals cannot produce in this way and these *essential amino acids* must be obtained directly from proteins in the diet. Animal proteins generally contain more essential amino acids than do plant pro-teins but since milk and eggs contain the highest proportion of all, a vegetarian who includes these in his diet should not lack essential amino acids.

If proteins are eaten in excess, there will be more amino acids in the body than are needed to produce or replace cells. The excess amino acids are converted in the liver to carbo-hydrates which are then oxidized for energy, or converted to glycogen and stored. The energy value of protein is 17 kJ/g.

Fats

Like carbohydrates, fats contain only carbon, hydrogen and oxygen but in different proportions. They are present in milk, butter and cheese, animal fat, egg-yolk and margarine.

Although fats are less easily digested and absorbed than carbohydrates, they have more than double the energy value, providing 39 kJ/g. Fat can be stored in the body.

Mineral salts

A wide variety of salts is essential for the chemical activities in the body and for the construction of certain tissues. The red pigment in the blood contains iron; bones and teeth contain calcium, magnesium and phosphorus; sodium and potassium are essential in nearly all cells, in the blood fluid and in nerves; iodine is necessary for the proper functioning of the *thyroid gland*. In addition to the sulphur and nitrogen in the body, traces of copper, cobalt and manganese are required.

Salts of these elements are present in small quantities in a normal diet and the body can absorb and concentrate them.

Vitamins

Vitamins are complex chemical compounds which, although they have no energy value, are essential in small quantities for the normal chemical activities of the body.

It has long been known that certain diseases could be pre-vented or cured by making alterations in the diet. In about 1750, James Lind, a naval doctor, cured scurvy in seamen by providing them with citrus fruits. Christiaan Eijkman, a Dutch doctor working in Java, in 1896, was searching for the "germ" which he thought transmitted the disease beri-beri, but found that the disease could be caused by feeding chickens with polished (de-husked) rice and cured by feeding them un-polished rice. Although Eijkman concluded from this result that the disease was caused by a poisonous substance in the polished rice, which was normally neutralized by something present in the husk of the unpolished grain, it was realized in the next ten years by Eijkman and other scientists, notably the

English biochemist Gowland Hopkins, that the husk contained an *accessory food factor* whose absence led to the disease.

Gowland Hopkins, experimenting with rats, originated the concept of essential amino acids and between 1906 and 1912 demonstrated the importance of vitamins. He fed young rats on a diet containing starch, sucrose, lard, salts and purified casein, a milk protein rich in essential amino acids. The rats ceased growing and soon died unless the diet also contained 3 cm³ milk per day (Fig. 17.1). Since the calorific value of 3 cm³ milk is very little and it contains no more amino acids than casein, the restorative properties of the milk must have been due to factors in the milk which did not provide calories or amino acids. Many similar accessory food factors have been identified and although their composition varies widely they are usually listed under the general heading of *vitamins*.

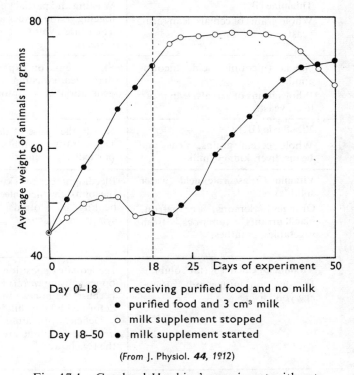

Day 0–18   o   receiving purified food and no milk
            ●   purified food and 3 cm³ milk
            o   milk supplement stopped
Day 18–50   ●   milk supplement started

(From J. Physiol. **44**, 1912)

Fig. 17.1   Gowland Hopkins' experiment with rats

Fifteen or more vitamins have been isolated and most of them seem to act as catalysts in essential chemical changes in the body, each one influencing a number of vital processes. Vitamins A, D, E and K are the *fat-soluble vitamins*, occurring mainly in animal fats and oils and absorbed along with the products of fat digestion. Vitamins B and C are the *water-soluble vitamins*.

If a diet is deficient in one or more vitamins, this results in a breakdown of normal bodily activities and produces symptoms of disease. Such diseases can usually be effectively remedied by including the necessary vitamins in the diet.

Plants can build up their vitamins from simple substances, but animals must obtain them "ready-made" directly or in-directly from plants.

Some of the important vitamins are set out in the Table overleaf together with their properties. It must be emphasized that nearly all normal, mixed diets will include adequate amounts of vitamins, and deficiency diseases are most likely to occur where, as in rice-eating countries, the bulk of the diet consists of only one or two kinds of food.

Vitamins and their Characteristics

NAME AND SOURCE OF VITAMIN	DISEASES AND SYMPTOMS CAUSED BY LACK OF VITAMIN	NOTES
Vitamin A; retinol (fat-soluble) Milk, butter, cheese, liver, cod-liver oil. Fresh green vegetables for *carotene* (water soluble).	Reduced resistance to disease, particularly those which gain access through the skin. Poor night vision. Skin and cornea of eyes become dry leading to *xerophthalmia*.	The yellow pigment, carotene, present in green leaves and carrots, is oxidized in the body to make retinol. Retinol contributes to a light-sensitive pigment in the retina. Retinol is stored in the liver.
Vitamin B complex Ten or more water-soluble vitamins usually occurring together. Three are described here.	Since the B vitamins are present in most unprocessed food, deficiency diseases usually arise only in populations living on restricted diets.	Many of the B vitamins act as catalysts (co-enzymes) in the normal oxidation of carbohydrates during respiration. The absence of these catalysts leads to metabolic disturbances.
Thiamine (B_1) Whole grains of cereals, lean meat, yeast.	Wasting and partial paralysis, or water-logging of the tissues and heart failure. These are symptoms of two forms of *beriberi*.	Rice husks contain both thiamine and niacin, so highly milled rice is deficient in both. Maize is deficient in niacin. Populations living largely on milled rice or maize are very prone to the deficiency diseases of beriberi and pellagra.
Niacin (nicotinic acid, nicotinamide) Whole grains of cereals, lean meat, liver, yeast.	Skin eczema on exposure to sunlight, diarrhoea, wasting and mental degeneration; all symptoms of *pellagra*.	White flour (72% extraction) is deficient in thiamine which must be added to white flour used in bread-making. Badly planned slimming diets may be deficient in thiamine.
Riboflavin (B_2) Whole cereal grains, peas and beans, liver, kidney, milk.	Rarely the cause of deficiency disease. Degenerative condition of the skin, particularly round the mouth.	
Vitamin C; ascorbic acid (water soluble) Oranges, lemons, grapefruit, blackcurrants, tomatoes, fresh vegetables, potatoes.	Bleeding under the skin, particularly at the joints. Swollen, bleeding gums, poor healing of wounds. These are all symptoms of *scurvy*.	Possibly acts as a catalyst in cell respiration. Scurvy is only likely to occur when fresh food is not available. Milk contains very little ascorbic acid, so babies need additional sources such as rose-hip syrup.
Vitamin D; calciferol (fat-soluble) Fish-liver oil, butter, milk, cheese, egg yolk, liver.	Inadequate deposition of calcium in the bones causing *rickets* in young children because the bones remain soft and are deformed by the child's weight. Deficiency in adults causes *osteomalacia*; the vertebrae are compressed and the legs bowed.	Calciferol helps the absorption of calcium from the intestine and the deposition of calcium salts in the bones. Natural fats beneath the skin are converted to a form of calciferol by the action of sunlight or ultra-violet radiation. Impaired absorption of fats in the intestine leads to rickets or osteomalacia.
Vitamin E; tocopherol (fat-soluble) Wheat germ (embryo), e.g. in wholemeal bread; dairy products, meat.	Few deficiency effects apparent in man. Severe deficiency in infants may lead to high rates of destruction of red blood cells and hence to anaemia.	Experimentally induced deficiencies in pregnant rats leads to reabsorption of the embryos and in males to degeneration of the testes. Other pathological symptoms have been demonstrated in guinea pigs, rabbits and monkeys.
Vitamin K; phylloquinone, menaquinone (fat-soluble) Green vegetables contain phylloquinone.	In its absence, blood takes longer to clot and may result in bleeding diseases. Dietary deficiency unlikely to cause disease because intestinal bacteria can synthesize menaquinone.	Vitamin K is needed for the manufacture of a blood-clotting factor in the liver. Anything which impairs absorption of fats, e.g. bile deficiency, may lead to symptoms of fat-soluble vitamin deficiency disease irrespective of the diet.

Water

Water makes up a large proportion of all the tissues in the body and is an essential constituent of normal protoplasm. It plays an important part in the digestion and transport of food material and all the chemical reactions of the body take place in solution.

Dietary fibre (roughage)

Fibre consists largely of plant-cell walls which cannot be digested by man but are digested by bacteria in the colon. The fibre is thought to be important in maintaining a healthy digestive system in a variety of ways. It adds bulk to the food and enables the muscles of the alimentary canal to grip it and keep it moving by peristalsis (p. 86).

Milk

The sole article of diet during the first few weeks or months of a mammal's life, milk is an almost ideal food since it contains proteins, fats, carbohydrates, mineral salts, particularly those of calcium and magnesium, and vitamins. For adults, however, it is less satisfactory because of its high water content and lack of iron. Large volumes would have to be consumed if it were the principal article of diet for an adult, and serious blood deficiencies would result from the lack of iron. In the body of the embryo mammal, iron is stored while the embryo develops inside its mother, and this supply must suffice until the young mammal begins to eat solid food.

PRACTICAL WORK

Food tests. Test for starch. A little starch powder is shaken in a test-tube with some cold water and then boiled to make a clear solution. When the solution is cold, 3 or 4 drops of *iodine solution** are added. The dark blue colour that results is characteristic of the reaction between starch and iodine. This test is very sensitive.

Test for glucose. A water-bath is prepared as shown in Fig. 10.2, p. 51. A little glucose is mixed with some Benedict's solution* in a test-tube, and the test-tube placed in the boiling water-bath. The solution changes from blue to opaque green, yellow, and finally a red precipitate of copper(I) oxide appears. Sucrose is recognized by its failure to react with Benedict's solution until after it has been boiled with dilute hydrochloric acid and neutralized with sodium hydrogencarbonate.

Test for protein. (Biuret test). To a one per cent solution of albumen is added 5 cm^3 dilute sodium hydroxide (CARE: this solution is caustic) and 5 cm^3 one per cent copper sulphate solution. A purple colour is indicative of protein.

Test for fats. Two drops of cooking oil are thoroughly shaken with about 5 cm^3 ethanol in a dry test-tube until the fat dissolves. The alcoholic solution is poured into a test-tube containing a few cm of water. A cloudy white emulsion is formed.

Application of the food tests. These tests can be applied to samples of food such as milk, raisins, potato, onion, beans, egg-yolk, ground almonds, to find out what food materials are present in them.

They can also be carried out in connexion with work on seeds and storage organs. The seeds are crushed in a mortar and shaken with warm water to extract soluble products. The solution and suspension of crushed seed is subjected to the tests given above.

[*Note:* these experiments and several others are fully detailed in the laboratory manual *Food Tests—see* p. 1.]

QUESTIONS

1. What principles must be observed when working out a diet in order to lose weight? What dangers are there if such diets are not scientifically planned?
2. Kwashiorkor is a protein deficiency disease which affects young children in the developing countries. Its onset is usually most severe when a child is weaned from the mother's milk to the starchy plant foods, e.g. yam, cassava, of the adults. Why should weaning mark the onset of the disease?
3. It is a common fallacy that manual workers need much more protein than sedentary workers. Suggest why this idea is fallacious.
4. Eating a large amount of protein at one sitting is wasteful. It is better to take in a little protein at each meal. Why do you think this is the case?
5. What dietary problems confront (*a*) people living predominantly off rice or maize, (*b*) strict vegetarians?

* Methods of preparing the reagents are given on p. 211.

18 The Digestion, Absorption and Metabolism of Food

To be of any value to the body, the food taken in through the mouth must enter the blood stream and be distributed to all the living regions.

Digestion is the process by which insoluble food, consisting of large molecules, is broken down into soluble compounds having smaller molecules. These smaller molecules, in solution, pass through the walls of the intestine and enter the blood stream. Digestion and absorption take place in the *alimentary canal* (Fig. 18.1 and Plate 18), digestion being brought about by means of active chemical compounds called *enzymes*. The alimentary canal is a muscular tube, with an internal glandular lining, running from mouth to anus. Some regions have particular functions and, accordingly, different structures. Juices are secreted in the alimentary canal from glands in its lining or are poured into it through ducts from glandular organs outside it. As the food passes through the alimentary canal it is broken down in stages until the digestible material is dissolved and absorbed. The indigestible residue is expelled through the anus.

Enzymes

Enzymes are chemical compounds, protein in nature, made in the cells of living organisms. They act as catalysts—substances which accelerate the rate of most chemical changes in the organism without altering the end-products. They occur in great numbers and varieties in all protoplasm and without them the chemical reactions would be too slow to maintain life. The vast majority of enzymes are *intracellular*, that is, they carry out their functions in the protoplasm of the cell in which they are made. Some enzymes, however, are secreted out of the cells in which they are made, to be used elsewhere. These are called *extracellular* enzymes. Bacteria (p. 73) and fungi (p. 75) secrete such extracellular enzymes into the medium in which they are growing. The higher organisms secrete extracellular enzymes into the alimentary tract to act on food taken into it.

These digestive enzymes accelerate the rate at which insoluble compounds are broken down into soluble ones. Enzymes which act on starch are called *amylases*, those acting on proteins are *proteinases*, and *lipases* act on fat.

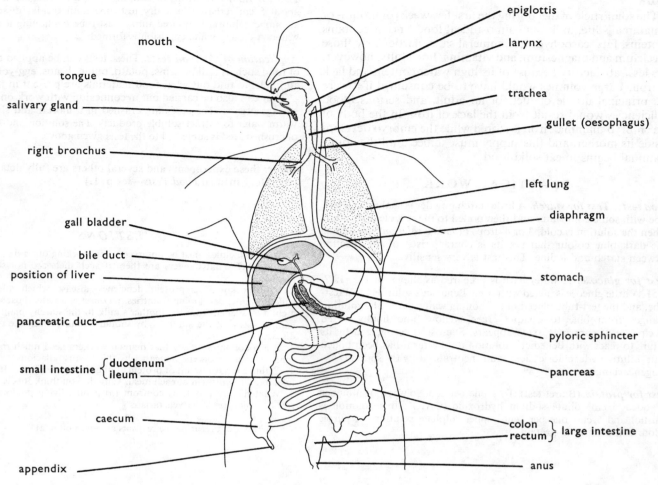

Fig. 18.1   The alimentary canal

Every enzyme has the following characteristics:
  (*a*) it is destroyed by heating, since it is a protein,
  (*b*) it acts best within a narrow temperature range,
  (*c*) it acts most rapidly in a particular degree of acidity or alkalinity (pH),
  (*d*) it acts on only one kind of substance,
  (*e*) it always forms the same end-product or products, since an enzyme affects only the rate of reaction.

Movement of food through the alimentary canal

Ingestion is the act of taking food into the alimentary canal through the mouth.

Swallowing (*see* Fig. 18.3). In swallowing, the following actions take place: (*a*) the tongue presses upwards and back against the roof of the mouth, forcing the pellet of food, called a *bolus*, to the back of the mouth, or *pharynx*; (*b*) the *soft palate* closes the opening between the nasal cavity and the pharynx; (*c*) the *laryngeal cartilage* round the top of the trachea, or windpipe, is pulled upwards by muscles so that the opening of the larynx lies beneath the back of the tongue, and the opening of the trachea is constricted by the contraction of a ring of muscle; and (*d*) the *epiglottis*, a flap of cartilage, directs food over the laryngeal orifice. In this way food is able to pass over the trachea without entering it. The beginning of this action is voluntary, but once the bolus of food reaches the pharynx swallowing becomes an automatic or reflex action. The food is forced into and down the oesophagus, or gullet, by peristalsis.

This takes about six seconds with fairly solid food, and then the food is admitted to the stomach. Liquid travels more rapidly down the gullet.

Peristalsis (Fig. 18.2). The walls of the alimentary canal contain circular and longitudinal muscle fibres. The circular muscles, by contracting and relaxing alternately, urge the food in a wave-like motion through the various regions of the alimentary canal.

Egestion. The expulsion from the alimentary canal of the undigested remains of food is called egestion.

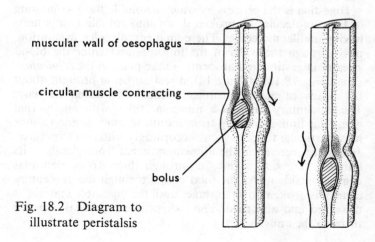

Fig. 18.2   Diagram to illustrate peristalsis

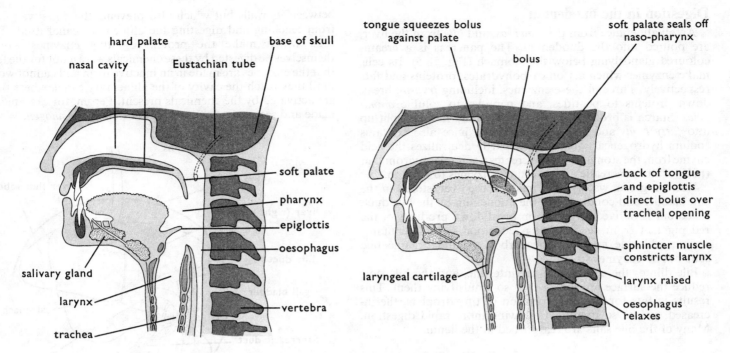

Fig. 18.3   Section through head to show swallowing action

Digestion in the mouth

In the mouth the food is chewed and mixed with saliva. Chewing reduces the food to suitable sizes for swallowing and increases the available surface for enzymes to act on.

Saliva is a digestive juice secreted by three pairs of glands the ducts of which lead into the mouth (Fig. 18.3). It is a watery fluid, not particularly acid or alkaline, containing a little mucus, which helps to lubricate the food and makes the particles adhere to one another. An adult may secrete from 1 to 1·5 litres of saliva per day. One enzyme, *salivary amylase*, is present in saliva. Salivary amylase acts on cooked starch and begins to break it down into *maltose*, a soluble sugar.

The longer food is retained in the mouth, the further this starch digestion proceeds and the more finely divided does the food become as a result of chewing. In fact even well-chewed food does not remain in the mouth long enough for much digestion of starch to take place, but saliva will continue to act for a time even when food is passed into the stomach.

Digestion in the stomach

This part of the alimentary canal has elastic walls and so can be extended as the food accumulates in it. This enables food from a particular meal to be stored for some time and released slowly to the rest of the alimentary canal.

Very little absorption takes place in the stomach, but its glandular lining (Fig. 18.4) produces *gastric juice* containing the enzyme *pepsin*, and it may also contain, in young children, an enzyme called *rennin*. Pepsin acts on proteins and breaks them down into more soluble compounds called *peptides*. Rennin, if present, clots protein in milk. The stomach wall also secretes *hydrochloric acid* which makes a 0·5 per cent solution in the gastric juice. The acid provides the best degree of acidity (optimum pH) for pepsin to work in, and probably also kills many of the bacteria taken in with the food. The salivary amylase from the mouth cannot digest starch in an acid atmosphere, but it seems likely that it continues to act within

the bolus of food until this is broken up and the hydrochloric acid reaches all its contents.

The rhythmic, peristaltic movements of the stomach, about every twenty seconds, help to mix the food and gastric juice to a creamy fluid called *chyme*. Each wave of peristalsis also pumps a little of the chyme from the stomach into the first part of the small intestine, called the *duodenum*. The pyloric sphincter is usually relaxed but contracts at the end of each wave of peristalsis, so limiting the amount of chyme which escapes. Even when relaxed, the pyloric opening is narrow and only liquid is allowed through. When the acid contents of the stomach enter the duodenum, they set off a reflex action (p. 139), which closes the pyloric sphincter until the duodenal contents have been partially neutralized.

A meal of carbohydrate such as porridge may be retained less than an hour, and a mixed meal containing protein and fat may be in the stomach for one or two hours.

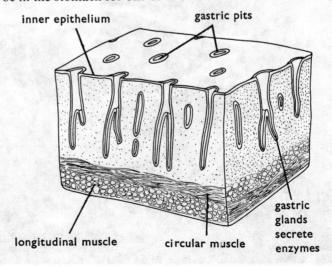

Fig. 18.4   Stereogram of section through stomach wall

Digestion in the duodenum

An alkaline juice from the *pancreas*, and *bile* from the liver, are poured into the duodenum. The pancreas is a cream-coloured gland lying below the stomach (Fig. 18.5). Its cells make enzymes which act on carbohydrates, proteins and fats respectively. Three of these enzymes, including *trypsin*, break down proteins to peptides, and peptides to soluble *amino acids*. Starch is broken down to *maltose* and fats are split up into *fatty acids* and *glycerol*. Pancreatic juice also contains sodium hydrogencarbonate which partly neutralizes the acid chyme from the stomach, and so creates a suitable environment (pH) for the pancreatic and intestinal enzymes to work in.

Bile is a green, watery fluid made in the liver, stored in the gall bladder and conducted to the duodenum by the bile duct. Its colour is derived largely from breakdown products of the red pigment from decomposing red blood cells. It contains sodium chloride and hydrogencarbonate, and organic bile salts but no enzymes.

Bile dilutes the contents of the intestine, and the bile salts reduce the surface tension of fats, so emulsifying them. This results in fats forming a suspension of tiny droplets, the increased surface so presented allowing more rapid digestion. Many of the bile salts are reabsorbed in the ileum.

Digestion in the ileum

In the small intestine, digestion is continued by the action of the pancreatic enzymes. All the digestible material is changed to soluble compounds which pass into the cells lining the ileum. In these cells, peptides are broken down to amino acids, and maltose and sucrose are broken down to glucose and fructose. These substances then enter the blood capillaries.

The glandular lining of the alimentary canal is continually secreting mucus which helps to lubricate the passage of food between its walls but which also prevents the digestive juices from reaching and digesting the alimentary canal itself. The cells which make the protein-digesting enzymes would themselves be digested by these chemicals were it not for the fact that the enzymes are made in an inactive form and cannot work until they reach the cavity of the alimentary canal, where they are activated by the chemicals present. Pepsin, for example, is made and secreted as an inactive substance, *pepsinogen*. When

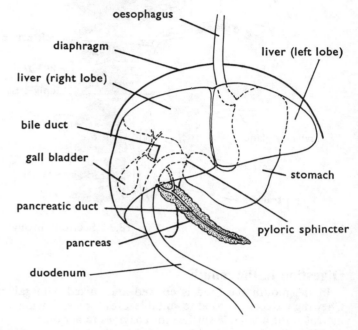

Fig. 18.5 Diagram to show relation of stomach, liver and pancreas

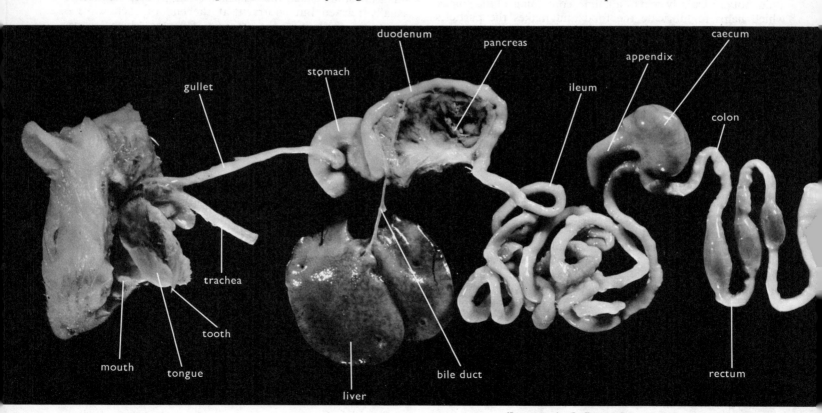

(Dissection by Griffin and George Ltd. Gerrard Biological Centre)

Plate 18. ALIMENTARY CANAL OF A RAT UNRAVELLED

pepsinogen is set free in the stomach the hydrochloric acid present converts it to active pepsin. This pepsin cannot now digest the stomach walls because of their protective coating of mucus.

Absorption in the ileum

Nearly all the absorption of digested food takes place in the ileum, and certain of its characteristics are important adaptations to its absorbing properties:

(a) it is usually fairly long and presents a large absorbing surface to the digested food,

(b) its internal surface is greatly increased by thousands of tiny, finger-like projections called *villi* (Fig. 18.6 and Plate 19),

(c) the lining epithelium is very thin and the fluids can pass fairly rapidly through it,

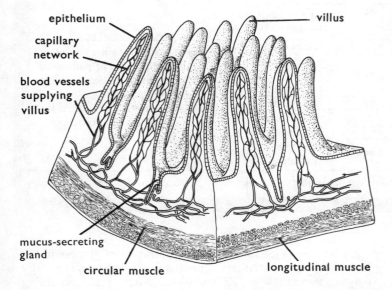

Fig. 18.6   Stereogram to show structure of ileum

(d) there is a dense network of blood capillaries in each villus (Fig. 18.7).

The small molecules of the digested food, principally amino acids and glucose, pass through the epithelium and the capillary walls and enter the blood plasma. They are then carried away in the capillaries which unite to form veins and eventually join up to form one large vein, the *hepatic portal* vein. This carries all the blood from the intestine to the liver, which may retain or alter any of the digestion products. The digested food then reaches the general circulation.

Some of the fatty acids, and glycerol from the digestion of fats, enter the blood capillaries of the villi but a large proportion may be recombined in the intestinal lining to form fats once again and then these fats pass into the *lacteals*. It may be that some of the finely emulsified fat is absorbed directly, i.e. without digestion, as minute droplets which subsequently enter the lacteals. The fluid in the lacteals enters the *lymphatic system* which forms a network all over the body and eventually empties its contents into the blood stream (p. 100).

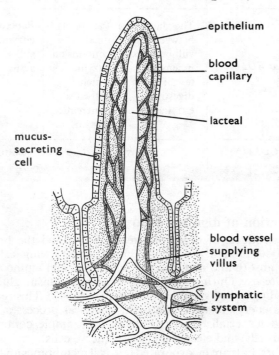

Fig. 18.7   Villus structure

The large intestine (colon and rectum)

The material passing into the large intestine consists of water with undigested matter, largely cellulose and vegetable fibres (roughage), mucus and dead cells from the lining of the alimentary canal. The large intestine secretes no enzymes but the bacteria in the colon digest part of the fibre. The colon absorbs much of the water from the undigested residues. This semi-solid waste, *the faeces*, is passed into the *rectum* by peristalsis and is expelled at intervals through the *anus*. The residues may spend from 12 to 24 hours in the intestine.

The caecum and appendix

These are relatively small, probably vestigial* structures in man. In herbivores like the rabbit and the horse they are much larger, and it is here that most of the cellulose digestion takes place, largely as a result of bacterial activity.

* i.e. structures which have become apparently functionless through disuse in the course of evolution.

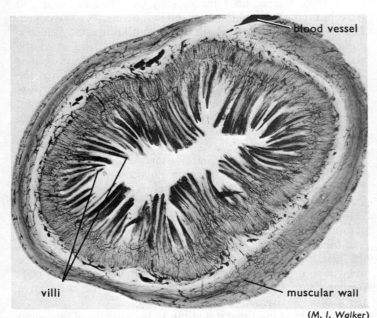

(M. I. Walker)

Plate 19.   TRANSVERSE SECTION THROUGH
ILEUM OF CAT, SHOWING VILLI (× 12)

Digestive Action

REGION OF ALIMENTARY CANAL	DIGESTIVE GLAND	DIGESTIVE JUICE PRODUCED	ENZYMES IN THE JUICE	CLASS OF FOOD ACTED UPON	SUBSTANCES PRODUCED	NOTES
MOUTH	Salivary glands	Saliva	Salivary amylase	Starch	Maltose	Slightly acid or neutral. Mucus helps form bolus. Water lubricates food.
STOMACH	Gastric glands (in stomach lining)	Gastric juice	Pepsin	Proteins	Peptides	0·5% hydrochloric acid also secreted, provides acid medium for pepsin and kills most bacteria. No absorption except of alcohol.
			(Rennin)	(Milk protein)	(Clots it)	
DUODE-NUM	Pancreas	Pancreatic juice	Trypsin	Proteins and peptides	Amino acids	Two other protein-digesting enzymes are present. Bile emulsifies fats and aids their absorption. Duodenum contents slightly acid.
			Amylase	Starch	Maltose	
	(Liver)	(Bile)	Lipase	Fats	Fatty acids and glycerol	
ILEUM	The glands between the villi produce mucus but few, if any, digestive enzymes	Intestinal juice contains an enzyme which activates pancreatic trypsin	Pancreatic enzymes still active	Peptides	Amino acids	These final stages of digestion take place in the ileum with the aid of pancreatic enzymes. Some digestion occurs in the epithelial cells of the villi. The main function of the ileum is the absorption of the digested products.
				Fats	Fatty acids and glycerol	
				Maltose	Glucose	
				Sucrose	Glucose and fructose	
				Lactose	Glucose and galactose	
COLON	Bacterial enzymes produce fatty acids from vegetable fibre					Absorption of water

Utilization of digested food

The products of digestion are carried round the body in the blood plasma. From the blood, most living cells are able to absorb and metabolize glucose, fats and amino acids.

(a) *Glucose.* During respiration in the protoplasm, glucose is oxidized to carbon dioxide and water (*see* p. 47). This reaction releases energy to drive the many chemical processes in the cell, and in specialized cells produces, for example, contraction (muscle cells) and electrical changes (nerve cells).

(b) *Fats.* Fats are incorporated into cell membranes and other structures in cells. Fatty acids are oxidized in muscles to provide energy for muscle contraction. Twice as much energy can be obtained from fats as from glucose.

(c) *Amino acids* are absorbed by cells and reassembled to make proteins (p. 82). These proteins may form visible structures such as the cell membrane and other components of the protoplasm or the proteins may be enzymes which control and co-ordinate the chemical activity within the cell.

Amino acids not required for building proteins are de-aminated in the liver, that is, their nitrogen is removed and the residue is used in the same way as carbohydrate, namely oxidized, or converted to glycogen and stored.

Storage of digested food

If the quantity of food taken in exceeds the energy requirements of the body or the demand for structural materials, it is stored in one of the following ways:

(a) *Glucose* (Fig. 18.8). The concentration of glucose in the blood of a person who has not eaten for eight hours is usually between 90 and 100 mg/100 cm^3 blood. After a meal containing carbohydrate, the blood sugar level may rise to 140 mg/100 cm^3 but two hours later, the level returns to about 95 mg.

The sugar not required immediately for the energy supply in the cells is converted in the liver and in the muscles to glycogen. The glycogen molecule is built up by combining many glucose molecules in a long branching chain rather similar to the starch molecule. About 100 g of this insoluble glycogen is stored in the liver and about 300 g in the muscles. When the blood sugar level falls below 80 mg/100 cm^3, the liver converts its glycogen back to glucose and releases it into the circulation. The muscle glycogen is not normally returned to the circulation but is used by active muscle as a source of energy in much the same way as glucose.

The glycogen in the liver is a "short-term" store, sufficient for about only six hours if no other glucose supply is available. Excess glucose not stored as glycogen is converted to fat and stored in the fat cells of the fat depots. (See below.)

(b) *Fats.* Certain cells can accumulate drops of fat in their cytoplasm. As these drops increase in size and number, they join together to form one large globule of fat in the middle of the cell, pushing the cytoplasm into a thin layer and the nucleus to one side (Figs. 18.9 and 18.10). Groups of fat cells form *adipose tissue* beneath the skin and in the connective tissue of most organs. (Plate 25, p. 110.)

Unlike glycogen, there is no limit to the amount of fat stored and because of its high energy value it is an important reserve of energy-giving food.

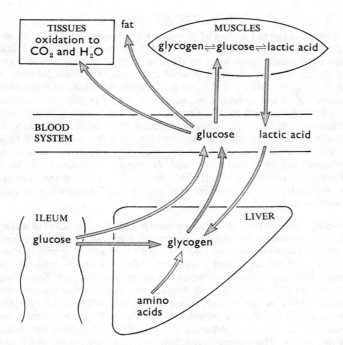

Fig. 18.8   Carbohydrate metabolism

(*Figs. 18.8 and 18.11 by kind permission of Bell, G. H., Davidson, J. N., Scarborough, H. (1959) Textbook of Physiology and Biochemistry, 4e, Edinburgh, Livingstone*)

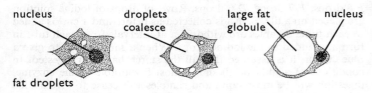

Fig. 18.9   Accumulation of fat in a fat cell

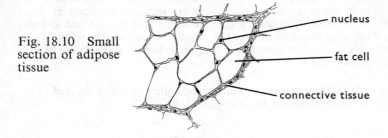

Fig. 18.10   Small section of adipose tissue

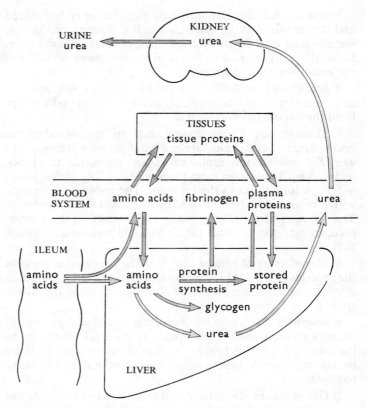

Fig. 18.11   Protein metabolism

(*c*) *Amino acids* (Fig. 18.11). Amino acids are not stored in the body. Those not used in protein formation are deaminated. The protein of the liver and tissues can act as a kind of protein store to maintain the protein level in the blood but absence of protein in the diet soon leads to serious disorders.

The rate of oxidation of glucose and its conversion to glycogen or fat is controlled by hormones (p. 144). When intake of carbohydrate and fat exceeds the energy requirements of the body, the excess will be stored mainly as fat. Some people never seem to get fat no matter how much they eat, while others start to lay down fat when their intake only marginally exceeds their needs. Putting on weight is unquestionably the result of eating more food than the body needs but it is not clear why individuals should differ so much in their reaction. The explanation probably lies in the hormonal balance which, to some extent, is determined by heredity. A slimming diet designed to reduce calorific intake must, nevertheless, always include the essential amino acids, vitamins, mineral salts and certain essential fatty acids.

The liver

The liver is a large, reddish-brown organ which lies just below the diaphragm and partly overlaps the stomach (Fig. 18.5). In addition to a supply of oxygenated blood from the *hepatic artery*, it receives all the blood which leaves the alimentary canal. It has a great many important functions, some of which are described below. (*See* also Figs. 18.8 and 18.11.)

1. **Regulation of blood sugar.** The liver is able to convert glucose, amino acids and other substances to an insoluble carbohydrate, *glycogen*. Some of the glucose so converted may be taken from the *hepatic portal* vein carrying blood, rich in digested food, from the ileum to the liver. About 100 g glycogen is stored in the liver of a healthy man. If the concentration of glucose in the blood falls below about 80 mg/100 cm³ blood, some of the glycogen stored in the liver is converted by enzyme action into glucose and it enters the circulation. If the blood sugar level rises above 160 mg/100 cm³, glucose is excreted by the kidneys. A blood glucose level below 40 mg/100 cm³ affects the brain cells adversely, leading to convulsions and coma. By helping to keep the glucose concentration between 80 and 150 mg the liver prevents these undesirable effects and so contributes to the homeostasis (*see* below) of the body. (*See* Fig. 19.2 for circulatory supply to liver.)

2. **Formation of bile.** Green and yellow pigments are formed when the red blood cells break down. These pigments are removed from the blood by the liver and excreted in the bile. The liver also produces bile salts which play an important part in the emulsification and subsequent absorption of fats (p. 88).

Bile is produced continuously by the liver cells, but stored and concentrated in the gall bladder. It is discharged through the bile duct into the duodenum when the acid chyme arrives there. Bile salts are reabsorbed with the fats they emulsify and eventually return to the liver.

3. **Storage of iron.** Millions of red blood cells break up every day. In the liver their decomposition is completed and the iron from the haemoglobin is stored.

4. **Deamination.** Excess amino acids are not stored in the body. Amino acids which are not built up into proteins and used for growth and replacement are converted to carbohydrates by the removal from the molecule of the *amino group*, $-NH_2$, which contains the nitrogen. The residue can be converted to glycogen, being stored or oxidized to release energy. The nitrogen of the amino group is converted in the liver to *urea*, an excretory product that is constantly eliminated by the kidneys.

5. **Manufacture of plasma proteins.** The liver makes most of the proteins found in blood plasma, including fibrinogen which plays an important part in the clotting action of the blood (p. 95).

6. **Body heat.** The above list shows that a great many chemical changes go on in the liver and many of them release energy in the form of heat. This heat is distributed throughout the body by the circulatory system and helps to maintain the body temperature.

7. **Use of fats in the body.** When fats stored in the body are required for use in providing energy, they travel in the blood stream from the fat depots. Some are used directly by the muscles and some are oxidized in the liver to substances which can be oxidized by other tissues to release energy.

8. **Detoxication.** Poisonous compounds, produced in the large intestine by the action of bacteria on amino acids, enter the blood, but on reaching the liver are converted to harmless substances, later excreted in the urine. Many other chemical substances normally present in the body or introduced as drugs are modified by the liver before being excreted by the kidneys. The hormones, for example, are converted to inactive compounds in the liver so limiting their period of activity in the body.

9. **Storage of vitamins.** The fat-soluble vitamins A and D are stored in the liver. This is the reason why animal liver is a valuable source of these vitamins in the diet. The liver also stores a product of the vitamin B_{12}. This product is necessary for the normal production of red cells in the bone marrow.

Homeostasis

A complete account of the functions of the liver would involve a very long list. It is most important, however, to realize that the one vital function of the liver, embodying all the details outlined above, is that it helps to maintain the concentration and composition of the body fluids, particularly the blood.

Within reason, a variation in the kind of food eaten will not produce changes in the composition of the blood.

If this *internal environment*, as it is called, were not so constant, the chemical changes that maintain life would become erratic and unpredictable so that with quite slight changes of diet or activity the whole organization might break down. The regulation of the internal environment is called *homeostasis* and is discussed again on pp. 94, 107, 108 and 145.

PRACTICAL WORK

To investigate the action of pepsin on egg-white. The white of one egg is stirred into 500 cm³ tap water. The mixture is boiled and filtered through glass wool to remove large particles. About 2 cm of the cloudy suspension is poured into each of 4 tubes labelled *A* to *D*. To *A* is added 1 cm³ 1 per cent pepsin solution, to *B* is added 3 drops of bench hydrochloric acid, to *C* 1 cm³ pepsin solution and 3 drops of acid and to *D* 3 drops of acid and 1 cm³ boiled pepsin solution. All four test-tubes are placed in a beaker of water at 35–40°C for 5–10 minutes, after which time the contents of tube *C* will be clear. The change from a cloudy suspension to a clear solution suggests that the solid egg-white particles have been digested to soluble products. This result, and those of the controls *A*, *B* and *D*, support the idea that pepsin digests egg-white in acid conditions.

The digestive action of saliva on starch. Saliva is collected in a test-tube after rinsing the mouth to remove traces of food, and about 1 cm³ is added to each of two test-tubes containing approximately 2 cm³ of 2 per cent starch solution. After 5 minutes, one tube is tested with iodine solution while the other is tested with Benedict's reagent as described on p. 85. Failure to obtain a blue colour with iodine indicates that starch is no longer present, while a red precipitate with Benedict's reagent shows that a sugar has been produced. A control is conducted by repeating the experiment using boiled saliva. The theory that saliva contains an enzyme which can change starch to sugar is supported by this experiment.

The effect of acidity and alkalinity on an enzyme reaction. Six test-tubes, labelled *A* to *F*, each have 5 cm³ 1 per cent starch solution placed in them. Acid or alkali is added as follows: 0.1 M sodium bicarbonate solution, *A* 10 drops, *B* 4 drops; 0.1 M hydrochloric acid, *F* 8 drops, *E* 7 drops, *D* 6 drops. Rows of drops of iodine solution are placed on a tile. Saliva is collected as before and 1 cm³ is added to each tube. Samples are withdrawn at intervals from each tube in turn and added to the iodine drops. When a sample *fails* to give a blue colour, it is assumed that all the starch has been digested. In conditions of acidity or alkalinity most favourable to the enzyme, digestion will be most rapid and samples will cease to give a blue colour after a short time.

The effect of temperature on an enzyme reaction. Three test-tubes each containing 10 cm³ 2 per cent starch solution are placed separately in beakers of ice-water, cold tap water, warm water at about 40°C. After 5 minutes, 1 cm³ saliva is added to each and samples tested as described above. The tube at the temperature most favourable to starch digestion will give the first samples which fail to turn iodine blue.

[*Note:* these experiments and several others are fully detailed in the laboratory manual *Enzymes—see* p. 1.]

QUESTIONS

1. List the chemical changes undergone by (*a*) a molecule of starch from the time it is placed in the mouth to its ultimate use in providing energy, (*b*) a molecule of protein from the time it is swallowed to the time when its components are used in a cell (other than in the liver).
   In each case, state where the changes are taking place.
2. Write down the menu for your breakfast and lunch (or supper); indicate the principal food substances present in each component of the meal and state the final digestion product of each and the use your body is likely to have made of them.
3. What advantage is it to an animal to take food into its alimentary canal for digestion rather than digest it externally as do the fungi?
4. Herbivorous animals have very long intestines with a large caecum and appendix but carnivorous animals have a relatively shorter intestine with small caecum and appendix. In what ways are these differences related to the differences in diet? (*See* also p. 125.)

19 Blood, its Composition, Function and Circulation

COMPOSITION

BLOOD consists of a suspension of cells in an aqueous solution. In an adult man there are five to six litres of blood in the body.

Cells

Red cells (erythrocytes) (Fig. 19.1a and Plate 20). Minute, biconcave discs, the red cells consist of spongy cytoplasm in an elastic membrane. They have no nuclei. In their cytoplasm is a red pigment, *haemoglobin*, which is a protein with iron in its molecule. It has an affinity for oxygen and readily combines

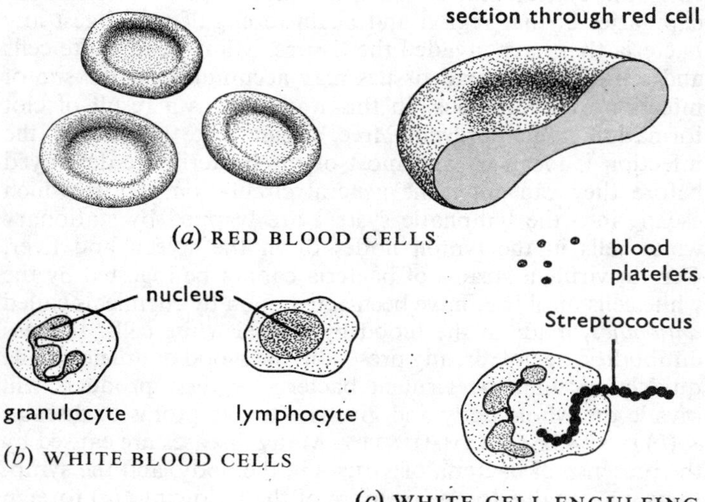

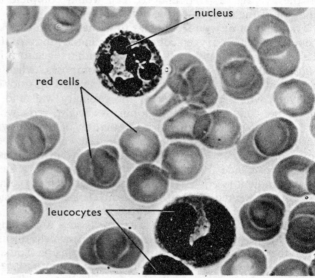

section through red cell

(a) RED BLOOD CELLS

nucleus

granulocyte   lymphocyte

(b) WHITE BLOOD CELLS

blood platelets

Streptococcus

(c) WHITE CELL ENGULFING A STREPTOCOCCUS

Fig. 19.1   Blood cells

with it in conditions of high oxygen concentration. It forms an unstable compound called *oxy-haemoglobin* which, however, in conditions of low oxygen concentration readily breaks down and releases the oxygen. This property makes it most efficient in transporting oxygen from the lungs to the tissues.

The red cells are made in the red bone-marrow of the short bones such as the sternum, ribs and vertebrae. There are about five-and-a-half million in a cubic millimetre of blood. A red blood cell lasts for about four months, after which it breaks down and is disintegrated in the liver or spleen. About 200,000,000,000 are formed and destroyed each day, which means that about 1 per cent of the total is replaced daily.

White cells (leucocytes) (Fig. 19.1b and Plate 20). There are about 600 red cells to every white cell. The actual numbers vary between 4000 and 13,000/mm^3. Various kinds of white cell occur; they are made in the bone marrow, the lymph nodes or the spleen. *Granulocytes* are irregular in shape, can change their form, and all have a lobed nucleus. They can move by a flowing action of their cytoplasm and can pass out of blood capillaries by squeezing between the cells of the capillary wall. They ingest and destroy bacteria and dead cells by flowing round, engulfing and digesting them (Fig. 19.1c). Granulocytes accumulate at

the site of an injury or infection and devour invading bacteria and damaged tissue, so preventing the spread of harmful bacteria as well as accelerating the healing of the infected region. *Lymphocytes* produce antibodies (p. 94).

Platelets are cell fragments budded off from special, very large cells in the red bone marrow and they play an important part in the clotting action of the blood. There are about 400,000 of them in a cubic millimetre of blood and they appear as tiny, round or oval structures.

Plate 20.   RED AND WHITE CELLS FROM THE HUMAN BLOOD (× 1500)

Plasma. The liquid part of the blood is called plasma, which is a solution in water of many compounds. Some of the most important of these compounds are sodium chloride, sodium bicarbonate, glucose, amino acids and proteins including *albumin*, *fibrinogen* and the *globulin antibodies*; *hormones*, *urea* and other nitrogenous compounds. In the plasma, digested food, carbon dioxide and excretory products are carried round the body.

Serum is blood plasma from which the fibrinogen has been removed.

FUNCTIONS OF THE BLOOD

It will be convenient at this point to distinguish between (a) the functions of the blood as the agent replenishing the tissue fluid surrounding the cells, and (b) the circulation of blood.

(a) Homeostatic functions

All the cells of the body are bathed by a fluid, tissue fluid, derived from plasma, which supplies them with the food and oxygen necessary for their living chemistry, and removes the products of their activities which, if they accumulated, would poison the cells.

The composition of the blood plasma is very precisely regulated by the liver and kidneys so that, within narrow limits, the living cells are soaked in a liquid of unvarying composition. This provides them with the environment they need and enables them to live and grow in the most favourable conditions. By delivering oxygen and nutrients to the tissue fluid and removing the excretory products, the blood fulfils a homeostatic function (p. 108), maintaining the constancy of the internal environment. (*See* p. 98 for further details.)

(*b*) Circulation

The movement of the blood in vessels round the body constantly changes the fluid surrounding the living cells so that fresh supplies of oxygen and food are brought in as fast as they are used up and poisonous end-products are not allowed to accumulate. The following account is concerned principally with the circulation as a transport system, rather than with the chemical properties of blood fluid as an internal environment.

On the average, a particular red cell would complete the circulation of the body in 45 seconds.

1. **Transport of oxygen from lungs to tissues.** When exposed to the relatively high oxygen concentration in the lungs (p. 103) the haemoglobin in the red cells combines with oxygen forming *oxy-haemoglobin.* Oxy-haemoglobin decomposes when it reaches an active tissue where oxygen is being used up, and sets free oxygen which diffuses out of the capillary wall and so reaches the cells. Oxy-haemoglobin is bright red, while haemoglobin is a dark red. The combination of oxygen with haemoglobin as soon as it enters the red cell, effectively removes oxygen from solution so that its concentration as a dissolved gas inside the cell is kept very low. Thus a steep diffusion gradient (*see* p. 62) is maintained between the source of oxygen and the red cell and, as a result, the rate of diffusion of oxygen into the erythrocyte is rapid.

2. **Transport of carbon dioxide from the tissues to the lungs.** Carbon dioxide produced from actively respiring cells diffuses through the capillary wall and dissolves in the plasma. Some of it enters the red cells and some of it forms sodium hydrogen carbonate in the plasma. In the lungs (p. 103) it is released, diffuses into the air sacs, and is expelled.

3. **Transport of nitrogenous waste from the liver to the kidneys.** When the liver changes amino acids into glycogen (p. 92), the amino ($-NH_2$) part of the molecules is changed into the nitrogenous waste product, *urea.* This substance is carried away in the blood circulation. When the blood passes through the kidneys, the urea is removed and excreted (p. 107).

4. **Transport of digested food from the ileum to the tissues.** The soluble products of digestion pass into the capillaries of the villi lining the ileum (p. 89). They are carried in solution by the plasma and after passing through the liver enter the general circulation. Glucose and amino acids diffuse out of the capillaries and into the cells of the body. Glucose may be oxidized in a muscle, for example, and provide the energy for contraction; amino acids will be built up into new proteins and make new cells and fresh tissues.

5. **Distribution of hormones.** Hormones are chemicals which affect the rate of vital processes in the body. They are carried in the blood plasma, from the glands which make them, all round the body. When they reach certain organs such as the heart they affect the rate at which these organs work (p. 144).

6. **Distribution of heat and temperature control.** Muscular and chemical activity release heat. These processes occur more rapidly in some parts of the body than others; for example, chemical activity in the liver and muscular action in the limbs. The heat so produced locally is distributed all round the body by the blood and in this way an even temperature is maintained in all regions.

The diversion of blood to or away from the skin also plays a part in keeping the temperature constant (*see* p. 110).

7. **Formation of clots.** When a blood vessel is cut open, or its lining damaged, the blood platelets and damaged tissue produce chemicals which help to convert the protein *fibrinogen* to *fibrin.* This makes a network of fibres across the wound within which red cells become entangled, forming a clot which stops further loss of blood and prevents entry of bacteria and poisons. The platelets also adhere to the damaged area and help to form a plug before causing the fibrin to precipitate. The dried clot eventually becomes a scab which protects the damaged area while new skin is forming.

8. **Prevention of infection.** (*a*) INFECTED WOUNDS. Normally the skin provides a barrier to the entry of any bacteria. The layer of dead cells on the skin provides a mechanical barrier while the mucus and chemicals of the alimentary canal offer a chemical defence. If the skin is broken, however, and bacteria enter the cut, certain of the white cells migrate through the capillaries in that region and begin to engulf and digest any bacteria that have invaded the tissues. Many dead white cells and self-digested, dead tissues may accumulate at the site of infection and form pus. In this way, and as a result of clot formation which prevents free circulation, the site of the infection is localized and most of the bacteria are destroyed before they can enter the general circulation. Those which escape into the lymphatic system are trapped by stationary white cells in the lymph nodes or in the spleen and liver. Certain virulent strains of bacteria cannot be ingested by the white cells until they have been acted upon by chemicals called *antibodies,* made in the blood by special white cells. If these antibodies are not already present in the blood or are not made quickly enough, the virulent bacteria or their products will invade the whole body and give rise to symptoms of disease.

(*b*) DISEASE AND IMMUNITY. Many diseases are caused by the presence of bacteria or viruses in the body, and the symptoms may be due to one or more of the following: (*a*) foreign proteins of the bacteria themselves; (*b*) the poisonous chemicals (usually proteins) called *toxins*, which are produced by the bacteria; (*c*) the breakdown products of the infected tissue. Recovery from the disease and subsequent immunity depend to a large extent on the production of *antibodies* in the blood. These antibodies are proteins released into the plasma and they may affect bacteria or their products in a number of ways:

(*a*) *opsonins* adhere to the outer surface of bacteria and so make it easier for the phagocytic white cells to ingest them,

(*b*) *agglutinins* cause bacteria to stick together in clumps; in this condition the bacteria cannot invade the tissues,

(*c*) *lysins* destroy bacteria by dissolving their outer coats, and

(*d*) *anti-toxins* combine with and so neutralize the poisonous toxins produced by bacteria.

The substances which stimulate the production of antibodies are called *antigens.*

When the organism recovers from the disease the antibodies remain for only a short time in the circulatory system but the ability to produce them is greatly increased so that a further invasion by bacteria or viruses is likely to be stopped at once and the person is said to be "immune" to the disease. People may possess this immunity from birth, they may acquire it after recovering from an attack as in measles, or it may be induced in them by vaccination or inoculation. Natural or acquired immunity may occur because disease bacteria are present in the body without being sufficiently numerous or suitably placed to produce disease symptoms.

A **vaccine** is a preparation of killed disease bacteria or viruses, or forms of these treated in such a way as to prevent their reproduction. When these are injected into the blood stream the organism undergoes a mild form of the disease and its cells manufacture an excess of antibodies. In this way immunity is artificially acquired. The period of immunity, during which antibodies can be produced rapidly, varies from a few months to many years, according to the nature of the infection.

Serum. The blood of a person or animal which has recently recovered from a disease will contain antitoxins and antibodies. If the cells and fibrinogen are removed from a sample of this blood a serum is obtained which, when injected into other people, may give temporary immunity or cure them if they already have the disease. Sera for treating tetanus and snake bites are prepared from horse's blood. The horse is injected with diluted poison which stimulates the formation of antitoxins in the blood. Samples of the blood are then taken from the horse, and serum prepared from the samples is used to treat cases.

CIRCULATORY SYSTEM (Figs. 19.2 and 19.3)

The blood is distributed round the body in vessels, most but not all of them tubular, and varying in size from about 1 cm to 0·001 mm in diameter. They form a continuous system, communicating with every living part of the body (Plate 21). Blood flows in them, always in the same direction, passing repeatedly through the heart, the muscular contractions of which maintain the circulation.

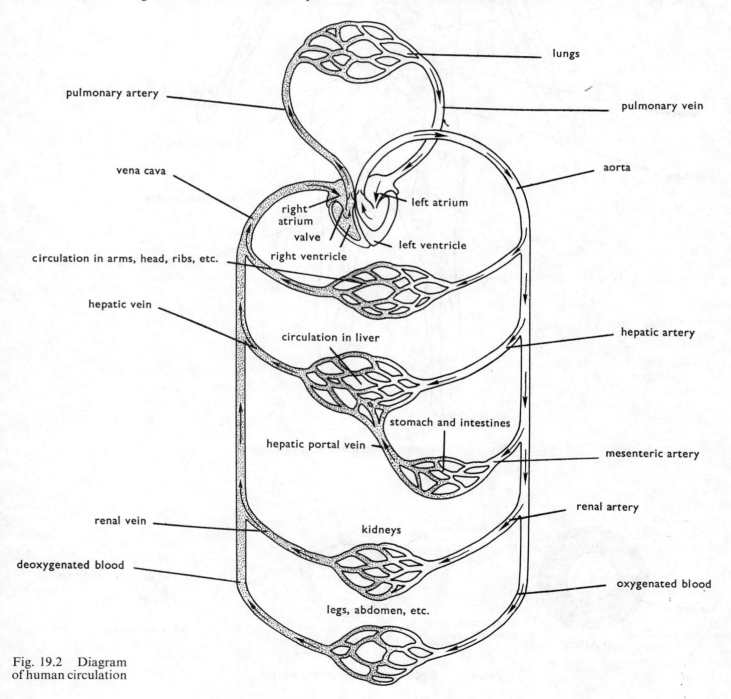

Fig. 19.2 Diagram of human circulation

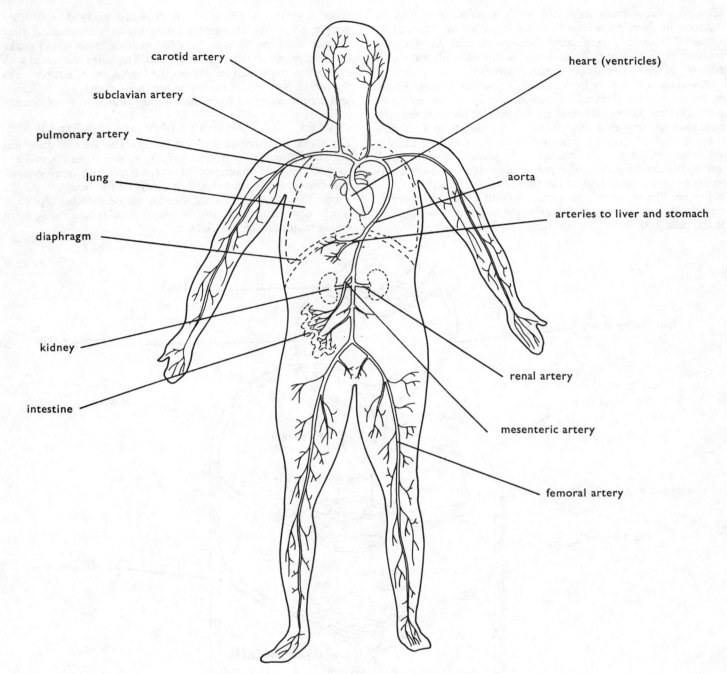

Fig. 19.3   Diagram of human arterial system

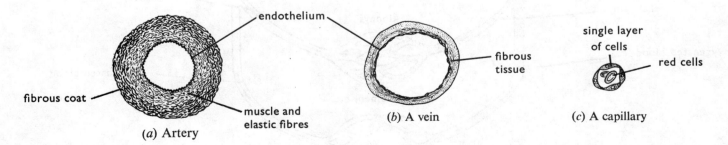

Fig. 19.4   Blood vessels, transverse sections

There are three types of blood vessel, arteries, veins and capillaries, connected to form a continuous system (Plate 22).

Arteries (Fig. 19.4a) are fairly wide vessels which carry blood from the heart to the limbs and organs of the body. They are thick-walled, muscular and elastic and must stand up to the surges of high pressure caused by the heart-beat. The arteries divide into smaller vessels, called *arterioles*, which themselves divide repeatedly until they form a dense network of microscopic vessels permeating between the cells of every living tissue. These final branches are called *capillaries*.

Capillaries (Fig. 19.4c and 19.5) are tiny vessels with walls often only one cell thick. Although the blood seems to be

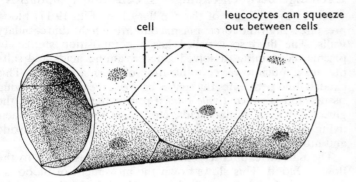

Fig. 19.5   Stereogram of blood capillary

physically confined within the capillary walls, the latter are permeable, allowing water and dissolved substances, other than proteins, to pass in and out. Through these thin walls, oxygen, carbon dioxide, dissolved food and excretory products are exchanged with the tissues round the capillary. The capillary network is so dense that no living cell is far from a supply of oxygen and food (p. 98). In the liver every cell is in direct contact with a capillary. Some capillaries are so narrow that the red cells are squeezed flat in passing through them. Eventually, the capillaries unite into larger vessels, *venules*, which join to form veins and these return blood to the heart.

Veins (Fig. 19.4b) return blood from the tissues to the heart. The blood pressure in them is steady and is less than in the arteries. They are wider and have thinner walls than the arteries. They also have valves (Fig. 19.6) in them which prevent blood

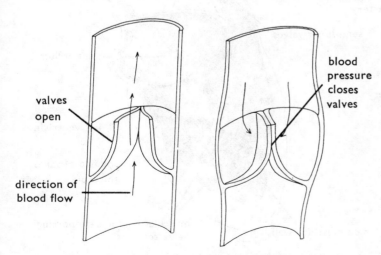

Fig. 19.6   Diagram to show the action of valves in a vein or the semilunar valves in the arteries leaving the heart

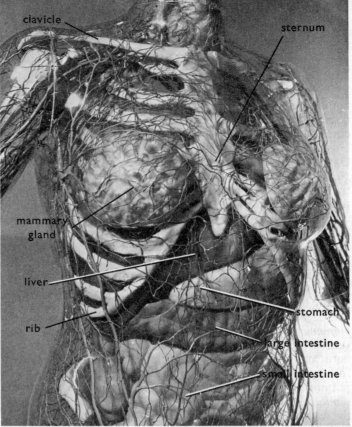

(A model in the Cleveland Health Museum, U.S.A.)

Plate 21.   "JUNO", THE TRANSPARENT WOMAN

flowing away from the heart. Contractions of skeletal muscles during activity compress the veins, so forcing blood along in a direction determined by the valves. This assists the return of blood to the heart. The blood in the veins will usually contain less oxygen and food, and more nitrogenous waste and carbon dioxide, while the arterial blood has a higher concentration of oxygen and dissolved food.

Exceptions to this are the *pulmonary artery* which carries deoxygenated blood to the lungs, the *pulmonary vein* which returns oxygenated blood to the heart, the *hepatic portal vein* to the liver from the alimentary canal which carries blood rich in glucose and amino acids, and the *renal vein* from the kidney where some water, salts and urea have been eliminated.

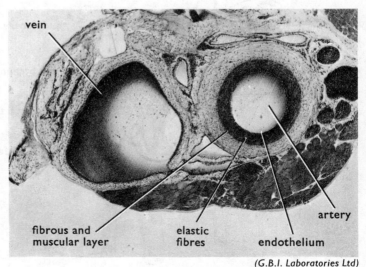

(G.B.I. Laboratories Ltd)

Plate 22.   TRANSVERSE SECTION THROUGH
AN ARTERY AND VEIN (×20)

The heart

The heart is a muscular pumping organ. It is thought that it has evolved from the highly muscular region of an artery. It is divided into four chambers; the left and right sides do not communicate. The upper chambers, the *atria*, which are relatively thin-walled, receive blood from the veins (Fig. 19.8). Oxygenated blood from the lungs enters the *left atrium* via the *pulmonary vein* and deoxygenated blood from the body enters the *right atrium* from the *venae cavae* (Fig. 19.9). Relaxation of the ventricular muscle allows the ventricles to expand and fill with blood which flows in from the atria and veins (Fig. 19.10). Simultaneous contraction of both atria forces the blood they contain into the corresponding ventricles and, about 0·1 sec. later, the ventricles contract simultaneously, expelling their blood into the arteries and round the body. Both ventricles have thick muscular walls but those of the left are thicker, having to pump blood all round the entire body via the *aorta*. The *right ventricle* pumps blood to the lungs through the *pulmonary artery*. When the ventricles contract, blood is prevented from returning to the atria and veins by the closure of parachute-like valves between the atria and ventricles. Powerful contraction of the ventricles forces blood into the aorta and pulmonary arteries. When the ventricles relax, the pocket-like *semilunar valves* in these two arteries are closed and prevent the return of blood to the ventricles. The heart contracts about 70 times a minute when an adult person is at rest, but this rate increases to 100 or more during activity or excitement. In a sparrow the rate is nearly 500 a minute. The heart's rhythmic muscular contraction is basically automatic and needs no nervous stimulation to bring it about. If kept in the right solution of salts a frog's heart will continue to beat for some hours after removal from the body, and the same is true of a mammalian heart if an artificial circulation to the heart muscle is maintained. Nervous stimulation is, however, superimposed on the heart's natural rhythm and helps to maintain and control its rate. An increased heartbeat increases the speed with which the blood is supplied to the tissues and so allows a greater rate of activity. The *coronary arteries* shown in Fig. 19.7 carry oxygenated blood to the ventricular muscle whose constant activity demands an unceasing supply of food and oxygen.

Blood pressure. To force blood through a capillary system and to overcome atmospheric pressure, which tends to flatten the vessels, a fairly high pressure must be developed by the heart. This pressure varies according to the part of the body considered and the age of the individual, but an average pressure produced in the ventricle when it contracts is equal to 130 mm of mercury.

Exchange between capillaries, cells and lymphatics

At the arterial end of the capillary bed (Fig. 19.11) blood pressure is high and forces plasma out through the thin capillary walls. The fluid so expelled has a composition similar to plasma, containing dissolved glucose, amino acids and salts but has a much lower concentration of plasma proteins. This exuded fluid permeates the spaces between the cells of all living tissues and is called *tissue fluid*. From it the cells extract the glucose, oxygen, amino acids, etc. which they need for their living processes and into it they excrete their carbon dioxide and nitrogenous waste.

The narrow capillaries offer considerable resistance to the flow of blood. This slows down the movement of blood, so facilitating the exchange of substances by diffusion between the plasma and the tissue fluid (Fig. 19.12). The capillary resistance also results in a drop of pressure so that at the venous end of a capillary bed the blood pressure is less than that of the tissue fluid and the latter passes back into the capillaries.

The fact that the plasma contains more proteins than the tissue fluid gives the blood a low osmotic potential (p. 63) which tends to cause water to pass from the tissue fluid into the capillary. At the arterial end of the capillary network, the blood pressure is greater than this osmotic pressure, so forcing water out, but at the venous end water from the tissue fluid enters the capillary by osmosis.

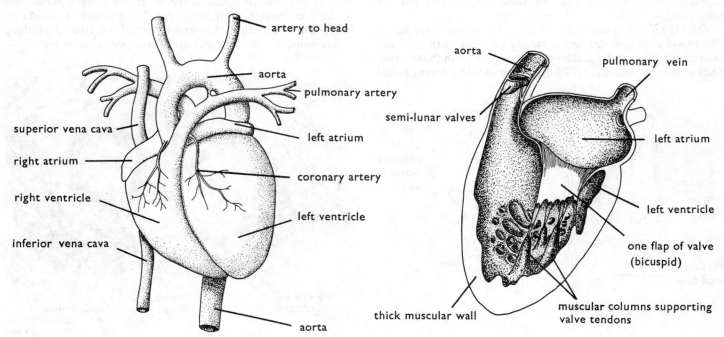

Fig. 19.7   External view of mammalian heart
(pulmonary veins not shown)

Fig. 19.8   Diagram of heart cut open

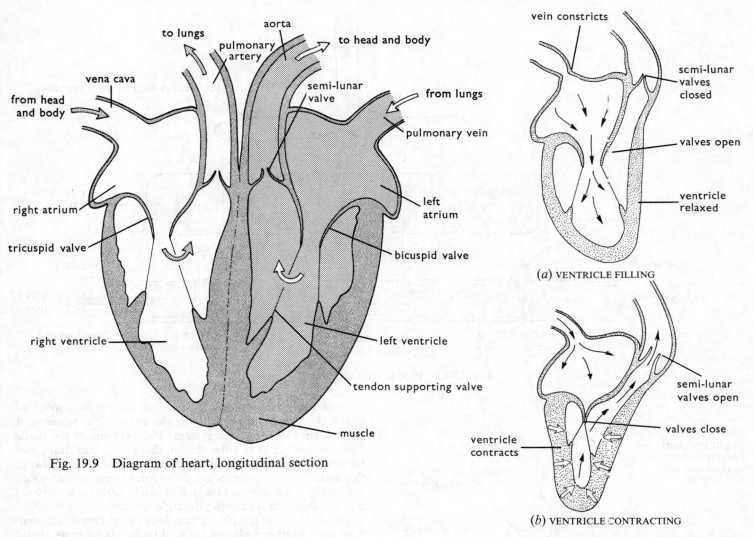

Fig. 19.9   Diagram of heart, longitudinal section

(a) VENTRICLE FILLING

(b) VENTRICLE CONTRACTING

Fig. 19.10   Diagram of heart beat
(only right side shown)

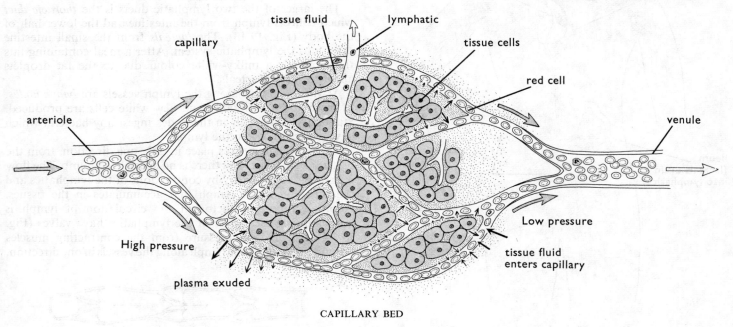

CAPILLARY BED

Fig. 19.11   Relationship between capillaries, cells and lymphatics

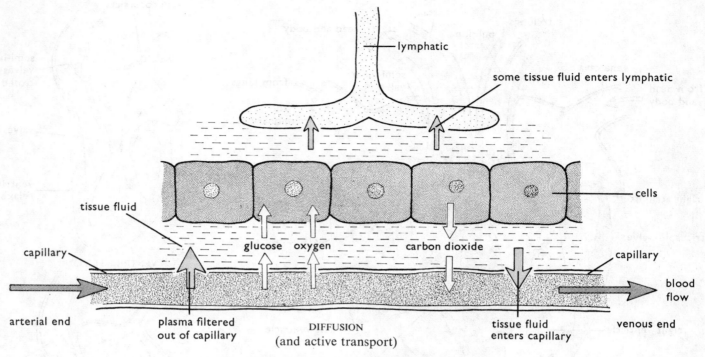

Fig. 19.12   Blood, tissue fluid and lymph

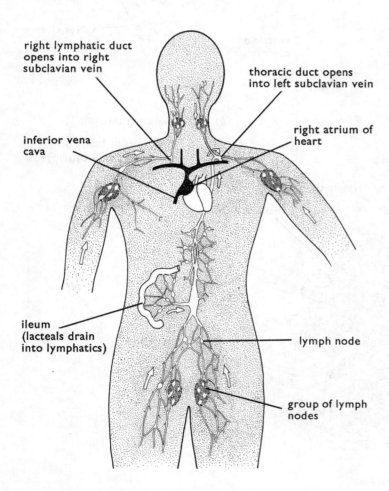

Fig. 19.13   Main drainage routes of
lymphatic system

Lymphatic system. The capillaries are not the only route by which the tissue fluid returns to the circulation. Some of it returns via the lymphatic system. The proteins in the tissue fluid are unable to re-enter the capillaries but can drain into blindly-ending, thin-walled vessels which are found between the cells. These *lymphatics* join up to form larger vessels which eventually unite into two main ducts and empty their contents into the large veins entering the right atrium.

The fluid in the lymphatic vessels is called *lymph*. Its composition is similar to plasma but it contains less proteins. It also contains a certain type of white cell, *lymphocyte*, which is made in the lymph glands and produces antibodies.

The larger of the two lymphatic ducts is the *thoracic duct* which collects lymph from the intestine and the lower half of the body (Fig. 19.13). The *lacteals* from the small intestine open into the lymphatic system. After a meal containing fats the lymph is a milky-white colour due to the fat droplets absorbed in the lacteals.

At various points along the lymph vessels are *lymph nodes*. In these nodes antibodies and new white cells are produced. Stationary white cells in the nodes ingest any bacteria which have gained access to the lymph.

The lymph flow takes place in only one direction, from the tissues to the heart, and there is no specialized pump. The flow is brought about partly by contractions of the lymphatics and the pressure of the lymph that accumulates in the tissues. Another important factor in the circulation of lymph is muscular exercise. Some of the lymphatics have valves (Fig. 19.14) in them and pressure from the contracting muscles around them forces the lymph along the vessels in one direction.

Fig. 19.14   Deep lymphatic vessel cut open to show valves

PRACTICAL WORK

1. **Blood smear.** To examine blood under the microscope a thin film must be spread on a slide. The finger is pricked with a sterile lancet which must be used by one person only. One technique is to squeeze the first finger between the middle-finger and thumb of the same hand and pierce the soft area at the tip. The capillaries are nearer the surface, however, on the side of the top joints of the fingers, just to one side of the insertion of the finger-nail.

The drop of blood is placed at one end of a clean, dry slide. A second slide is placed as shown in Fig. 19.15 so that the blood drop spreads across the region of contact by capillary attraction. By

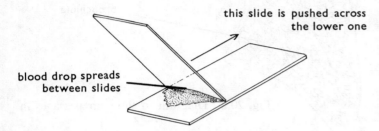

this slide is pushed across the lower one

blood drop spreads between slides

Fig. 19.15   Making a blood smear

pushing the top slide across the lower one, a thin film of blood is made which will show red cells quite clearly under the microscope.

To see white cells, it is best to study a prepared slide.

2. **Valves in the veins.** If a light tourniquet is applied to the upper arm the veins in the forearm can be made to stand out. The lower end of one of these is blocked off near the wrist by pressing it with a finger. The blood can be expelled from the vein by running a finger with light pressure along its length towards the elbow. When this has been done the vein will remain collapsed up to a certain point; above this the vein will fill up and swell once more. The boundary between the filled and collapsed regions indicates the position of a non-return valve.

3. **Effect of gravity on circulation.** Allow the left arm to hang straight down at the side of the body. Open and clench the hand repeatedly between once and twice a second. It should be possible to continue these movements for 3 or 4 minutes or up to 500 times without feeling acute discomfort. After a period of rest, hold the arm straight up and repeat the exercises. After about one minute, or 100 closures, the movement becomes almost impossible. One reason for this is the reduced blood supply resulting from the retarding effect of gravity on the circulation. It is interesting to speculate on which particular aspect of circulation, i.e. oxygen transport, waste removal, etc., is responsible for the fatigue.

4. **Capillaries.** These are best seen in the web of a frog's foot, tadpole's tail or newt larva's external gills where the red cells can be seen streaming through the narrow vessels. (*See* p. 164, Fig. 30.5.)

Our own capillaries can be seen by soaking the back of the top joint of a finger in a clearing agent such as cedar-wood oil and examining, by reflected light under a microscope, the area below the nail cuticle. Capillary loops can usually be seen even with a good hand lens.

5. **Pulse rate.** The swelling of the arteries as a result of the surge of pressure from the heart can be felt in certain places and gives an indication of the rate of the heart's contractions. The pulse in the wrist is the most usual region for this. Count the number of pulsations over a period of 30 seconds and make a note of it. Then take some form of exercise, e.g. standing up and getting off a stool once in two seconds for about half a minute, and take the pulse rate again. Find out how long it takes to return to its original rate.

6. **Blood pressure.** A very effective but simply constructed sphygmomanometer is described in *The Science Masters' Book*, Series III, Part 3, p. 191 (John Murray).

QUESTIONS

1. Although the walls of the left ventricle are thicker than those of the right ventricle, the volume of the ventricles is the same. Why is this necessary?
2. State in detail the course taken by (*a*) a glucose molecule and a fat molecule from the time they are ready for absorption in the ileum, and (*b*) a molecule of oxygen absorbed in the lungs, to the time when all three reach a muscle cell in the leg.
3. Why is a person whose heart valves are damaged by disease unable to participate in active sport?
4. A system for transporting substances in solution might as well be filled with water. What advantages has the blood circulatory system over a water circulatory system?
5. What is the advantage to an animal of having capillaries which are (i) very narrow, (ii) repeatedly branched and (iii) very thin-walled?
6. How do you think microscopic animals can survive without having a circulatory system?

20   Breathing

THE various processes carried out by the body, e.g. movement, growth and reproduction, require the expenditure of energy. In animals this energy can be obtained only from the food they eat. Before the energy can be used by the cells of the body it must be set free from the chemicals of the food. This process of liberating energy is called respiration (Chapter 9) and involves the use of oxygen and the production of carbon dioxide.

Oxygen enters the animal's body from the air or water surrounding it. In the less complex animals the oxygen is absorbed by the entire exposed surface of the body, but in the higher animals there are special respiratory areas such as lungs or gills. Excess carbon dioxide is usually eliminated from the same area. In the respiratory organ oxygen combines with the haemoglobin in the blood and is so carried to all living parts of the body where it is used in tissue respiration.

An efficient respiratory organ has a large surface area, a dense capillary network or similar blood supply, a very thin *epithelium* separating the air or water from the blood vessels and, in land-dwelling animals, a layer of moisture over the absorbing surface. In many animals there is also a mechanism which renews the air or water in contact with or near the respiratory surface, a process called *ventilation*. In mammals, the respiratory organs are lungs.

Lungs (Fig. 20.1)

The lungs have a spongy, elastic texture and are enclosed in the *thorax*. They can be expanded or compressed by movements of the thorax in such a way that air is repeatedly taken in and expelled. They communicate with the atmosphere through the wind-pipe or *trachea*, which opens into the *pharynx* (Fig. 18.3). In the lungs, a gaseous exchange takes place; some of the atmospheric oxygen is absorbed and carbon dioxide from the blood is released into the lung cavities.

Lung structure. The trachea divides into two *bronchi* which enter the lungs and divide into smaller branches (Fig. 20.2). These divide further into *bronchioles* which terminate in a mass of little thin-walled, pouch-like air sacs or *alveoli* (Figs. 20.3, 20.4 and Plate 23).

(*a*) AIR PASSAGES. Rings of cartilage keep the trachea and bronchi open and prevent their closing up when the pressure inside them falls during inspiration. The lining of the air passages is covered with numerous *cilia*. These are minute, cytoplasmic hairs which constantly flick to and fro. Mucus is

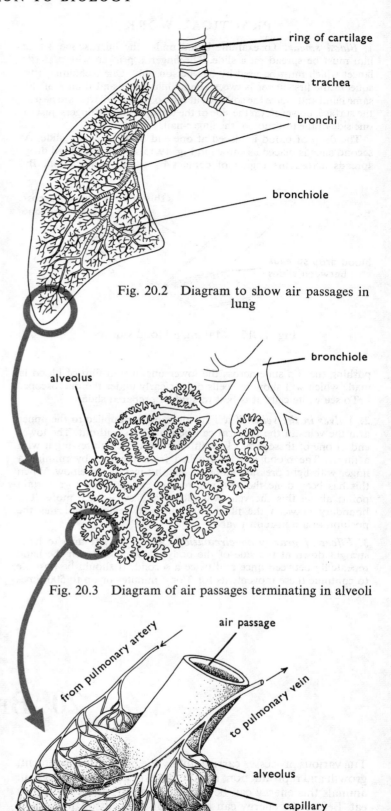

Fig. 20.2   Diagram to show air passages in lung

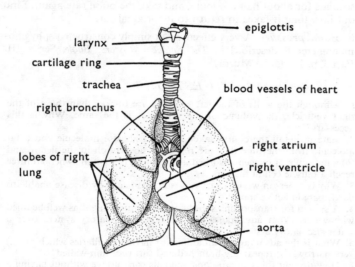

Fig. 20.1   Diagram of lungs showing position of heart

Fig. 20.3   Diagram of air passages terminating in alveoli

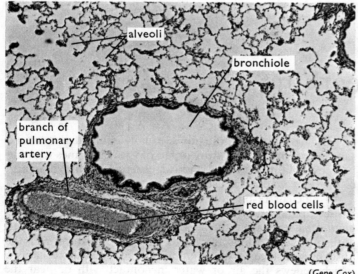

(Gene Cox)

Plate 23.   MICROSCOPIC STRUCTURE OF LUNG TISSUE SEEN IN SECTION (× 100)

Fig. 20.4   Diagram to show relation of blood vessels to alveoli

secreted by glandular cells, also in the lining. Dust particles, bacteria, etc., which are carried in with the air during inspiration become trapped in the mucus film and, by the movements of the cilia, are swept away in it up to the larynx and into the oesophagus where they are swallowed.

The epiglottis and other structures at the top of the trachea prevent food and drink from entering the air passages, particularly during swallowing. Choking and coughing are reflex actions which tend to remove any foreign particles which accidentally enter the trachea or bronchi.

(b) ALVEOLI. The alveoli have thin, elastic walls consisting internally of a single cell layer, or epithelium, and beneath this, a dense network of capillaries (Fig. 20.4) supplied with de-oxygenated blood pumped from the right ventricle through the pulmonary artery. In one human lung there are about 350 million alveoli with a total absorbing surface of about 90 m².

Gaseous exchange (Fig. 20.5)

The lining of the alveoli is covered with a film of moisture. The oxygen concentration in the blood is lower than in the alveolus, hence oxygen in the air space dissolves in the film of moisture and diffuses through the epithelium, the capillary wall, the plasma and into a red cell, where it combines with the haemoglobin (see p. 94). The capillaries reunite and eventually form the pulmonary veins which return the oxygenated blood to the left atrium of the heart. The low concentration of carbon dioxide in the alveoli stimulates the enzyme, *carbonic anhydrase*, in the blood to break down the bicarbonate salts and liberate carbon dioxide. This gas diffuses into the alveoli and is eventually expelled.

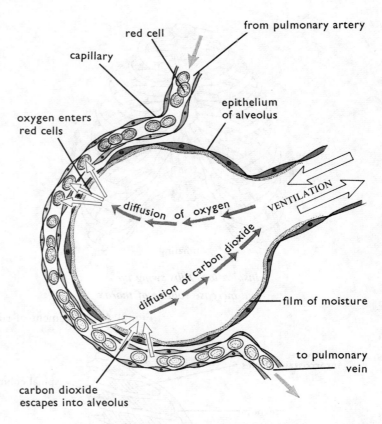

Fig. 20.5   Gaseous exchange in the alveolus

Approximate Composition of Inspired and Expired Air

	Inspired%	Expired%
Oxygen	21	16
Carbon dioxide	0·04	4
Nitrogen	79	79
Water vapour	varies	saturated

Although nitrogen does dissolve in blood plasma, it plays no part in the chemical reactions of the body so the rates of diffusion into and out of the blood are the same.

Diffusion gradient. A steep diffusion gradient (p. 62) of oxygen is maintained by (i) replenishment of air in the air passages by ventilation, (ii) the very short distance between the alveolar lining and the blood, (iii) the combination of oxygen with haemoglobin, so removing oxygen from solution, (iv) the blood flow which constantly replenishes oxygenated blood with deoxygenated blood. Similar factors work in the reverse direction for the diffusion of carbon dioxide. The conversion of bicarbonate to carbon dioxide by carbonic anhydrase raises the concentration of carbon dioxide in the blood above that in the alveolus.

Rate of breathing

The rhythmical breathing movements are usually carried out quite unconsciously about 16 times a minute. They are controlled by a region of the brain which is very sensitive to the carbon dioxide concentration in the blood. If there is a rise in the carbon dioxide concentration of the blood reaching this region of the brain, nerve impulses are automatically sent to the diaphragm and rib muscles which increase the rate and depth of breathing. The concentration of carbon dioxide in the blood is most likely to rise during vigorous activity, and the accelerated rate of breathing helps to expel the rapidly accumulating carbon dioxide and to increase the amount of oxygen in the

blood, so meeting the demands of increased tissue respiration. By regulating the oxygen and carbon dioxide levels in the blood, the lungs are fulfilling a homeostatic function (p. 108). At most times the rate of breathing can be controlled voluntarily, as in singing or when playing a wind instrument.

Lung capacity

The total capacity of the lungs, when fully inflated in an adult man, is about 5½ litres, but during quiet breathing only about 500 cm³ of air is exchanged. This is called *tidal air*. During activity the thoracic movements are more extensive, and deep inspiration can take in another 2 litres while vigorous expiration can expel an additional 1½ litres. The thorax cannot collapse completely, so that 1½ litres of air can never be expelled. This *residual air*, which remains stationary in the alveoli, exchanges carbon dioxide and oxygen by diffusion with the tidal air that sweeps into the bronchi and air passages.

The nose

The ciliated epithelium and film of mucus which line the nasal passages help to trap dust and bacteria. The air is also warmed slightly before it enters the lungs. In addition, in the linings of the nasal cavity there are sensory organs which respond to chemicals in the air and confer a sense of smell.

Voice

The *vocal cords* are two folds protruding from the lining of the larynx. They contain ligaments which are controlled by muscles. When air is passed over them in a certain way they vibrate and produce sounds. The controlling muscles can alter the tension in the cords and the distance between them. In this way they vary the pitch and quality of the sounds produced.

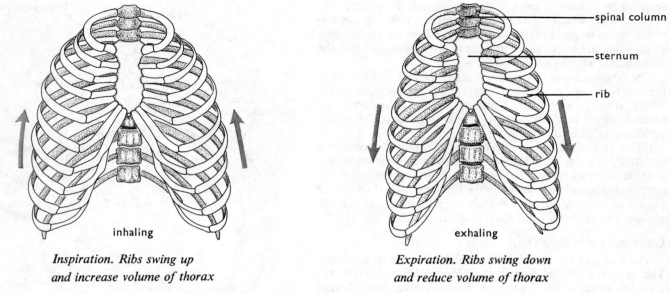

inhaling

*Inspiration. Ribs swing up
and increase volume of thorax*

spinal column

sternum

rib

exhaling

*Expiration. Ribs swing down
and reduce volume of thorax*

Fig. 20.6   Movement of rib cage during breathing

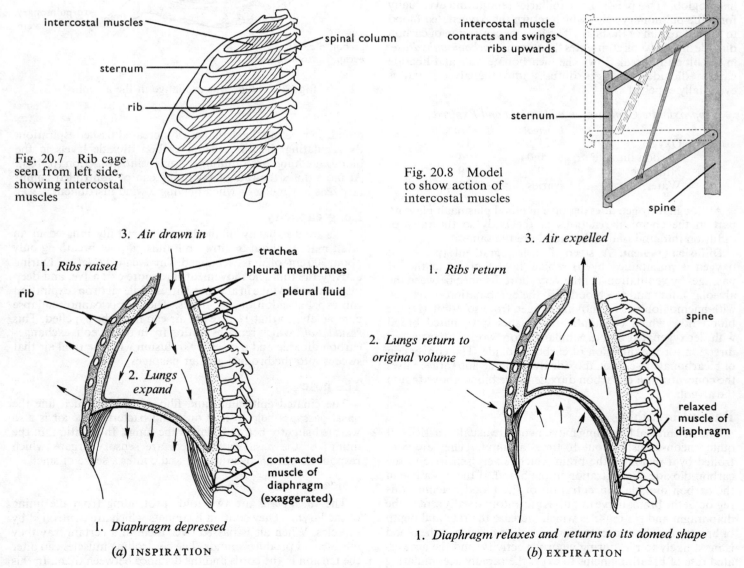

intercostal muscles

spinal column

sternum

rib

Fig. 20.7   Rib cage
seen from left side,
showing intercostal
muscles

intercostal muscle
contracts and swings
ribs upwards

sternum

Fig. 20.8   Model
to show action of
intercostal muscles

spine

3. *Air drawn in*

1. *Ribs raised*

rib

trachea
pleural membranes
pleural fluid

2. *Lungs
expand*

contracted
muscle of
diaphragm
(exaggerated)

1. *Diaphragm depressed*

(*a*) INSPIRATION

3. *Air expelled*

1. *Ribs return*

2. *Lungs return to
original volume*

spine

relaxed
muscle of
diaphragm

1. *Diaphragm relaxes and returns to its domed shape*

(*b*) EXPIRATION

Fig. 20.9   Diagrams of thorax to show mechanism of breathing

Ventilation of the lungs

The exchange of air in the lungs is brought about by muscular movements of the thorax which alter its volume. The thorax is an airtight cavity enclosed by the ribs at the sides and the *diaphragm* below. The diaphragm is a muscular sheet of tissue extending across the body cavity between the thorax and abdomen. At rest, it is dome-shaped, extending upwards into the thoracic cavity, with the liver and stomach immediately below it. Any change in the volume of the thorax is followed by the lungs, which are too thin to oppose the movements.

INSPIRATION. During inspiration the volume of the thorax is increased by two movements (Fig. 20.9).

(*a*) The muscles of the diaphragm contract and cause it to flatten from its domed position.

(*b*) The lower ribs are raised upwards and outwards (Fig. 20.6) by contraction of the intercostal muscles which run obliquely from one rib to the next (Figs. 20.7 and 20.8).

Both these movements increase the volume of the thorax and, consequently, the volume of the lungs which follow the movements. The increase in volume raises the capacity of the lungs so that atmospheric pressure forces air into them through the nose and trachea.

EXPIRATION. Expiration, or breathing out, results mainly from a relaxation of the muscles of the ribs and diaphragm. The ribs move down under their own weight, and the organs below the diaphragm, under pressure from the muscular walls of the abdomen, push the relaxed diaphragm back into its domed position. The lungs, as a result of these movements and by virtue of their elasticity, return to their original volume.

In this way air containing less oxygen and more carbon dioxide and water vapour than when it entered the lungs is expelled from them. Usually, in quiet breathing the movements of the diaphragm alone are responsible for the ventilation of the lungs.

Pleural membranes (Fig. 20.9). The pleural membrane is the lining which covers the outside of the lungs and the inside of the thorax. It produces pleural fluid which lubricates the surfaces in the regions of contact between the lungs and thorax. As a result, they can slide freely over one another with very little friction during the breathing movements.

Gaseous exchange in other organisms

(*a*) **Green plants.** The leaves and stem of a plant exchange oxygen and carbon dioxide with the atmosphere by diffusion (*see* pp. 16, 18, 51 and 62). Roots obtain their oxygen from the air dissolved in soil-water or in air spaces.

(*b*) **Micro-organisms.** The surface area of microscopic plants and animals is large in comparison with their volume and the distance from the cell surface to the centre of the protoplasm is very small. Consequently, simple diffusion of gases is rapid enough to meet the respiratory needs of the organism and the diffusion gradients are maintained by the consumption of oxygen and production of carbon dioxide in the protoplasm (*see* p. 178).

(*c*) **Insects** use their tracheal system for gaseous exchange (*see* p. 146).

(*d*) **Fish.** The respiratory surface of a fish is provided by the gills, and ventilation is achieved by passing a current of water over them (*see* p. 161).

(*e*) **Frog and tadpole.** Gaseous exchange takes place through the skin, gills and lungs at various stages of the life cycle and in different situations (*see* pp. 164–167).

PRACTICAL WORK

1. *Composition of exhaled air.* (Fig. 20.10). By placing tube T in the mouth and breathing gently in and out, air is made to pass into the lungs via test-tube A and out via B. After a few seconds the difference in the lime water of each test-tube will indicate one of the differences between the composition of inhaled and exhaled air.

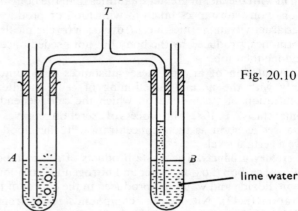

Fig. 20.10

lime water

2. *Oxygen concentration.* Exhaled air is collected in a gas jar by downward displacement of water. A lighted splint placed in the gas jar will give some indication of the oxygen concentration. The air exhaled first (tracheal and bronchial) and last (alveolar) should be collected separately and compared.

3. *Lung capacity.* A large bottle is calibrated up to 5 litres by filling it with water 1 litre at a time and marking the levels. The bottle, full of water, is inverted in a trough or bowl of water, the stopper removed, and a rubber tube inserted through the neck. The experimenter takes a deep breath and exhales through the tube so that the exhaled air collects in the bottle, displacing the water. The level of water left in the bottle will give a measure of the lung capacity.

QUESTIONS

1. Outline the events which take place in the course of vigorous exercise which lead to a change in the rate and depth of breathing both during and after the activity. (*See* also Chapter 9.)

2. The lungs and ileum are adapted for absorption. Point out the features they have in common which facilitate absorption.

	Inhaled air	Exhaled air	Alveolar air
% Oxygen	21	16	14
% Carbon dioxide	0.03	4	5.5

3. The table above gives the approximate percentage volume composition of air inhaled, exhaled or retained in the lungs. Explain how these differences in composition are brought about by events in the lungs.

4. An artificial pneumothorax is a method of resting an infected lung. Air is injected into the pleural cavity and the lung collapses. After a few months, the air is absorbed and the lung works normally again. Try to explain why the introduction of air into the pleural cavity should cause the lung to stop working and say why it is possible for a person with a collapsed lung to lead a normal life.

21 Excretion

EXCRETION means getting rid of unwanted substances from the body. These substances, called *excretory products*, may be (*a*) by-products from the chemical reactions in cells; (*b*) substances taken in with the diet in greater quantities than the body needs; (*c*) poisonous substances taken in with food or produced by bacterial activity in the intestine, (*d*) drugs; or (*e*) chemicals such as hormones, produced by the body but not needed once they have done their job.

Accumulation of any of these substances could interfere directly with the normal functioning of cells, or alter the concentration of tissue fluid on which the cells depend (*see* "Homeostasis", p. 108). The process of excretion removes these substances as soon as their concentration in the blood rises above a certain level.

Excretory products. The waste products of cell metabolism (p. 50) are carbon dioxide, water and nitrogenous compounds. Carbon dioxide and water are produced in the course of tissue respiration (p.47). Nitrogenous compounds result from the breakdown of proteins and amino acids. Deamination in the liver removes the $-NH_2$ from excess amino acids (p. 92). The ammonia (NH_3) which results from this reaction is highly toxic and is converted at once by the liver into a less harmful compound called *urea*. Bacteria in the colon also produce some ammonia and this is also converted to urea by the liver. The nitrogenous compound, *uric acid*, is produced in small quantities and a very small amount of ammonia remains in the blood.

Water and salts taken in with the food are usually in excess of the body's needs and are removed by excretion. Hormones are changed into less active substances and are excreted in the urine. Drugs and toxic substances such as alcohol may be excreted unchanged, or altered by chemical reactions in the body. Tests on urine are used to estimate the level of hormones or drugs in the blood.

Excretory organs. In man, the excretory organs are the lungs, the liver and the kidneys. The lungs excrete carbon dioxide; the liver excretes bile pigments derived from the decomposition of haemoglobin; the kidneys remove nitrogenous compounds from the blood and eliminate excess water and salts.

The kidneys

Gross structure. The two kidneys are fairly solid, oval structures, with an indentation on their innermost sides. They are red-brown, enclosed in a transparent membrane, and attached to the back of the abdominal cavity (Fig. 21.1). The *renal artery*, branching from the aorta, brings oxygenated blood to them, and the *renal vein* takes deoxygenated blood away to the vena cava. A tube, the *ureter*, runs from each kidney to the base of the *bladder* in the lower abdomen (*see* Plate 27).

The kidney tissue consists of many capillaries and tiny tubes, called *renal tubules*, held together with connective tissue. A section through a kidney shows a darker, outer region, the *cortex*, and a lighter inner zone, the *medulla*. Where the ureter leaves the kidney is a space called the *pelvis* and into this project cones or *pyramids* of kidney tissue (Fig. 21.2).

Detailed structure. The renal artery divides up into a great many arterioles and capillaries (Fig. 21.3), mostly in the cortex. Each arteriole leads to a *glomerulus*, which is a capillary repeatedly divided and coiled, making a little knot of vessels (Fig. 21.5). The glomerulus is almost entirely surrounded by a cup-shaped organ called a *Bowman's capsule*, which leads to a coiled renal tubule. This tubule, after a series of coils and loops, joins other tubules and passes through the medulla to

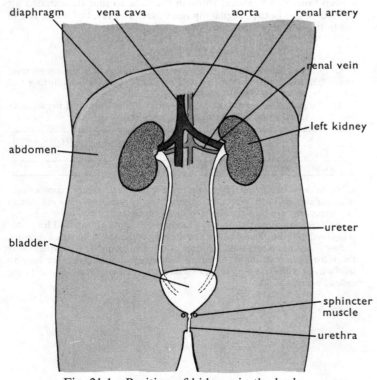

Fig. 21.1 Position of kidneys in the body

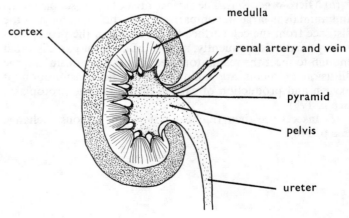

Fig. 21.2 Section through kidney to show regions

open into the pelvis at the apex of a pyramid (Fig. 21.4 and Plate 24).

Mechanism of excretion in the kidney. The tortuous capillaries of the glomerulus offer resistance to the flow of blood, so that a high pressure is set up. This pressure causes fluid to filter out through the capillary walls and collect in the Bowman's capsule. The filtered fluid, *glomerular filtrate*, contains a solution of glucose, salts, amino acids and urea, but

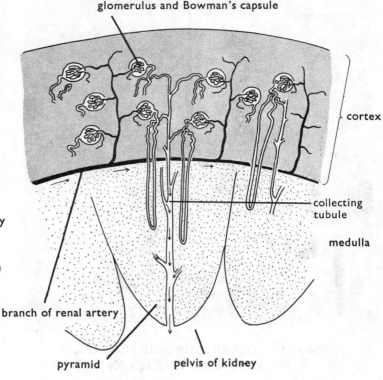

Fig. 21.4   Section through cortex and medulla

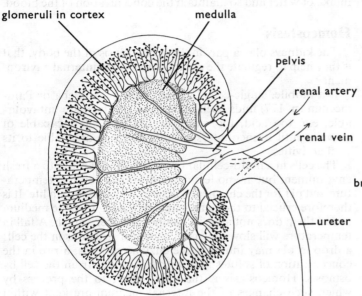

Fig. 21.3   Section through kidney to show distribution of glomeruli

fibrinogen and other proteins remain in the blood. In man, 180 litres per day of this filtrate, carrying 145 g glucose and 1100 g sodium chloride pass into the Bowman's capsule. As the filtered serum passes down the renal tubule, all the glucose and amino acids, some of the salts and much of the water are absorbed back into a network of capillaries surrounding the tubule (Fig. 21.5). This selective reabsorption prevents the loss of useful substances from the blood serum. The remaining liquid, now called *urine*, contains only the waste products such as inactive hormones, urea and excess salts and water. This liquid passes down the *collecting tubule* where more water is reabsorbed and the concentration of the blood is regulated. If the blood is too dilute, e.g. after drinking a great deal, less water is absorbed back into the blood and the urine is dilute. If the blood is too concentrated, e.g. after sweating profusely, more water is reabsorbed from the collecting tubule, making the urine more concentrated. From the collecting tubes, the urine enters the pelvis of the kidney where it collects and continues down the ureter to the bladder as the result of waves of contraction in the ureter.

The capillaries from the glomeruli and the renal tubules unite to form the renal vein. It is the cells of the kidney tubules which selectively reabsorb substances from the glomerular filtrate. They do this often against a diffusion gradient by methods which are not fully understood but which certainly need energy supplied by respiration within the cells. In consequence, the blood leaving the kidneys in the renal vein contains less oxygen and glucose, more carbon dioxide and, as a result of excretion, less water, salts and nitrogenous waste.

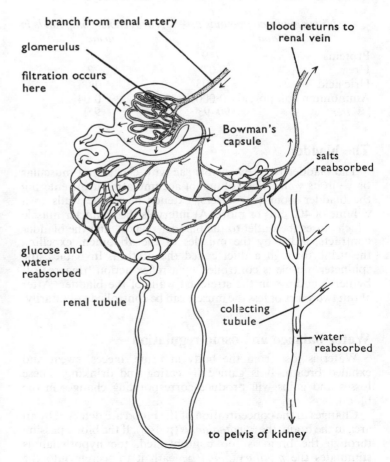

Fig. 21.5   Diagram of glomerulus and Bowman's capsule

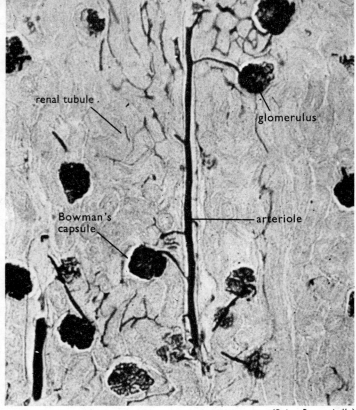

renal tubule

glomerulus

Bowman's capsule

arteriole

(Brian Bracegirdle)

Plate 24. SECTION THROUGH CORTEX
TO SHOW GLOMERULI (×80)

The following Table shows the main nitrogenous substances which are removed from the blood by the kidneys.

	Nitrogenous compounds in blood %	Nitrogenous compounds in urine %
Proteins	7–9	0
Urea	0·03	2
Uric acid	0·005	0·05
Ammonium compounds	0·0001	0·04
[Water	90–93	95]

The bladder

The bladder is an extensible sac with elastic and muscular tissue in its walls. The volume of accumulating urine entering the bladder from the ureters, extends its elastic walls to a volume of 400 cm³ or more. At intervals the sphincter muscle which closes the outlet to the bladder relaxes, and the bladder contracts, aided by the muscles of the abdomen, expelling the urine through a duct called the urethra. In babies, the sphincter muscle is controlled by a reflex action triggered off by nerve endings in the stretched walls of the bladder. After about two years or less the muscle can be controlled voluntarily.

Water balance and osmo-regulation

Water is lost from the body in urine, faeces, sweat and exhaled breath. It is gained by eating and drinking. These losses and gains will produce corresponding changes in the blood.

Changes in the concentration of the blood are detected by an area in the brain, the *hypothalamus* (p. 143). If the blood passing through the brain is too concentrated, the hypothalamus stimulates the *pituitary gland* beneath it to secrete into the blood a hormone (p. 144) called *anti-diuretic* hormone (ADH).

When this hormone reaches the kidneys, it causes the kidney tubules to absorb more water from the glomerular filtrate back into the blood. Thus the urine becomes more concentrated and the further loss of water from the blood is reduced. If blood passing through the hypothalamus is too dilute, production of ADH from the pituitary is suppressed and less water is absorbed from the glomerular filtrate.

The mechanism which produces the sensation of thirst is not well understood but it undoubtedly serves to regulate the intake of water and so maintain the concentration of the blood.

Homeostasis

The kidneys play a part in the homeostasis of the body, that is they help to regulate the composition of the internal environment.

If a mobile, single-celled animal such as Amoeba or Paramecium (p. 177) finds itself in conditions which are unfavourable, e.g. too acid, too warm or too light, it is capable of moving until it encounters conditions more amenable to its vital activities.

The cells in a multi-cellular organism cannot move to a fresh environment but are no less dependent on a suitable temperature and pH for the chemical reactions which maintain life. It is therefore crucial to their efficient functioning that the medium round them does not alter its composition very much. A fall in temperature will slow down the chemical reactions in the cell; a drop in pH may inhibit some enzyme systems; a rise in the concentration of solutes may withdraw water from the cell by osmosis. Homeostasis is the name given to the process by which such changes of the internal medium are kept within narrow limits and many organ systems of the body contribute to this control.

The internal medium of most animals is the tissue fluid (p. 98) which is in contact with all living cells in the body. The constitution of the tissue fluid depends on the composition of the blood from which it is derived and, therefore, the homeostatic mechanisms of many animals act by adjusting the composition of the blood.

The skin helps to regulate blood temperature (p. 110), the liver adjusts its glucose concentration (p. 91), the lungs keep the carbon dioxide concentration down to a certain level (p. 103) and the kidneys control its composition in three principal ways: (a) they eliminate harmful compounds such as urea, (b) they remove excess water and (c) they expel salts above a certain concentration. These activities are both excretory, in that they remove the unwanted products of metabolism, and osmo-regulatory, in that they keep the osmotic potential of the blood more or less constant.

QUESTIONS

1. In an experiment, a man drank a litre of water. His urine output increased so that after two hours he had eliminated the extra water. When he drank a litre of 0·9 per cent sodium chloride solution, there was little or no increase in urine production. Explain the difference in these results.
2. In cold weather one may need to urinate frequently, producing a fairly colourless urine. In hot weather, urination is infrequent and the urine is often coloured. Explain these observations.
3. Consult pp. 89, 91 and 144 and then explain briefly why glucose does not normally appear in the urine.
4. Study the introduction to Chapter 12 and Experiment 3 (p. 63). In the artificial kidney a patient's blood is circulated through dialysis tubing immersed in a warm solution of sugar and salts. Explain how this results in the elimination from his blood of nitrogenous wastes without loss of essential glucose and salts.
5. Explain why the elimination of water by the kidneys may be considered to be both excretion and osmo-regulation.

22 Skin and Temperature Control

THE skin forms a continuous layer over the surface of the body. It has three principal functions:

(a) it protects the tissues beneath from mechanical injury, ultra-violet rays in sunlight, bacterial infection and desiccation;
(b) it contains numerous sense organs which are sensitive to temperature, touch and pain and so make the organism aware of changes in its surroundings (p. 128);
(c) it helps to keep the body temperature constant.

SKIN STRUCTURE

The skin consists of two main layers, (1) an outer *epidermis* and (2) an inner *dermis* (Fig 22.1 and Plate 25). The relative thickness of the layers and abundance of structures within the dermis varies with the position on the body. For example, the skin on the soles of the feet has a very thick epidermis and no hair follicles. The account given below is a generalized one.

1. Epidermis

(a) The **Malpighian** layer is a continuous layer of cells which can divide actively and so produce new epidermis. Also in this layer are the pigment granules, melanin, that determine the skin colour and act as a screen against ultra-violet light.

(b) The **granular layer** contains living cells but towards the outside it gives way gradually to the *cornified layer*.

(c) **Cornified layer.** In this region the cells are dead and form a tough outer coat which offers resistance to damage and bacterial invasion and reduces the loss of water by evaporation. The cells of the cornified layer are continually being worn away and replaced from beneath. On the palms of the hand and soles of the feet it may become very thick, particularly when the hands are used for heavy manual work (Plate 26).

2. Dermis

The dermis is a thicker layer of connective tissue with many elastic fibres in it. There are also blood capillaries, nerve endings or sensory organs, lymphatics, sweat glands and hair follicles.

Capillaries. The capillaries supply the skin with the necessary food and oxygen and remove its excretory products. The sweat glands and hair follicles have a network of capillaries supplying

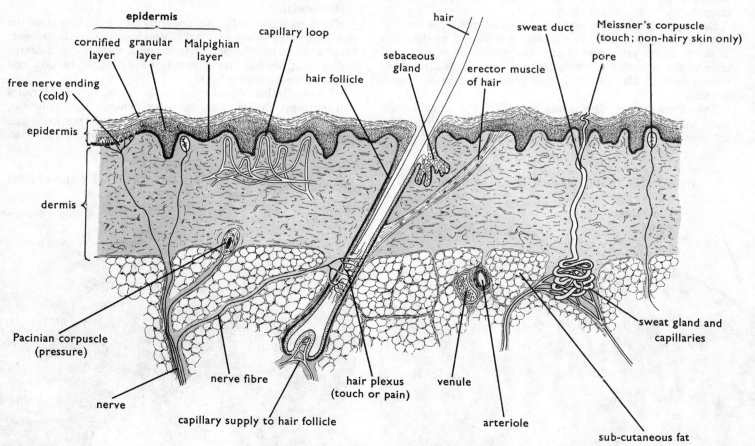

Fig. 22.1   Generalized section through skin

them. The capillaries beneath the epidermis play an important part in temperature control.

Sweat glands. The sweat gland is a coiled tube consisting of secretory cells which absorb fluid from the surrounding cells and capillaries and pass it into the duct through which it reaches the skin surface. The fluid is water with some salts, notably sodium chloride, dissolved in it and also small quantities of urea and lactic acid.

Although the body loses water vapour through the skin fairly constantly at most normal temperatures, the 2–3 million sweat glands do not operate until the body temperature rises about 0·2–0·5°C above normal. In a hot climate a man at work may lose about 1 kg per hour in sweat. Since salts, particularly sodium chloride and other solids, constitute up to 0·5 per cent of the sweat, these must be replaced by the food or in the drink of workers who lose much sweat; otherwise, if water alone is taken to replace that lost by sweating, the salt and water balance of the blood and tissues is upset leading to "heat cramp".

Hair follicle. The hair follicle is a deep pit of granular and Malpighian layers the cells of which multiply and build up a hair inside the follicle. The cells of the hair become impregnated with a horny substance, *keratin*, and die. The constant adding of new cells to the base of the hair causes it to grow. Growth continues for about four years; the hair then falls out and a new period of growth begins. The hairs of the body form a protective and heat-insulating layer in the regions where they grow thickly. The layer of stationary air held between the hairs reduces evaporation and heat loss. The follicle is supplied with sensory nerve-endings which respond to movements of the hair. This sensory function of the hair is well developed in the whiskers or *vibrissae* that grow on the sides of the face in mammals such as the cat or mouse.

Sebaceous glands. The sebaceous glands open into the hair follicles and produce an oily secretion which gives the hairs water-repelling properties, keeps the epidermis supple, and reduces the tendency for it to become too dry as a result of evaporation. It also has antiseptic properties against certain bacteria.

Sub-cutaneous fat. The layers beneath the dermis contain numerous fat cells (Fig. 18.9, p. 91) where fat is stored. The fat may also act as a heat-insulating layer.

TEMPERATURE CONTROL

Fish, amphibia, reptiles and all the invertebrates are *poikilothermic* (*see* p. 159), that is, their body temperature is the same as or only a few degrees above their surroundings, and varies accordingly. This makes them very dependent on temperature changes; for example, in cold conditions their low body temperature slows down all chemical changes and reduces the organism to a state of inactivity. Insects can be immobilized by a sudden fall in temperature.

Homoiothermic or constant-temperature animals are more independent of their surroundings because their body temperature is higher and does not alter with fluctuations in external temperature.

Heat loss and gain

Many of the chemical activities in living protoplasm release heat energy. Chemical changes in the liver and in contracting muscle produce a good deal of the body's heat which the circulatory system distributes round the body. At the same time the body loses heat from its surface to the atmosphere, mainly by convection and radiation. Evaporation of water from the surface of the skin also removes heat from the body. An outer layer of fur, feathers or clothing reduces the heat losses.

Normally a balance is maintained so that the rates of heat loss and gain are the same; hence man's body temperature, although varying in different parts of the body, remains at about 36·8°C, as shown by readings taken from under the tongue. There are regulating mechanisms in the body, under the control of the brain, that compensate for over-heating or cooling.

Over-heating

Vigorous activity, disease, absorption of radiation from the sun, and many other external causes may bring about over-heating. If the blood reaching the brain is a fraction of a degree higher than normal, nerve impulses are sent to the skin and produce two marked effects.

(1) **Vasodilation.** The dilating, or widening of the arterioles which supply the capillary network beneath the epidermis, causes more blood to flow near the surface. In consequence, more heat escapes into the air by convection and radiation

---

Plate 25.   SECTION THROUGH HAIRY SKIN (×30)

(Brian Bracegirdle)

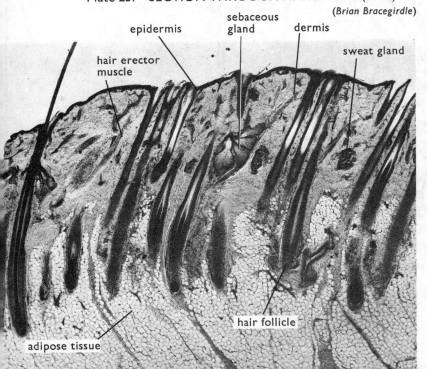

Plate 26.   SECTION THROUGH NON-HAIRY SKIN (×80)
(The sweat ducts are contorted in
passing through the cornified layer) *(Brian Bracegirdle)*

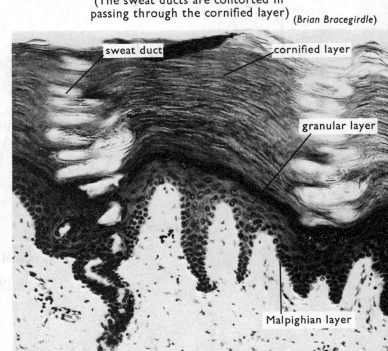

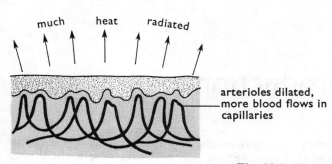

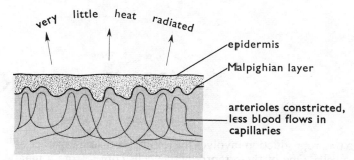

Fig. 22.2   Vasodilation and vasoconstriction

(Fig. 22.2). Vasodilation causes flushing of the skin because of the increased volume of blood beneath the epidermis.

(2) **Sweating.** Nerve impulses, starting mostly in the brain, increase the rate of sweat production so that a continuous layer of moisture may be produced on the skin surface. The latent heat absorbed by the sweat as it evaporates is taken from the body, so reducing the body's temperature.

Any air movement over the body helps to speed up the evaporation of sweat which is why fans, while not necessarily reducing the temperature of a room, have a cooling effect on the body.

In humid conditions, the air contains so much water vapour that the sweat may not evaporate rapidly enough to produce an adequate cooling effect, and may lead to *heat stagnation* in which the body temperature rises to over 41°C, causing collapse and sometimes death. *Heat-stroke* is a similar result of extreme over-heating when, after prolonged sweating due to vigorous activity at high temperatures, sweat production ceases and the body temperature rises to a lethal level. Both conditions may be called "*sun-stroke*" but it is not the effect of direct sunlight on the body so much as the high temperatures produced.

Over-cooling

If the body tends to lose more heat than it is generating the following compensatory changes may take place:

(1) **Decrease in sweat production,** thus minimizing heat lost by evaporation.

(2) **Vasoconstriction.** Constriction of the arterioles which supply the surface capillaries reduces the volume of blood flowing near the surface and hence diminishes heat losses. Vasoconstriction makes a person look pale or blue.

(3) **Increased metabolism.** An increase in the rate of chemical changes in the liver and certain special fat depots (brown adipose tissue), releases more heat.

(4) **Shivering.** This reflex action operates when the body temperature begins to drop. It is a spasmodic contraction of the muscles. These contractions produce heat which helps to raise the body temperature.

Furry mammals and birds can fluff out their fur or feathers by contraction of the erector muscles which are attached to them in the skin. This increases the layer of trapped air round the skin and so improves insulation by reducing convection and conduction. In man, a similar contraction of the muscles of the hairs only produces "goose-pimples".

Hibernation

When a mammal hibernates it falls asleep in some specially prepared burrow or nest and its body temperature falls well below normal, so that it may be only a few degrees above that of its surroundings. Breathing is often imperceptible, and all the chemical activities in the body go on very slowly, using food stored as fat and glycogen.

The animal is quite insensible and cannot be awakened by touching it, in fact it is likely to die if such attempts are made. At the end of its hibernation period its temperature rises spontaneously to normal, starting from the innermost regions of the body. Hedgehogs and dormice are examples of hibernating mammals.

Hibernation allows the small mammal to survive the period when, because of the heat losses from its body, its energy requirements are high, but at the same time, food is scarce.

Surface area and heat loss

Consider the cube drawn in Fig. 22.3a. If it is cut in half as shown in 22.3b each portion has half the volume of the original cube but more than half its surface area because an extra surface "X" has been added to each half. Each time the solid is cut into smaller parts, the ratio $\frac{surface\ area}{volume}$ increases. This means that small animals have a relatively larger surface area per unit of volume than have larger animals and so the former lose heat more rapidly to the surroundings. This is thought to be one reason why the smallest homoiothermic vertebrates, e.g. humming birds, are restricted to areas with warm climates and why the polar regions are populated with relatively large mammals and birds.

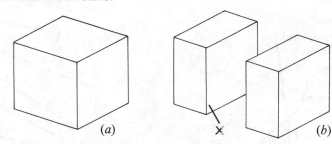

Fig. 22.3

QUESTIONS

1. Why is it more accurate to describe fish as "variable-temperatured" animals rather than "cold-blooded"?

2. When a dog is hot, it hangs its tongue out and pants. Why should this have a cooling effect?

3. Why do you think we experience more discomfort in hot humid weather than we do in hot dry weather?

4. You may "feel hot" after exercise or "feel cold" without your overcoat and yet your body temperature is not likely to differ by more than 0·5°C on the two occasions. Explain this apparent contradiction. (*See* also p. 128.)

5. Draw up a balance sheet showing all the possible ways in which the human body can gain and lose heat.

23 Sexual Reproduction

SEXUAL reproduction involves the joining or fusing together of two cells. One of these reproductive cells comes from a male animal and the other comes from a female. The reproductive cells are called *gametes*. The fusing of two gametes is called fertilization and the resulting, composite cell is called a *zygote*. The most important aspect of fertilization is the fusion of the nuclei of the male and female gametes, because the factors which determine the characteristics of the individual that grows from the zygote are in these nuclei.

Fertilization, in short, is the fusion of the nuclei of male and female gametes to form a zygote, from which can develop a new individual.

In animals, the male gametes are *sperms*, which are produced in the reproductive organs called *testes*. The female gamete is an *ovum* which is produced in a reproductive organ called an *ovary*.

Some animals such as earthworms and snails are *hermaphrodite*, that is, they have both testes and ovaries, but in most animals the sexes are separate.

Internal and external fertilization. In most fish and amphibia fertilization is external. The female lays the eggs first and the male fertilizes them by placing sperms on them afterwards. A behaviour pattern which brings the sexes into proximity usually ensures that sperm is shed near the eggs and so increases the chances of fertilization.

In reptiles and birds, the eggs are fertilized inside the body of the female by the male's passing sperms into the egg ducts. A sperm meets the ovum and fertilizes it before it is laid. Very little development of the egg takes place before laying, however, and the embryo grows in the egg *after* it has left its mother's body.

In mammals, sperms are placed in the body of the female and the eggs are fertilized internally. They are not laid after fertilization but retained in the female's body while they develop to quite an advanced stage, after which the young are born more or less fully formed, being fed on a secretion of milk from the mammary glands and protected by their parents until they become independent.

Sexual reproduction in man

Female reproductive organs (Fig. 23.1 and Plate 27). The female reproductive organs are the *ovaries*, two cream-coloured, oval bodies lying in the lower part of the abdomen. They are attached by a membrane to the uterus and supplied with blood vessels. Close to each ovary is the expanded, funnel-shaped opening of the *oviduct*, the tube down which the ova pass when they are released from the ovary.

The oviducts are narrow tubes that open into a wider tube, the *uterus or womb*, lower down in the abdomen. When there is no embryo developing in it the uterus is only about 80 mm long. It communicates with the outside through a muscular tube, the *vagina*. The *cervix* is a ring of muscle closing the lower end of the uterus where it joins the vagina. There is normally only a very small aperture connecting these two organs at this point. The *urethra*, from the bladder, opens into the vulva just in front of the vagina.

Male reproductive organs (Figs. 23.4 and 23.5). The two testes lie outside the abdominal cavity in man, in a special sac called the *scrotum*, consequently the testes remain at a temperature rather below that of the rest of the body, which is favourable to sperm production. The testes consist of a radiating mass of sperm-producing tubes. These tubes meet and join to form

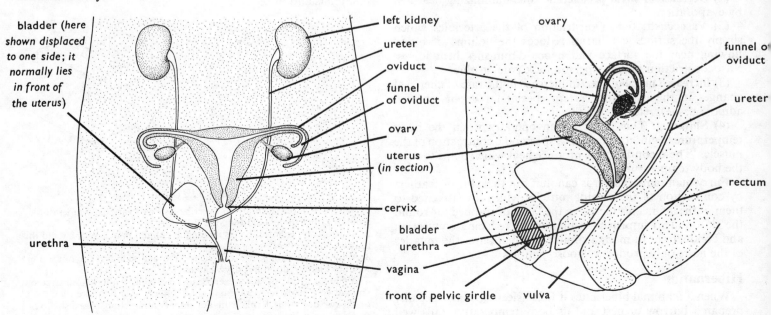

Fig. 23.1   Female reproductive organs

Fig. 23.2   Female reproductive organs (vertical section)

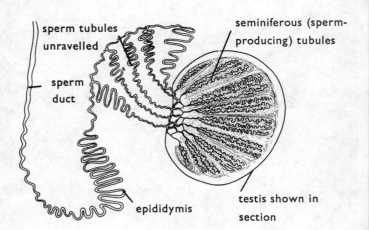

Fig. 23.3   Diagram to show relation of sperm ducts and testis

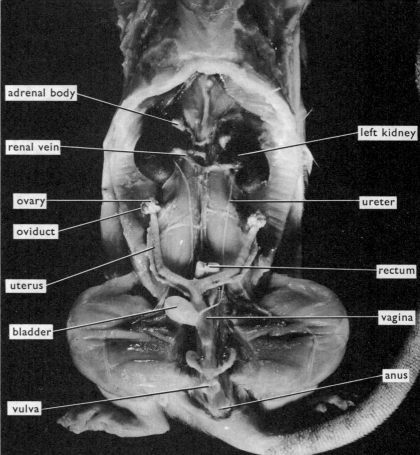

(Dissection by Griffin and George Ltd. Gerrard Biological Centre.)

Plate 27.   DISSECTION OF FEMALE REPRODUCTIVE ORGANS OF A RAT

(Note that there are two uteri in the rat)

ducts leading to the *epididymis*, a coiled tube about 6 m long on the outside of each testis. The epididymis, in turn, leads into a muscular sperm duct. The two muscular sperm ducts, one from each testis, open into the top of the urethra just after it leaves the bladder. A short, coiled tube, the *seminal vesicle*, branches from each sperm duct just before the latter enters the prostate gland. Surrounding the urethra at this point, is the *prostate gland* and farther down, *Cowper's gland*. The urethra conducts, at different times, either urine or sperms. In males the urethra is prolonged into a *penis*, which consists of connective tissue with numerous small blood spaces in it.

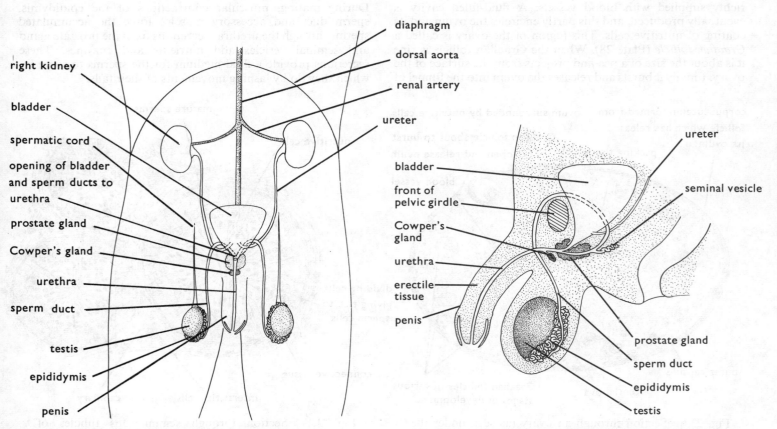

Fig. 23.4   Male reproductive organs

Fig. 23.5   Male reproductive organs (vertical section)

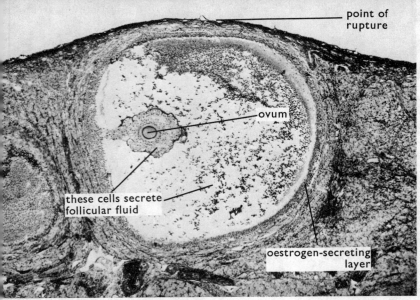

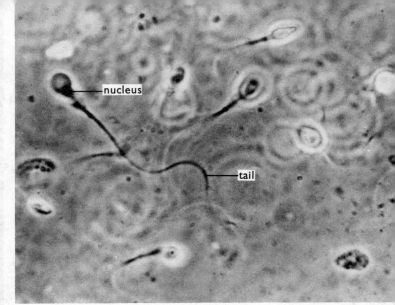

(Brian Bracegirdle)

(Brian Bracegirdle)

Plate 28.   MATURE GRAAFIAN FOLLICLE (×40)

Plate 29.   HUMAN SPERMS (×1000)

Ovulation. The ovary consists of connective tissues, blood vessels and potential egg cells (Fig. 23.6). It is thought that 70,000 potential egg cells are already present at birth; they are not manufactured by the ovary during the lifetime. Of these 70,000 potential egg cells only about 500 will ever become mature ova (eggs). The ovaries also secrete hormones, *oestrogens*, which control the secondary sexual characters (p. 117) and initiate the thickening of the uterine lining which occurs each month.

Between the ages of about 11 and 16 years the ovaries become active and begin to produce mature eggs. The beginning of this period of life is called *puberty*. In the ovary some of the ova start to grow; the cells around them divide rapidly and become richly supplied with blood vessels. A fluid-filled cavity is eventually produced, and this partly encircles the ovum and its coating of nutritive cells. This region of the ovary is called a *Graafian follicle* (Plate 28). When the Graafian follicle is ripe it is about the size of a pea and projects from the surface of the ovary. Finally it bursts and releases the ovum into the funnel of the oviduct, the ciliated cells of which waft it into the tube. At this stage the ovum (Fig. 23.8) is a spherical mass of protoplasm about 0·13 mm in diameter, with a central nucleus and some of the follicle cells still adhering to it. An ovum is produced more or less alternately from the two ovaries every four weeks and it may spend about three days travelling down the oviduct to the uterus.

Sperm production (Fig. 23.3). The lining of the tubes making up the testis consists of actively dividing cells which give rise ultimately to sperms (Fig. 23.7 and Plate 29). A sperm is a nucleus surrounded by a little cytoplasm which extends into a long tail. Sperms are quite immobile when first produced; they pass into the epididymis where they are stored. During mating, muscular contractions of the epididymis, sperm duct and accessory muscles force the accumulated sperms through the urethra. Secretions from the prostate gland and seminal vesicles add nutrients and enzymes. These secretions provide a fluid medium for the sperms to swim in, which they do by lashing movements of their tails.

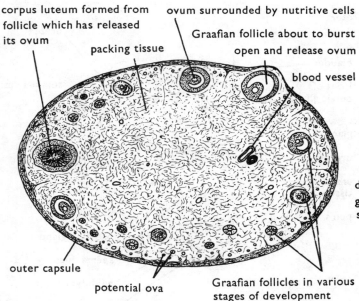

Fig. 23.6   Section through an ovary (as seen under the low power of a microscope)

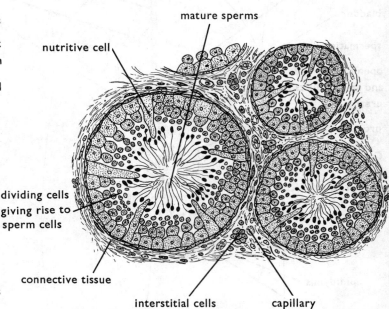

Fig. 23.7   Section through seminiferous tubules of mammalian testis (greatly enlarged)

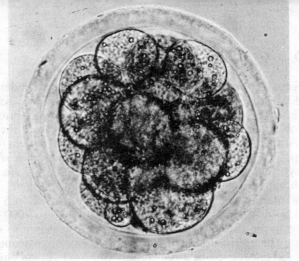

(Prof. W. J. Hamilton)

Plate 30. EARLY STAGES OF CELL DIVISION
IN ZYGOTE OF SHEEP (×400)

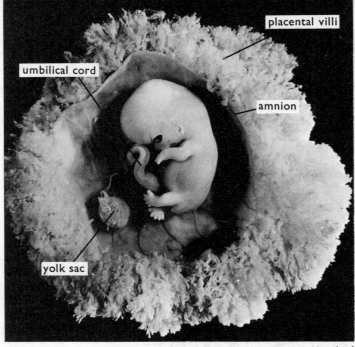

(Prof. W. J. Hamilton)

Plate 31. HUMAN FOETUS, 7 WEEKS (×1·5)

Fertilization. Fertilization occurs internally when the sperms meet the ovum as it passes down the oviduct. They are introduced into the female through the penis which is placed in the vagina. To facilitate this action the penis becomes erect, largely as a result of blood flowing into the blood spaces more rapidly than it escapes, so increasing the turgidity of the tissues round the urethra. The stimulation of the sensory organs in the penis sets off a reflex action which results in the accumulated sperms, together with the secretions of the prostate and Cowper's glands, being ejaculated into the vagina. The action is called copulation and may result in fertilization. The sperms deposited in the vagina swim through the cervix into the uterus and travel (it is not known exactly how) to the oviduct. If an ovum is present in the oviduct, one of the sperms will eventually collide with it and the head of the sperm sticks to the ovum. The sperm's nucleus passes into the cytoplasm and fuses with the female nucleus there (Fig. 23.8). (*See* also p. 191.)

Although a single ejaculation may contain two or three hundred million sperms, only one will actually fertilize the ovum, though the others may assist in the fertilization as a

result of an enzyme they produce which helps to disperse the remaining follicle cells adhering to the surface of the ovum (not confirmed in man so far). In some mammals such as the rabbit, several ova are released from the ovary at the same time and will be fertilized by the corresponding number of sperms.

In man, the released ovum is thought to live for 8–24 hours, while the sperm might be able to fertilize the ovum for up to 3 days. Thus there is only a relatively short period of 3–4 days each month when fertilization can occur (Fig. 23.12).

The cells of the follicle from which the ovum has been released continue to divide and grow, forming a solid body, the *corpus luteum*. If fertilization does not occur, the corpus luteum degenerates. If fertilization does occur, the corpus luteum enlarges further and continues to produce a hormone, *progesterone*, which stimulates the further thickening and increased blood supply of the uterine lining.

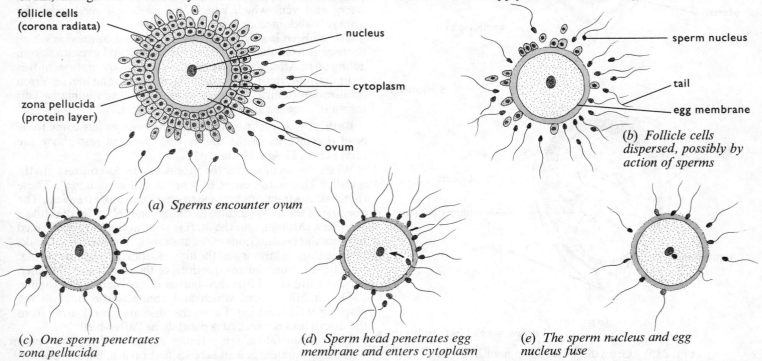

Fig. 23.8 Fertilization of human ovum (the diagrams show what is thought to happen but the events are not known for certain)

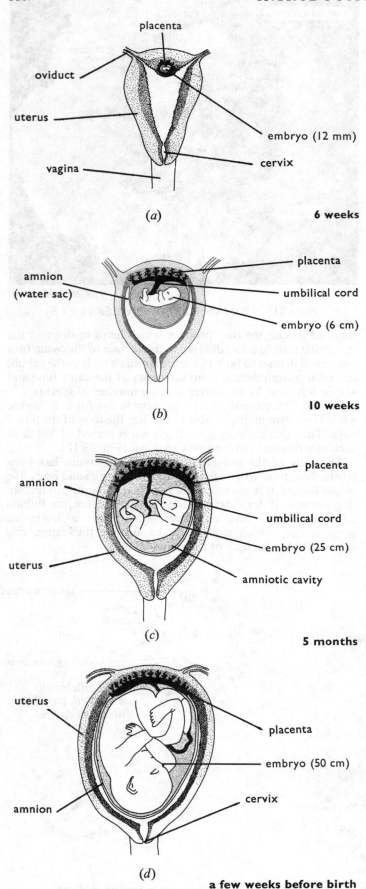

placenta

oviduct

uterus

embryo (12 mm)

cervix

vagina

(a)   6 weeks

amnion
(water sac)

placenta

umbilical cord

embryo (6 cm)

(b)   10 weeks

amnion

placenta

umbilical cord

embryo (25 cm)

uterus

amniotic cavity

(c)   5 months

uterus

placenta

embryo (50 cm)

cervix

amnion

(d)

a few weeks before birth

Fig. 23.9   Growth and development in uterus
(not to scale)

Pregnancy and development. The fertilized egg undergoes rapid cell division (Plate 30) as it passes down the oviduct and into the uterus where it adheres to and sinks into the uterine lining. Finger-like processes, *villi*, grow from the embryo into the uterine lining and absorb nourishment (Fig. 23.9*a*). The region bearing the *villi* does not form part of the embryo but develops into a special organ called the *placenta* which supplies the embryo with both food and oxygen.

The uterus, which at first has a volume of only 2 to 5 cm^3, extends with the growth of the embryo to 5000 to 7000 cm^3, enlarging the abdomen and displacing the organs in it to some extent. The uterine lining, under the influence of oestrone and progesterone, develops a rich supply of blood vessels and the walls become increasingly muscular.

The cells of the embryo divide repeatedly to form tissues. The tissues swell, roll, extend, etc., and so form organs of the body. The embryo's heart and circulatory system are formed quite early, after about one month.

Although the embryo depends on the mother's blood for its food and oxygen its circulatory system is never directly connected with the maternal blood vessels. If this did occur, the adult's blood pressure would burst the delicate capillaries forming in the embryo, and many substances in the mother's circulation would be poisonous to the embryo.

The placenta. The placenta becomes a large disc of tissue adhering closely to the uterine lining (Fig. 23.10 *b* and *c*). From the placenta villi protrude into the uterine lining which has thickened and developed a rich blood supply. The membranes separating the maternal and embryonic blood vessels are very thin; hence dissolved substances can pass across in both directions (Fig. 23.11). Dissolved oxygen, glucose, amino acids and salts in the mother's blood pass from the uterine vessels into those of the embryo, while carbon dioxide and nitrogenous waste from the embryo pass across in the opposite direction (Fig. 23.12). The membrane exerts some selective influence over the substances that pass into the placenta and so prevents harmful material from reaching the embryo. The capillaries in the placenta are connected to an artery and vein which run in the *umbilical cord* from the embryo's abdomen to the placenta.

The embryo is surrounded by a *water sac* or *amnion* which protects it from damage and prevents unequal pressures from acting on it. After about five month's growth the embryo moves its limbs quite vigorously inside the water sac and uterus. From fertilization to birth takes about nine months in humans. This length of time is called the *gestation* period.

Birth. A few weeks before birth the embryo has come to lie head downwards in the uterus with its head just above the cervix (Fig. 23.9*d* and Plate 32).

When the birth starts, the uterus begins to contract rhythmically. This is the onset of what is called "labour". These rhythmic contractions become stronger and more frequent. The opening of the cervix gradually dilates enough to let the child's head pass through, and the uterine contractions are reinforced by muscular contractions of the abdomen. The water sac breaks at some stage in labour and the fluid escapes through the vagina. Finally, the muscular contractions of the uterus and abdomen expel the child head first through the dilated cervix and vagina.

The umbilical cord which still connects the child to the placenta is tied and cut. Later, the placenta breaks away from the uterus and is expelled separately as "after-birth".

The sudden fall in temperature experienced by the newly born baby stimulates it to take its first breath, usually accompanied by crying. In a few days the remains of the umbilical cord attached to the baby's abdomen shrivel and fall away, leaving a scar in the abdominal wall, called the navel. The average birth weight of babies is 3 kg.

Shortly after its birth the baby starts to suckle. During pregnancy the mammary glands of the breast have developed and are stimulated to secrete milk by the first sucklings.

Parental care. All mammals suckle their young and protect them in various ways until they are old enough to move about efficiently and obtain their own food. Most mammals prepare a nest in which to bear their young, and in this way the babies are protected from animals of prey and from temperature changes. The nest reduces the chances of their wandering and getting lost and prevents them from injuring themselves. The young mammals are often born without fur but the presence of the mother's body prevents them from losing heat. The food at first is entirely milk, suckled from the mother. The milk contains nearly all the food, vitamins and salts that the young need for their energy requirements and tissue building but there is no iron present for the manufacture of haemoglobin. All the iron needed for the first weeks or months is stored in the body of the embryo during the period of gestation in the uterus. The parent's milk supply increases as the young animals grow larger and is supplemented by solid food. In carnivorous mammals the prey is brought back to the nest site and torn into pieces small enough for the young to swallow. Often the parents are more aggressive when they have young and react violently to intruders. When the young animals are old enough to obtain their own food and escape from predators, they leave the nest site and disperse. In humans, the period of dependence on parents is lengthy.

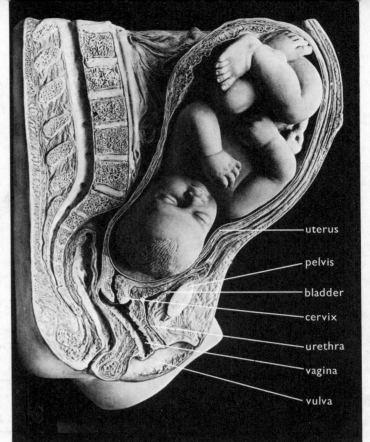

(Reproduced with permission from the Birth Atlas published by Maternity Centre Association, New York)

Plate 32. MODEL OF HUMAN FOETUS JUST BEFORE BIRTH

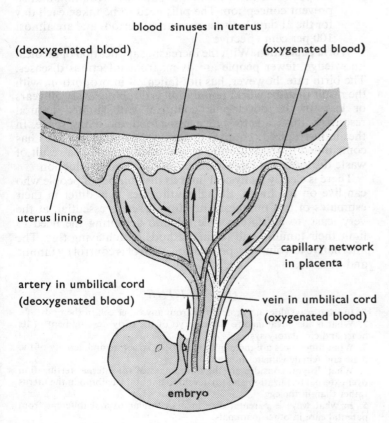

Fig. 23.10 Diagram to show relationship between blood supply of embryo, placenta and uterus

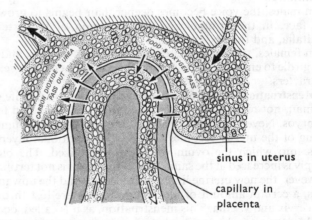

Fig. 23.11 Diagram to show exchange of oxygen and food from uterus to placenta

Twins. IDENTICAL TWINS. Sometimes a fertilized ovum divides into two parts at an early stage of cell division and each part develops separately into a normal embryo. Such "one egg" twins are the same sex and are identical in nearly every physical respect, although differences in position and blood supply while in the uterus may cause them to differ initially in weight and vigour.

FRATERNAL TWINS. If two ova are released from the ovary and fertilized simultaneously, twins will result. These twins may be different sexes and are not necessarily any more alike than other brothers and sisters of the same family.

Secondary sexual characters. In addition to producing gametes, the ovaries and testes make chemicals called hormones. At puberty these hormones are released into the blood stream, and as they circulate round the body they give rise to physical

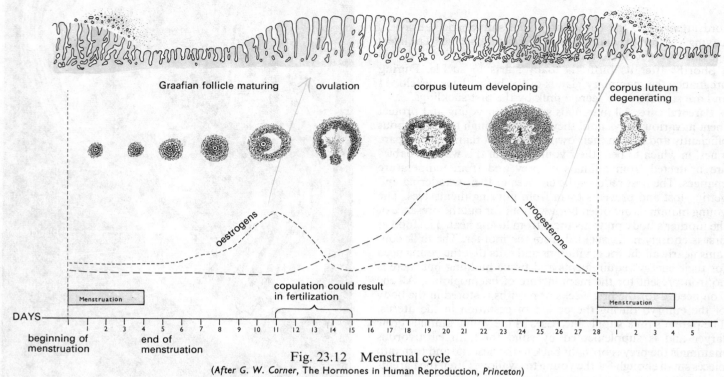

uterine lining thickening   becoming highly vascular   breaking down

Graafian follicle maturing   ovulation   corpus luteum developing   corpus luteum degenerating

oestrogens

progesterone

copulation could result in fertilization

Menstruation   Menstruation

DAYS

1 2 3 4 5 6 7 8 9 10 11 12 13 14 15 16 17 18 19 20 21 22 23 24 25 26 27 28 1 2 3 4 5

beginning of menstruation   end of menstruation

Fig. 23.12   Menstrual cycle

(*After G. W. Corner*, The Hormones in Human Reproduction, *Princeton*)

and mental changes which we associate with masculinity or femininity.

In males, the voice becomes deeper, hair begins to grow on the face, in the armpits and in the region of the external genitalia, and the body becomes more muscular.

In females, the hormones cause the breasts to grow and the hip girdle to enlarge. These features are called secondary sexual characters.

Menstruation. Of the 500 or so ova produced in the life of a woman, not more than 12 are likely to be fertilized and form embryos. Nevertheless, at the time of release of each ovum the lining of the uterus becomes thicker with additional layers of cells into which the ovum will sink if fertilized. The blood supply is increased at the same time. If the ovum is not fertilized, however, the new uterine lining disintegrates and the unwanted cells, a certain amount of blood, and mucus are lost through the cervix and vagina. This menstruation, as it is called, occurs 12–14 days after the egg is released, once in about four weeks.

Birth control. When people wish to limit the size of their families they usually make use of one or other form of birth control. This may involve either a contraceptive practice or restricting copulation to a period in which fertilization is unlikely to occur, i.e. from 5 to 10 or 18 to 28 days after the onset of menstruation (*see* Fig. 23.12). This is not a very reliable method because there is considerable individual variation in the time of ovulation, the time interval between menstrual periods and the regularity of their occurrence.

The principal methods of contraception are as follows:

(*a*) *The sheath or diaphragm.* A thin rubber sheath worn on the penis, or a small rubber diaphragm inserted in the vagina prevents the sperms from reaching the cervix.

(*b*) *Intra-uterine loop.* A small plastic strip bent into a loop or coil is inserted and retained in the uterus. Whether it interferes with fertilization or implantation is not certain but it is very effective.

(*c*) *The contraceptive pill.* The pill contains chemicals which have the same effect on the body as the hormones oestrogen and progesterone. When mixed in suitable proportions, these hormones suppress ovulation and so prevent conception. The pills need to be taken each day for the 21 days between menstrual periods and are almost 100 per cent effective.

World population. With the increasing application of medical knowledge, fewer people are dying from infectious diseases. The birth rate, however, has not fallen off in proportion, with the result that the world population is doubling every 50 years or less. In the developing countries · with limited natural resources, this leads to shortages of food and living space. In the industrialized countries, the population increase has contributed to the pollution of the environment as a result of waste disposal and intensive methods of food production.

There is clearly a physical limit to the number of people who can live on the Earth, though authorities may differ in their estimates of this number. Therefore it seems essential, at the very least, to educate people (*a*) into accepting the need to limit their families and (*b*) in methods of achieving this. The alternative to voluntary population control is control by famine and disaster.

QUESTIONS

1. In what ways does a zygote differ from any other cell in the body?
2. What is the advantage to the embryo of the early development of its heart and circulatory system?
3. What differences are there in the numbers, structure and activity of the male and female gametes in man?
4. What do you consider are the advantages of (*a*) internal fertilization over external fertilization and (*b*) development of the embryo in the uterus rather than in the egg?
5. In what ways is parental care in man similar to and different from parental care in other mammals?
6. List the changes in the composition of the maternal blood which are likely to occur when it passes through the placenta.
7. Explain why there are only a few days in each menstrual cycle when fertilization is likely to occur.

24 The Skeleton, Muscles and Movement

SKELETAL tissues are hard substances formed by living cells. Frequently they contain non-living mineral matter such as calcium salts. The structures made of such non-living material can nevertheless grow and change as a result of the activities of living cells which dissolve away and replace the hard materials.

Exoskeletons. Where the hard material is formed mainly on the outside of the body it is often called an exoskeleton. Insects and crustaceans such as crabs have exoskeletons or cuticles, though there are projections from the exoskeleton into the body cavity for muscle attachment (p. 150). Animals with exoskeletons increase their size periodically by dissolving and absorbing most of the cuticle, splitting and shedding the outermost layers, and forming a new cuticle on the exposed surface. This is called *ecdysis* (p. 146).

Endoskeletons (Plate 33). Vertebrate animals have bony skeletons within their bodies. These animals can grow by a continuous increase in size and not by a series of ecdyses.

Functions of the skeleton

The functions can be grouped conveniently under the headings, support, protection, movement and locomotion, and muscle attachment.

Support. There are many invertebrate animals which have no skeleton. Those living in water may become fairly large because the water supports them and buoys them up to a certain extent. In others, e.g. the earthworm, they are supported by the pressure of fluid in their body cavities acting outwards against a muscular body wall. In larger, land-dwelling animals, a rigid skeletal support raises the body from the ground and allows rapid movement, it suspends some of the vital organs and prevents them from crushing each other, and maintains the shape of the body despite vigorous muscular activity. Examination of the skeleton of the rabbit (Fig. 24.1) will show the backbone as a bridge-like arch or span from which the organs of the body are suspended.

Protection. Certain delicate and important organs of the body are protected by a casing of bone. The brain is enclosed in the skull, the spinal cord in the backbone, while the heart and lungs are surrounded by a cage of ribs between the *sternum* and spine. The organs are thus protected from distortion resulting from pressure, or injury resulting from impact. The rib cage plays a positive part in the breathing mechanism (p. 104) in addition to protecting the organs of the thorax. In animals with exoskeletons the entire body is protected by the cuticle.

Movement. Many bones of the skeleton act as levers. When muscles pull on these levers they produce movements such as the chewing action of the jaws, the breathing movements of the ribs and the flexing of the arms. Locomotion is the result of the co-ordinated action of muscles on the limb bones and is discussed more fully on p. 124. Movements of the skeleton require a system of joints and muscle attachments.

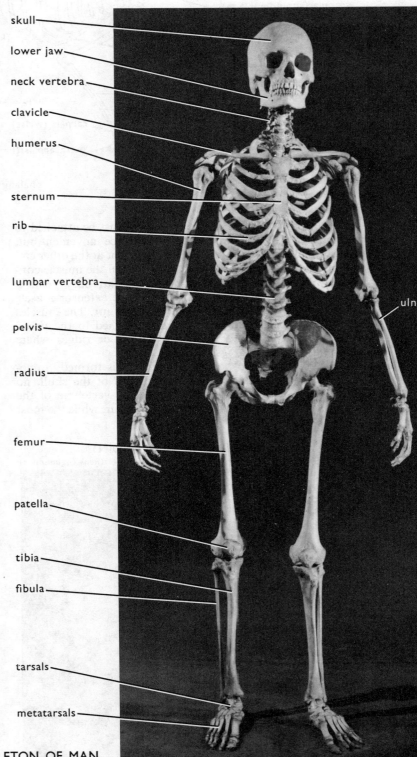

skull
lower jaw
neck vertebra
clavicle
humerus
sternum
rib
lumbar vertebra
ulna
pelvis
radius
femur
patella
tibia
fibula
tarsals
metatarsals

· Plate 33. SKELETON OF MAN
(Rank Organisation)

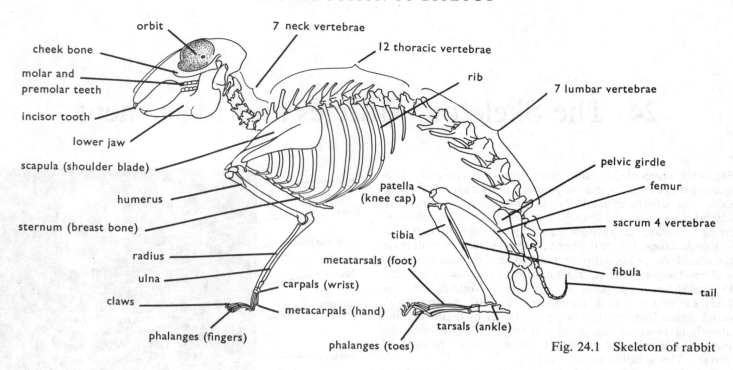

Fig. 24.1   Skeleton of rabbit

MUSCLE ATTACHMENT. The muscles must be attached to the limb bones at one end in order to produce movement but, in addition, they must have a rigid attachment at the other end so that only one part of the limb moves when the muscle contracts. (*See* Fig. 24.9.) Sometimes the "stationary" end is attached to the upper half of the limb. The extensor muscle which extends the foot is attached to the femur. The muscles which move the femur, however, are attached to the pelvic girdle. Bones frequently have projections or ridges where muscles are attached (Figs. 24.2 and 24.3).

JOINTS. Where two bones meet, a joint is formed. Sometimes, as in the sutures between the bones of the skull, no movement is permitted; in others, e.g. the vertebrae of the spine, only a very limited movement can occur, while the most

familiar joints, *synovial joints*, allow a considerable degree of movement. The ball-and-socket joints of the *humerus* and *scapula* (Plate 34) or *femur* and *pelvis* (Plate 35) allow movement in three planes (Fig. 24.5 *a* and *b*). The hinge joints of the elbow and knee allow movement in only one plane (Fig. 24.4).

The surfaces at the heads of the bones which move over each other are covered with a tough cartilage which is slippery and smooth. This, together with a liquid called *synovial fluid* which is formed in the joint, allows friction-free movement (Fig. 24.5a). The relevant bones of the joint are held together by strong *ligaments* (Fig. 24.5 *a* and *b*) which prevent dislocation during normal movement. Surrounding the joint is a capsule of fibrous material whose inner lining, the *synovial membrane*, secretes the synovial fluid.

Plate 34.   THE SHOULDER JOINT
(*Rank Organisation*)

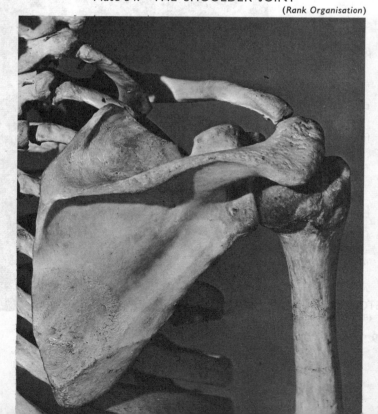

Plate 35.   THE HIP JOINT
(*Rank Organisation*)

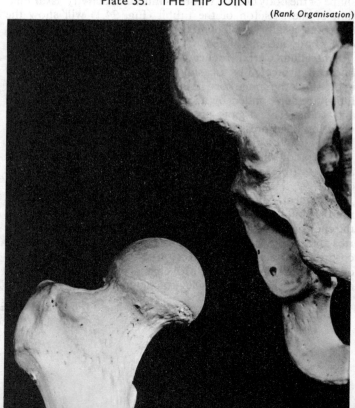

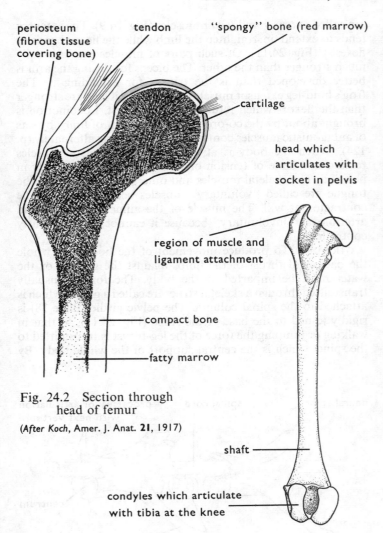

periosteum (fibrous tissue covering bone)

tendon

"spongy" bone (red marrow)

cartilage

head which articulates with socket in pelvis

region of muscle and ligament attachment

compact bone

fatty marrow

Fig. 24.2 Section through head of femur

(*After Koch, Amer. J. Anat.* **21**, 1917)

shaft

condyles which articulate with tibia at the knee

Fig. 24.3 Femur (rabbit)

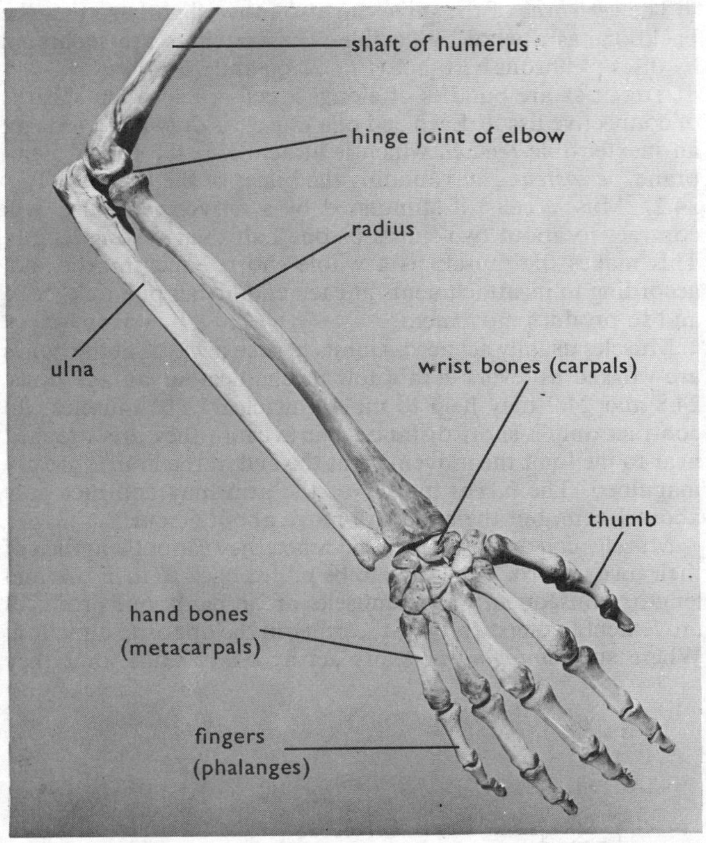

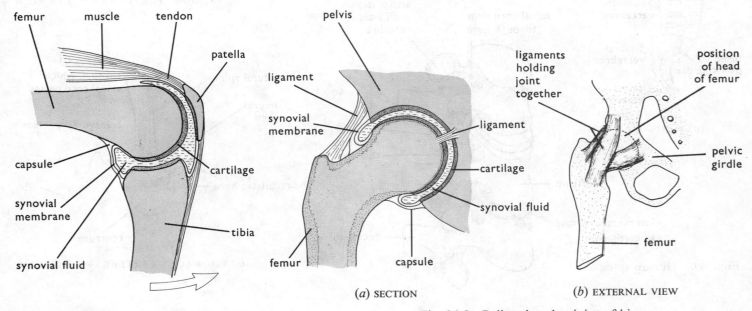

shaft of humerus

hinge joint of elbow

radius

ulna

wrist bones (carpals)

thumb

hand bones (metacarpals)

fingers (phalanges)

(*Rank Organisation*)

Plate 36. SKELETON OF THE FOREARM

femur   muscle   tendon

patella

capsule

cartilage

synovial membrane

synovial fluid

tibia

Fig. 24.4 Hinge joint of knee (section)

pelvis

ligament

synovial membrane

ligament

cartilage

synovial fluid

femur

capsule

(*a*) SECTION

ligaments holding joint together

position of head of femur

pelvic girdle

femur

(*b*) EXTERNAL VIEW

Fig. 24.5 Ball and socket joint of hip

The *vertebrae* of the spine can also move slightly so that the backbone as a whole is flexible. The vertebrae are separated by discs of fibrous cartilage (Fig. 24.6*b* and 24.7*a*).

MUSCLES are bundles of elongate cells enclosed in sheaths of connective tissue. Each end of a muscle is drawn out to form an inextensible *tendon*, which is attached to the tough membrane, *periosteum*, surrounding the bones of the skeleton (Fig. 24.2). Muscle cells, if stimulated by a nervous impulse, will contract to about two-thirds or one-half their resting length. This makes the muscle as a whole shorter and thicker and, according to its attachments at each end, it can pull on a bone and so produce movement.

Muscles usually act across joints in such a way that the bones are worked as levers with a low mechanical advantage. Figs. 24.8 and 24.9 may help to make this clear. The muscles can contract only a short distance, but because they are attached near to the joint the movement at the end of the limb is greatly magnified. The *biceps* muscle of the arm may contract only about 10 cm, but the hand will move about 60 cm.

Muscles can only contract and relax, they cannot lengthen of their own accord. They have to be pulled back to their original length. Consequently most muscles are in pairs: one produces movement in one direction, the other in the opposite direction. Where such *antagonistic* pairs act across a hinge joint they are called *extensor* and *flexor* muscles (Fig. 24.9). The extensor tends to extend or straighten the limb while the flexor bends or flexes it (Fig. 24.11). Of such pairs of muscles one is usually much stronger than the other. The biceps for flexing the arm is better developed than is the triceps for extending it. The frog's hind-leg extensor muscles which make it leap are stronger than the flexors which return the limb to rest. Locomotion is brought about by the co-ordinated movement of limbs by sets of antagonistic muscles contracting and relaxing alternately (p. 124). When the body is at rest, both antagonistic muscles remain in a state of tension or *tone* and so hold the body in position. The skeletal muscles and others such as those in the tongue are called "voluntary" muscles because they can be contracted at will. The muscle of the alimentary canal and arterioles is "involuntary" because it cannot be consciously controlled.

GIRDLES. To produce movement of the body as a whole the backward thrust of the limbs against the ground or the water must be imparted to the body. The force is usually transmitted through a skeletal structure called a girdle, which is attached to the spinal column. The pelvic girdle (Plate 38) is rigidly joined to the base of the spine (Fig. 24.11), so that in walking or jumping the force of the leg-thrust is transmitted to the spine, which is the central support of the whole body. By

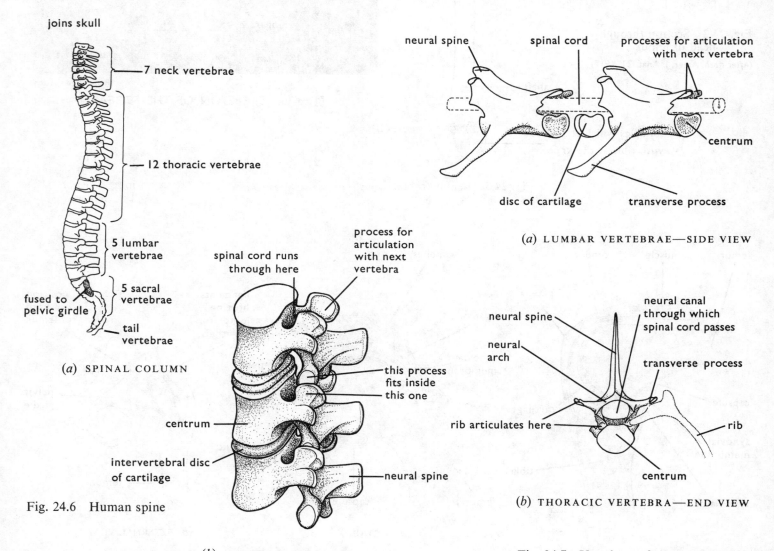

joins skull

— 7 neck vertebrae

— 12 thoracic vertebrae

5 lumbar vertebrae

fused to pelvic girdle

5 sacral vertebrae

tail vertebrae

(*a*) SPINAL COLUMN

Fig. 24.6   Human spine

spinal cord runs through here

process for articulation with next vertebra

this process fits inside this one

centrum

intervertebral disc of cartilage

neural spine

(*b*) LUMBAR VERTEBRAE

neural spine   spinal cord   processes for articulation with next vertebra

centrum

disc of cartilage   transverse process

(*a*) LUMBAR VERTEBRAE—SIDE VIEW

neural canal through which spinal cord passes

neural spine

neural arch

transverse process

rib articulates here

rib

centrum

(*b*) THORACIC VERTEBRA—END VIEW

Fig. 24.7   Vertebrae of mammal (rabbit)

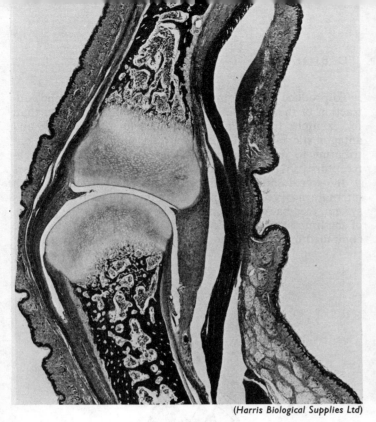

(Harris Biological Supplies Ltd)

Plate 37.   FINGER JOINT OF HUMAN FOETUS

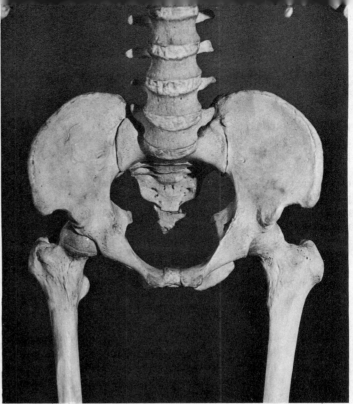

(Rank Organisation)

Plate 38.   THE PELVIC GIRDLE

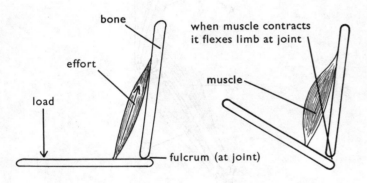

Fig. 24.8   The limb as a lever

this means also, the weight of the body is supported when at rest. The shoulder blades, which form part of the pectoral girdle (Plate 34) are not fused to the spine in mammals, but bound by muscles to the back of the thorax. The fusion of the pelvic girdle to the spine is very effective in transmitting force from the legs to the body. The muscular attachment of the shoulder blades to the spine is less effective in transmitting force from the arms to the body, but this function is not so necessary in the fore- as in the hind-limbs; and the free movement of the shoulders allows greater mobility of the arms. The shoulder in man is more mobile than in most mammals, and the *clavicle* (Plate 33) acts as a radius, limiting the movement of the scapula.

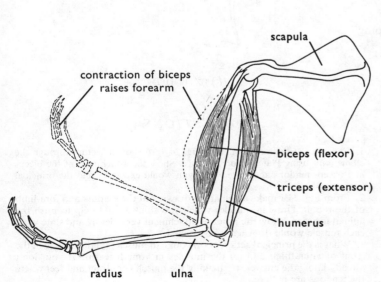

Fig. 24.9   Antagonistic muscles of the forearm

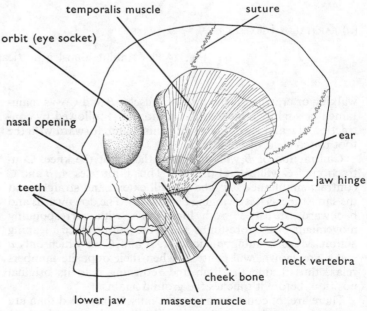

Fig. 24.10   Human skull and chewing muscles

Locomotion

The limbs are moved in a co-ordinated sequence, each one thrusting backwards on the ground and so propelling the animal forwards. While the limb is recovering its forward position it must be removed from contact with the ground.

In Fig. 24.11 some of the muscles of the hind-limb of the rabbit are shown in diagram form. If muscle A contracts it will pull the femur backwards. Friction between the ground and the toes prevents the foot sliding back so that a forward thrust is transmitted through the pelvic girdle to the spine and thence to the whole animal. Such a contraction would contribute to

To produce effective movement it is essential that the contraction of the many sets of muscles is co-ordinated so that, for example, the legs are moved in a logical sequence and antagonistic muscles do not contract simultaneously. Contributing to this co-ordination there is a system of stretch-receptors (p. 128) in the muscles. These fire nervous impulses to the spinal cord when the muscle is being stretched. Such internal sensory organs or proprioceptors, linked to the nervous system, feed back information to the brain about the position of the limbs and enable a pattern of muscular activity to be computed by the brain, so producing effective movement.

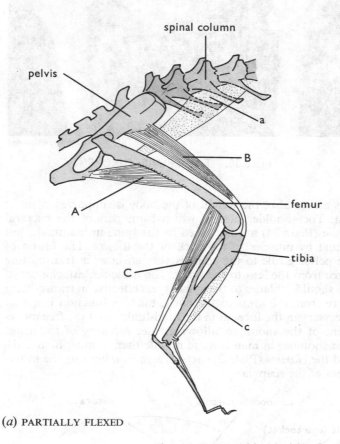

(a) PARTIALLY FLEXED

Fig. 24.11   Rabbit hind-leg in "leaping"

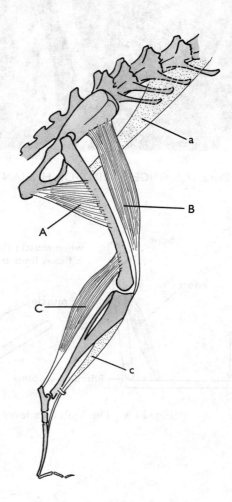

(b) EXTENDED

walking or crawling if the tone of muscles B and C was maintained or slightly adjusted. Relaxation of A followed by contraction of a and c would bring the hind-limb forward with the foot raised clear of the ground.

Contraction of B will straighten the leg at the knee. Contraction of C will extend the foot. Thus if muscles A, B and C contract simultaneously, the leg will extend and straighten at the same time as it is swinging backwards. The downwards and backwards thrust of both hind-limbs, with corresponding movements of the fore-limbs and spine will produce a leaping action. Muscles antagonistic to A, B and C, of which only a and c are shown, will contract when their opposite numbers relax, thus flexing the limb and returning it to its original position, before it touches the ground again.

There are, of course, many more muscles involved than are shown in Fig. 24.11 and their co-ordinated action is more subtle and complex than can be described here.

QUESTIONS

1. Construct a diagram similar to Fig. 24.4, to show a section through the elbow joint using Plate 36 for guidance. Show the attachment of the biceps and triceps tendons and state where you would expect to find the principal ligaments.

2. From Fig. 24.1 make an enlarged drawing of the scapula and fore-limb of the rabbit. Draw in muscles which you think would help to push the animal forward. Show the point of attachment very clearly and state what each muscle would do when it contracted.

3. What is the principal action of (a) your calf muscle, (b) the muscle in the front of your thigh and (c) the muscles in your forearm? If you don't already know the answer, try making the muscles contract and feel where the tendons are pulling.

4. Unlike most mammals, man stands upright on his hind-legs. What differences do you think this has made to his skeleton and musculature?

25 Teeth

TEETH are produced by the skin where it covers the jaws. Their roots become enclosed in the developing jaw bone, and their crowns break through the skin into the mouth cavity (Plate 39). They are thought to have arisen in the course of evolution from the scales of fish such as the dogfish. Where the skin bearing the scales stretched over the jaws, the scales became enlarged, pointed and specialized for holding prey or breaking up food.

Tooth structure (Fig. 25.1)

Enamel is the hardest substance made by animals. It is deposited on the outside of the crown of the tooth by cells in the gum before the tooth reaches the surface. The enamel is a non-living substance containing calcium salts; it forms an efficient, hard, biting surface.

Dentine is more like bone in its structure. It is hard but not so brittle as enamel and running through it are strands of cytoplasm from cells in the pulp. These cells are able to add more dentine to the inside of the tooth.

Pulp. In the centre of the tooth is soft connective tissue called pulp. From this the living strands of cytoplasm in the dentine derive their food and oxygen. The pulp contains sensory nerve-endings and blood capillaries. Oxygen and food brought by the blood enable the tooth to live and grow. The nerve-endings are particularly sensitive to heat and cold but produce only the sensation of pain.

The **root** is not set rigidly in the jaw bone, but is held by tough fibres so that it moves slightly in its socket as a result of chewing and biting movements.

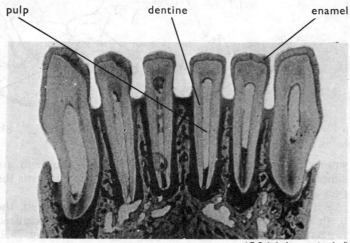

pulp   dentine   enamel

(G.B.I. Laboratories Ltd)

Plate 39. SECTION THROUGH TEETH AND JAW OF CAT

Cement is a thin layer of bone-like material covering the dentine at the root of the tooth. The fibres which hold the tooth in the jaw are embedded in the cement at one end and in the jaw bone at the other.

Specialization of teeth: carnivores

Where all the teeth serve the same purpose, such as merely holding the prey to prevent its escaping, they are all very

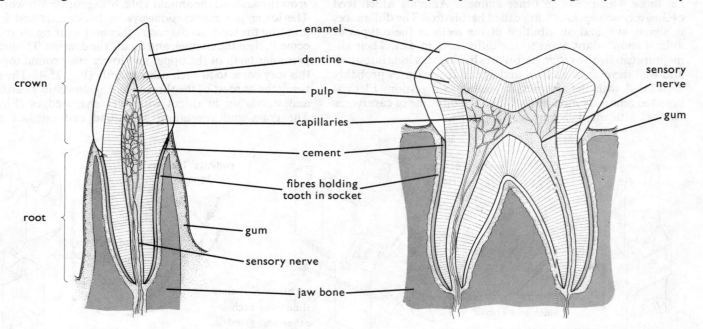

crown
root
enamel
dentine
pulp
capillaries
cement
fibres holding tooth in socket
gum
sensory nerve
jaw bone
sensory nerve
gum

(a) INCISOR (Section taken *across* line of jaw)

(b) MOLAR (Section taken *along* line of jaw)

Fig. 25.1   Sections through incisor and molar teeth

125

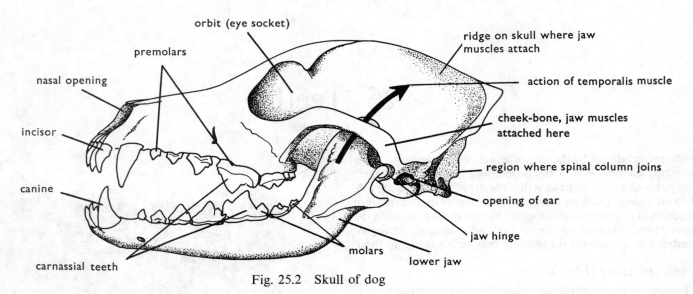

Fig. 25.2   Skull of dog

similar in structure. In most carnivorous fish the teeth are simply sharp pegs projecting backwards. In such animals the prey is often swallowed whole without chewing.

Where the food is captured, broken up and chewed, the teeth in different regions of the mouth may become specialized to a particular function. For example, a dog's *incisors* (Fig. 25.2) in the front of its mouth meet and can grip and strip off small pieces of flesh close to the bone. The long pointed *canines* are near the front and penetrate the prey, preventing its escape and often killing it. The massive *carnassial* teeth have sharp cutting edges. They pass each other and act like shears or snips, slicing off flesh and cracking bones (*see* Fig. 25.3). The *molars* have more flattened surfaces and meet each other, so crushing the bones and flesh to smaller particles before swallowing. The molars and premolars develop near the back of the jaw where the mechanical advantage is greater.

The dog's dentition is characteristic of carnivorous mammals, i.e. those which feed on other animals. Animals which feed exclusively on vegetation are called herbivores. The differences in shape, size and distribution of the teeth in these types of animal show adaptations to the differences in diet. Their alimentary canals also differ, herbivores having very long intestines and well developed caecum and appendix, features probably associated with the slowness of cellulose digestion. Flesh is digested relatively more rapidly and the intestine of carnivores is short with small caecum and appendix.

Growth of teeth

The teeth of herbivores grow throughout the life of the animal and are worn down at the same rate (see below). In carnivorous and omnivorous mammals, including man, the teeth cease to grow beyond a certain size; they are not worn away to any great extent.

Most mammals have two sets of teeth during their lives. The first set, or milk teeth, fall out as a result of their roots being dissolved away in the jaw, and they are replaced by the permanent teeth. A human two-year-old baby has 20 milk teeth which are replaced after the age of 5 years by 32 permanent teeth. These may not all appear until the age of 17 years or more.

Herbivores

The permanent teeth of herbivorous mammals continue to grow throughout the animal's life, being constantly worn down. The lower jaw moves sideways or backwards and forwards grinding the teeth across each other and wearing away first the cement, then the enamel, and finally the dentine. The molar and premolar teeth of the upper and lower jaw ground together in this way come to fit each other exactly (Fig. 25.4). The enamel, being the hardest of the three layers, is slower in wearing down and stands up in ridges with very sharp edges (Fig. 25.5). The grass and vegetation is ground and crushed between

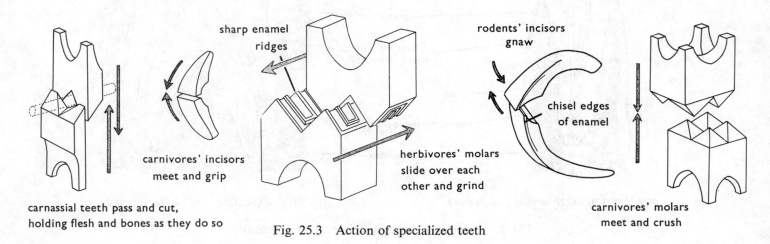

Fig. 25.3   Action of specialized teeth

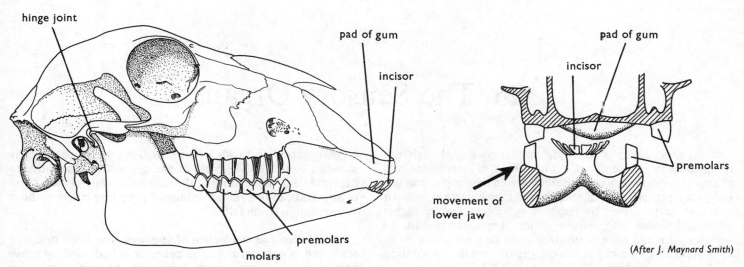

Fig. 25.4   Sheep's skull

Fig. 25.6   Section through herbivore's skull
to show action of teeth

(After J. Maynard Smith)

these exactly fitting edges, a high proportion of the cellulose cell walls are broken down so providing a greatly increased surface. Vertebrates, having no enzyme for digesting cellulose, depend on bacterial action for its digestion. An increased surface accelerates this process and the easily digestible protoplasm and cell sap are released from the crushed cells.

Herbivores' canine teeth, if present at all, are usually indistinguishable from incisors and the toothless gap between the incisors and premolars allows the tongue to manipulate the food (*see* Fig. 25.4). The premolars and molars are almost identical in shape and size, as might be expected from the fact that they have the same functions. Many herbivores have no incisors or canines in the top jaw. The grass or other vegetation is gripped between the bottom incisors and the gum of the upper jaw (Fig. 25.6).

Jaw articulation. In the herbivores, the joints between the lower jaw and skull are fairly loose, allowing the sideways or back and forth movement of the lower jaw (Fig. 25.6). In carnivores the jaw muscles, particularly the temporales (Fig. 25.2), are very powerful and the hinge joint allows only up

and down movement of the lower jaw, an adaptation which probably prevents dislocation of the jaw by (*a*) the strong chewing muscles and (*b*) the struggles of the prey.

Dentition in man

In man, the top and bottom incisors can pass each other when the jaws close and so cut off manageable portions of food. The molar and premolar surfaces meet and serve to crush the food. The canines are little different in shape or size from the incisors. Human dentition does not show the same degree of specialization as seen in the carnivorous and herbivorous mammals (Fig. 25.7).

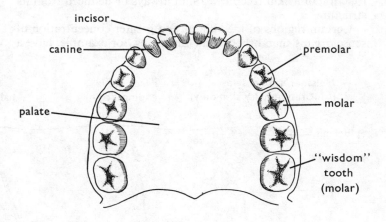

Fig. 25.7   Arrangement of teeth in man's upper jaw

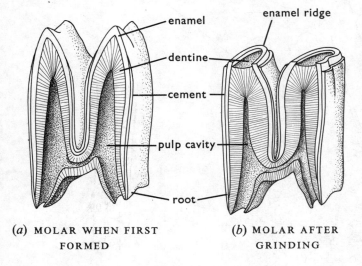

(*a*) MOLAR WHEN FIRST
FORMED

(*b*) MOLAR AFTER
GRINDING

Fig. 25.5   Sections of herbivore's molar to show how it
is worn down

QUESTIONS

1. Make a list of the differences in structure and position between the incisor, canine, premolar and molar teeth of a dog and a sheep. Relate each difference to the normal diet of each animal.
2. Study Fig. 25.1 and suggest why the correct method of brushing the teeth is to start with the brush on the gum and sweep it upwards over the crown rather than brushing across or up and down equally.
3. What general aspects of the diet of civilized man differ from the diets of other mammals? What effects might these differences have on the way man uses his teeth?

26 The Sensory Organs

THE sensory system makes the animal aware, though not necessarily in the sense of "conscious", of conditions and changes both outside and inside its body. In the simpler animals only very general stimuli such as light or darkness, heat or cold can be perceived by the sensory system. In the higher animals detailed information about surroundings such as distance, size and colours of objects can be gained as a result of the specialization of the sensory organs and the elaboration of the nervous system.

The general sensory system

Included in the general sensory system are the organs which are fairly evenly distributed through the dermis of the skin; hence any part of the skin is sensitive to touch, heat, pain and pressure. Examples of such sense organs are given in Fig. 26.1 and Plate 40.

In general, a particular sensory cell can respond to only one kind of stimulus. A light-sensitive cell in the eye does not respond to sound waves and a pressure-sensitive organ in the skin is not affected by the stimulus of heat. However, some sensory endings are less specific than this. The sensory nerve endings in the skin, which produce the sensation of pain, respond to a variety of stimuli, including pressure, heat and cold. The ear-lobe, which contains only free-nerve endings and hair plexuses, can detect touch, heat, cold and pressure. The function of a skin receptor cannot always be deduced from its structure.

Certain regions of the skin have a greater concentration of sense organs than others. The finger tips, for example, have a large number of touch organs, making them particularly sensitive to touch. The front of the upper arm is sensitive to heat and cold. Some areas of the skin have relatively few sense organs and can be pricked or burned in certain places without any sensation being felt.

Stimulation and conduction of impulses. The sense organ or sense cell is connected to the brain or spinal cord by nerve fibres. When the sense organ receives an appropriate stimulus it sets off an electrical impulse which travels along the nerve fibre to the brain or spinal cord. When the impulse reaches one of these centres it may produce an automatic or reflex action, or record an impression by which the animal feels the nature of the stimulus and where it was applied.

The sense organs of one kind and in a definite area are connected with one particular region of the brain. It is the region of the brain to which the impulse comes that gives rise to the knowledge about the nature of the stimulus and where it was received. For example, if the regions of the brain receiving impulses from the right leg were eliminated or suppressed by drugs, no amount of stimulation of the sensory-endings would produce any sensation at all, although the sense organs would still be functioning normally. On the other hand, if a region of the brain dealing with impulses from sense organs in the leg is stimulated by any means, the sensations produced seem to be from the leg.

Another important consideration is the fact that the impulses transmitted along the nerve fibres are fundamentally all exactly alike. It is not the sensations themselves that are carried but

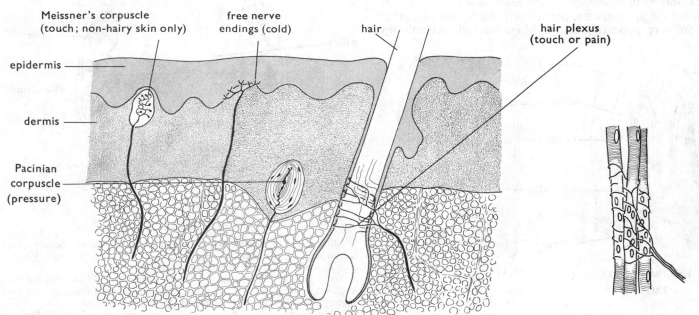

Fig. 26.1(a) Sense organs of the skin (generalized diagram)

Fig. 26.1 (b) Stretch receptor sensitive to tension in muscle fibre simply a surge of electricity, and this is so whether it is a heat organ or a touch organ that sets off the impulse. It is only in the brain that the stimulus is identified, according to the region of the brain which the impulse enters. For example, if the nerves from the arm and leg were changed over just before they entered the brain, stubbing one's toe would produce a sensation of pain in the arm or hand.

Intensity of sensation. A strong stimulus usually produces a more pronounced sensation than a weak stimulus. This is probably due to (*a*) the greater number of sense organs stimulated in the area, and (*b*) the stimulation of a number of sensory cells which do not respond at all unless the stimulus is intense. Many sensory organs are groups of cells, some of which are triggered off by the slightest stimulus while others need a powerful stimulus to affect them. When these latter are activated, the stimulus is recognized as being stronger than usual.

Vigorous stimulation does not affect the quality or intensity of the electrical impulse travelling in the nerve fibres but increases the total number of these impulses reaching the brain.

Pain. Although we tend to regard sensations of pain as inconvenient and alarming, they have important biological advantages. By making animals respond quickly or automatically by reflex action they tend to remove the animal or the affected part from danger. Our response when touching something unexpectedly hot affords a good example. If there were no sensations of pain, untold damage to tissues could result before one was aware of it. A sensation of pain is not essential for an effective reflex but it probably helps the animal to learn to avoid the same situation. Where pain occurs without producing a reflex action, as in tooth-ache, it serves as a warning that all is not well in that region and gives an opportunity to seek advice or treatment.

Internal sense organs (*proprioceptors*). The tissues of the body also have sense organs. One kind, occurring within the muscles, responds to the degree of stretching (Fig. 26.1*b*). These sense organs enable an animal to learn to place its limbs accurately in movement and to know their exact position without having to watch them. Sensory pain-endings occur in many internal organs, but not in the brain.

Special senses

Sight, hearing and balance, smell and taste are called the special senses. The relevant sense organs each consist of a great concentration of cells which are sensitive to one kind of stimulus. These sensory cells may be associated with structures that direct the stimulus on to the sensitive region.

Taste. In the lining of the nasal cavity and on the tongue are groups of sensory cells that can be stimulated by chemicals

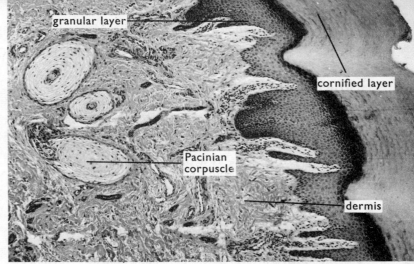

(Brian Bracegirdle)

Plate 40. PACINIAN CORPUSCLE IN HUMAN SKIN (× 60)

which dissolve in the moisture overlying them. On the tongue these groups are called *taste-buds*; they lie mostly in the grooves round the bases of the little projections on the upper surface of the tongue (Fig. 26.2).

Taste buds vary in their sensitivity to groups of chemicals which we describe as sweet, sour, salt or bitter. The chemicals producing a similar taste sensation often have little in common. A sour taste, however, usually indicates an acid. It can be seen that the sense of taste is very limited and probably serves to distinguish only between food suitable and unsuitable for eating. The sensation of flavour has a much greater range and comes from the sense of smell. If the nose is blocked it is difficult to distinguish between many kinds of food which normally have quite distinctive flavours.

Smell. In the epithelium lining the top of the nasal cavity there are spindle-shaped cells with processes extending out into the mucus film that spreads over the epithelium. From these cells nerve fibres pass into the brain. The cells are stimulated by substances which dissolve in the moist lining of this region of the nasal cavity and so produce the sensation of smell. No satisfactory classification of smells, or explanation of how they are distinguished, has yet been made.

The sense of smell is easily fatigued, that is, a smell experienced for a long period ceases to give any sensation and we become unaware of it, though a newcomer may detect it at once.

Sight. The eyes are the organs of sight. They are spherical organs housed in deep depressions of the skull, called *orbits*, and are attached to the wall of the orbit by six muscles which can also move the eye-ball (Fig. 26.3). The structure is best seen in a horizontal section as shown in Fig. 26.4.

Hearing and balance. *See* page 134.

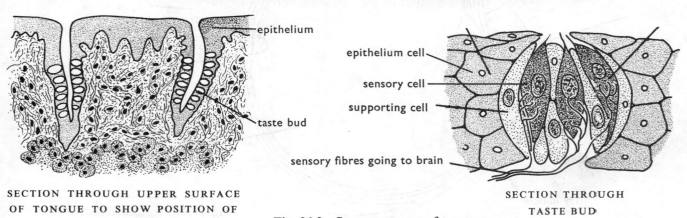

SECTION THROUGH UPPER SURFACE
OF TONGUE TO SHOW POSITION OF
TASTE BUDS

Fig. 26.2 Sensory system of tongue

SECTION THROUGH
TASTE BUD

Structure and functions of the parts of the eye

The EYELIDS can cover and so protect the eye. Closing the eyelids can be a voluntary or reflex action. Regular blinking serves to distribute fluid over the surface of the eye and prevents its drying up.

The CONJUNCTIVA is a thin epithelium which lines the inside of the eyelids and the front of the sclera, and is continuous with the epithelium of the cornea.

The TEAR GLANDS open under the top eyelids. They secrete a solution of sodium hydrogencarbonate and sodium chloride and keep the exposed surface of the conjunctiva and cornea moist. They also wash away dust and other particles. An enzyme which is present in tear fluid has a destructive action on bacteria. Excess fluid is drained into the nasal cavity through the *lachrymal duct* which opens at the inside corner of the eyes.

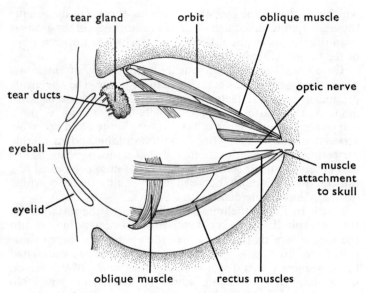

Fig. 26.3   Diagram to show eye muscle attachment
(left eye seen from side)

The EYE MUSCLES are attached to the *sclera* at one end and to the wall of the orbit at the other. Their contractions can make the eye move from side to side and up and down (Fig. 26.3).

The SCLERA is a tough, non-elastic, fibrous coat round the eye-ball.

The CORNEA is the transparent disc in the front part of the sclera. Light passes through the cornea into the eye. Since the cornea is a curved surface the light is refracted and the rays begin to converge.

The CHOROID is a layer of tissue lining the inside of the sclera. It contains a network of blood vessels supplying food and oxygen to the eye. It is also deeply pigmented, the black pigment reducing the reflection of light within the eye.

The AQUEOUS AND VITREOUS HUMOURS are solutions of salts, sugars and proteins in water. The aqueous humour is quite fluid, the vitreous jelly-like. These liquids help to refract the light and produce an image on the retina. Their pressure outwards on the sclera maintains the shape of the eye.

The crystalline lens, the cornea and the conjunctiva are made of living cells which are quite transparent. They contain no blood vessels and must absorb their food and oxygen from the aqueous humour.

The CRYSTALLINE LENS. The cornea begins and the crystalline lens continues the refraction of light so producing an image on the retina. The lens is held in position by the fibres of the *suspensory ligament* which radiate from its edge and attach it to the *ciliary body*. The shape of the lens can be altered by contraction or relaxation of muscles of the ciliary body.

The CILIARY BODY is the thickened edge of the choroid in the region round the lens. It contains blood vessels and muscle fibres some of which run in a circular direction, that is, parallel to the outer edge of the lens.

The IRIS consists of an opaque disc of tissue. At its outer edges it is continuous with the choroid. In the centre is a hole, the pupil, through which passes the light that will produce an image on the retina. The contraction or relaxation of opposing sets of circular and radial muscle fibres in the iris increases or decreases the size of the pupil, so controlling the intensity of

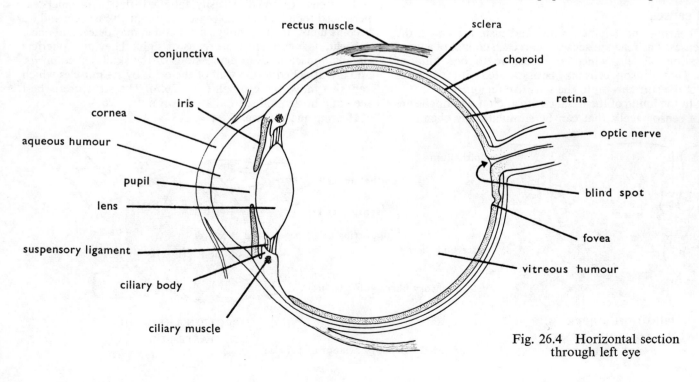

Fig. 26.4   Horizontal section
through left eye

light entering the eye. The iris contains blood vessels and sometimes a pigment layer that determines what is usually called the "colour" of the eyes. Blue eyes have no pigment, the colour being produced by a combination of the black backing, the blood capillaries and the white outer layers on the iris.

The RETINA. This is a layer of cells sensitive to light. There are two kinds of light-sensitive cell, called, according to their shape, *rods* and *cones*. Only the cones are sensitive to coloured light but the rods are more responsive to light of low intensity.

The nerve fibres from these cells pass across the front of the retina and all leave at one point to form the optic nerve which passes through the skull into the brain.

The BLIND SPOT. In the region where the nerve fibres leave the eye to enter the optic nerve there are no light-sensitive cells. If part of an image falls on that region no impression is recorded in the brain. We are not normally aware of this "blank" in our field of vision, partly because it is compensated by the use of two eyes scanning the same field and partly because it never coincides with the image of an object on which we are concentrating. (*See* Experiment 6, p. 136.)

The FOVEA is a small depression in the centre of the retina. It contains only cones, and it is the region of the retina with the greatest concentration of sensory cells and, therefore, gives the most accurate interpretation of an image. When an observer concentrates on an object, or part of an object, its image is thrown on to the fovea. Only in this region is there detailed appreciation of form and colour.

Image formation and vision. Light from an external object enters the eye. The curved surface of the cornea, the lens and the humours, refract the light and focus it so that "points" of light from the object produce points of light on the retina. The image thrown on to the retina is real, upside-down, and smaller than the object (Figs. 26.5 and 26.6). The light-sensitive cells are stimulated by the light falling on them, and impulses are fired off in the nerve fibres. These impulses pass along the optic nerve to the brain where, as a result, an impression is formed of the nature, size, colour and distance of the object. The inversion of the image on the retina is corrected in the optical centre of the brain to form the impression of an upright object.

The accuracy of the impression that the brain gains of the image depends on how numerous and how closely packed are the light-receiving cells of the retina, since each one can only record the presence or absence of a point of light and, in the cones, its colour. If there were only ten such cells, the image of a house falling on five of them would record an impression of its size, the fact that it was differently coloured at the top and bottom, and a vague representation of its shape. In a hawk's fovea there are about a million cones per square millimetre.

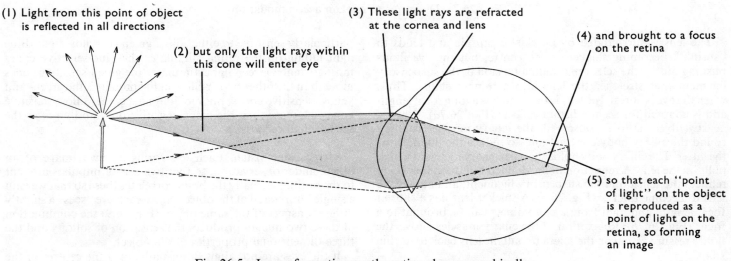

(1) Light from this point of object is reflected in all directions

(2) but only the light rays within this cone will enter eye

(3) These light rays are refracted at the cornea and lens

(4) and brought to a focus on the retina

(5) so that each "point of light" on the object is reproduced as a point of light on the retina, so forming an image

Fig. 26.5   Image formation on the retina shown graphically

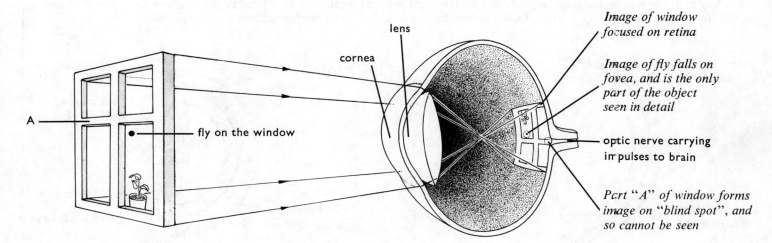

A

fly on the window

cornea

lens

Image of window focused on retina

Image of fly falls on fovea, and is the only part of the object seen in detail

optic nerve carrying impulses to brain

Part "A" of window forms image on "blind spot", and so cannot be seen

Fig. 26.6   Image formation in the eye

Accommodation. With a rigid lens of definite focal length placed at a fixed distance from a screen it is possible to obtain a sharply focused image of an object only if the object is at a certain distance from the lens. The focal length of the lens in the eye can be altered by making it thicker or thinner. In this way light from objects from about 25 cm to the limits of visibility can be brought to a focus. This ability of the eye to alter its focal length is called accommodation.

so admitting more light. This is a reflex action set off by changes in the intensity of light. In poor light the pupils are wide open; in bright light the pupils are contracted. In this way the retina is protected from damage by light of high intensity, and in poor light the wider aperture of the pupil helps to increase the brightness of the image.

Colour vision. There are three kinds of cone in the human retina. All three respond to more than one colour but each is

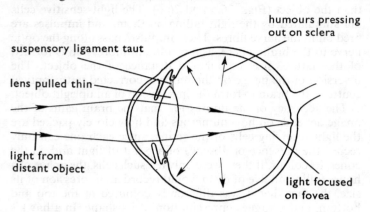

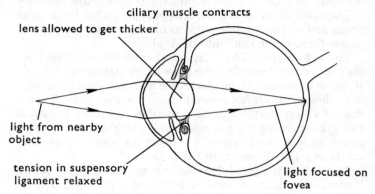

(a) EYE RELAXED (*distant accommodation*)

(b) EYE FOCUSED ON NEAR OBJECT

Fig. 26.7   Diagrams to explain accommodation

The lens is surrounded by an elastic capsule and tends to "shrink", becoming thicker in the centre, but the eye fluids pushing out on the sclera maintain a tension in the suspensory ligament that stretches the lens into a thinner shape. Thus, when the eye is at rest, the lens is thin and has a long focal length and is adapted for seeing distant objects (Fig. 26.7a). When a nearby object is to be observed, the ciliary muscles running round the ciliary body contract and so reduce the diameter of the latter. The ciliary body holds the suspensory ligament which pulls on the lens. A reduction in the diameter of the ciliary body reduces the tension in the suspensory ligament, and allows the lens to become thicker (Fig. 26.8). A thicker lens has a shorter focal length, and light from a close object can be brought to a focus (Fig. 26.7b). Relaxation of the ciliary muscles allows the fluid pressure acting on the sclera to pull the lens back to its thin shape.

Control of light intensity. When the circular muscles of the iris contract, the size of the pupil is reduced and less light is admitted. Contraction of the radial muscles widens the pupil,

particularly sensitive to either blue, green or yellow light. Blue light falling on the retina stimulates the blue-sensitive cones most strongly. Green light stimulates the green-sensitive cones more than the other two. Yellow light stimulates the green- and yellow-sensitive cones, but red light affects the yellow-sensitive cones far more than the green-sensitive ones and gives us the sensation of redness.

Stereoscopic vision. Each eye forms its own image of an object under observation, so that two sets of impulses are sent to the brain. Normally the brain correlates these so that we gain a single impression of the object. Since each eye "sees" a slightly different aspect of the same object (Fig. 26.9) the combination of these two images produces the sensation of solidity and the three-dimensional properties of the object.

If the eyes are not aligned normally, or if the centres of the brain dealing with sight impressions are dulled by alcohol, for example, the two sensory impressions from the eyes are not properly correlated and we "see double".

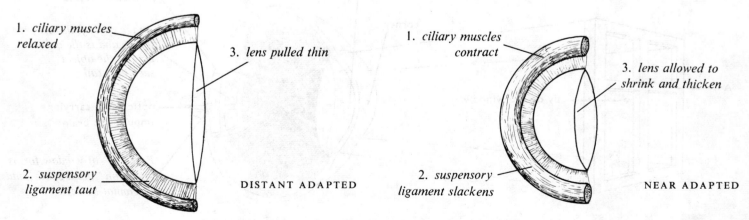

Fig. 26.8   Diagram to show how accommodation is brought about

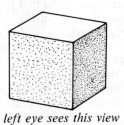
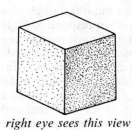

left eye sees this view   *right eye sees this view*

Fig. 26.9   Different views of a cube seen by left and right eyes

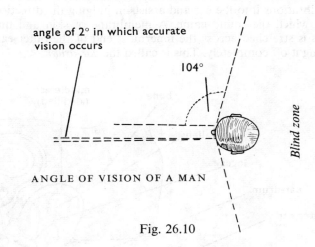

No blind zones

ANGLE OF VISION OF A HARE

angle of 2° in which accurate vision occurs

104°

Blind zone

ANGLE OF VISION OF A MAN

Fig. 26.10

Judgment of distance. Our ability to judge distance probably depends on a number of factors; the apparent size of the object, overlapping of objects, parallax effects (see below). When we concentrate on close objects, our eye muscles must contract to rotate the eyeballs slightly inwards, and the stretch receptors in these muscles may send impulses to the brain so providing information about distance. The stereoscopic vision described above probably helps one to judge distance. It is more difficult to estimate distance accurately using only one eye.

Animals with their eyes set in the front of their heads and directed forwards have stereoscopic vision. This occurs chiefly in animals of prey and is an advantage in judging the distance of their prey before they leap or dive. Examples are lions and tigers among the mammals, hawks, owls and gannets among the birds, and pike among the fish. The judgment of distance in the apes is probably associated with their tree-dwelling habits.

Most of the animals with eyes at the sides of their heads can judge distance only by the apparent size of objects and by parallax. Parallax is the name given to the apparent movement of nearby objects against a background of distant objects when the head is turned from side to side. In most birds the overlapping fields of vision of the two eyes could give stereoscopic vision within an angle of 6–10° from the head.

Animals with their eyes in the sides of their heads can usually see nearly all round (Fig. 26.10), including objects directly

behind them. The eyes of most mammals seem particularly sensitive to movement. These last two factors favour the rapid escape of animals such as deer and rabbits likely to be preyed upon by others.

Animals with eyes facing forward have a more limited field of vision, but even a man is aware of objects within an angle of about 200°, although only those included in an angle of 2° from the eye will form an image on the fovea and so be observed accurately. This is considerably less than most people imagine and means, for example, that only about two letters in any word on this page can be studied in detail.

Long and short sight. Its causes and corrections are explained diagrammatically in Fig. 26.11.

LONG SIGHT

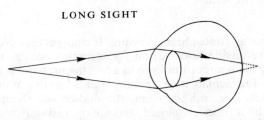

Long sight is caused by small eyeballs or "weak" lenses. Light from a distant object is brought to a focus on the retina but from a close object its focus is behind the retina

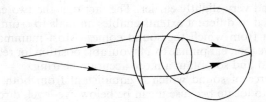

Long sight can be corrected by wearing converging lenses

SHORT SIGHT

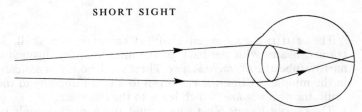

Short sight is usually caused by large eyeballs. Light from a distant object is focused in front of the retina

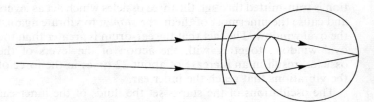

Short sight can be corrected by wearing diverging lenses

Fig. 26.11   Long and short sight

The ear

The ear contains receptors which are sensitive to sound vibrations in the air between the frequencies of 30 per second and 20,000 per second.

Structure (Fig. 26.12). The OUTER EAR is a tube opening on the side of the head and leading inwards to the ear-drum. At the outside end there may be an extension of skin and cartilage, the *pinna*, which in some mammals helps to concentrate and direct the vibrations into the ear and assists in judging the direction from which the sound came. A membrane of skin and fine fibres is stretched across the innermost end of the outer ear, closing it off completely. This is called the *ear-drum*.

worked out it is thought that the short fibres in the first part of the cochlea respond to high-frequency vibrations and the long fibres in the last part to low-frequency vibrations, with a continuous range of intermediate fibres which respond to other frequencies. When the transverse fibres vibrate they stimulate the sensory cells resting upon them. According to this theory the pitch of a note could be determined by the brain, owing to the fact that only a particular group of fibres is stimulated by a certain frequency and will send impulses through the auditory nerve to the brain, while the others will be unaffected.

Eustachian tubes. Air pressure in the middle ear is usually

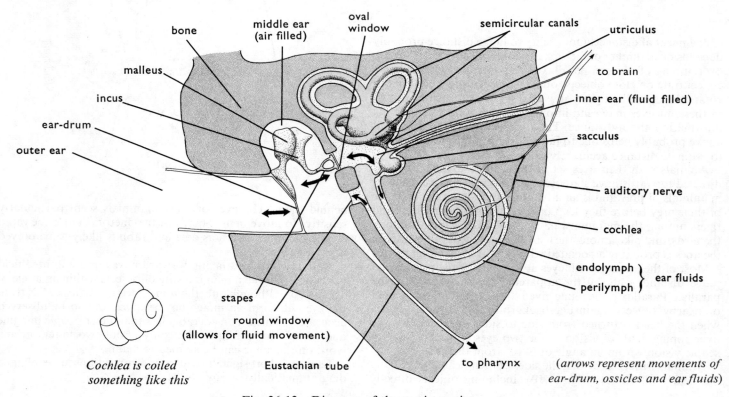

Cochlea is coiled something like this

(*arrows represent movements of ear-drum, ossicles and ear fluids*)

Fig. 26.12   Diagram of the ear in section

The MIDDLE EAR is an air-filled cavity in the skull. It communicates with the back of the mouth cavity through a narrow tube, the *Eustachian tube*. Three small bones, or *ossicles*, in the middle ear link the ear-drum to a small opening in the skull, the *oval window*, which leads to the *inner ear*.

The INNER EAR is filled with a fluid and consists mainly of a coiled tube, the *cochlea*, with sensory endings in it. It is here that the sound vibrations are converted to nervous impulses.

Hearing. Vibrations in the air that constitute sound waves enter the outer ear and set the ear-drum vibrating. This vibration is transmitted through the three ossicles which act as levers and cause the innermost of them, the *stapes*, to vibrate against the oval window. The fact that the ear-drum is greater than the oval window, together with the action of the levers of the ossicles, result in an increase of about 22 times in the force of the vibrations that reach the inner ear.

The oscillations of the stapes set the fluids of the inner ear vibrating, particularly in the cochlea. A membrane consisting of transverse fibres runs the length of the cochlea; in the first part of this membrane the fibres are short, and in the last part longer. Although details of the mechanism are not yet fully

the same as atmospheric pressure. If changes take place in the pressure outside the ear-drum, for example when gaining height rapidly in an aircraft or even in a lift, the pressure is equalized by the Eustachian tube opening and admitting more air to, or releasing excess air from, the middle ear. Normally the Eustachian tubes are closed, and are opened only when one is swallowing or yawning, when a "popping" sound may be heard in the ears.

Sense of direction of sound. With two ears the sound from a single source will be heard more loudly in one ear than the other, and very slightly earlier. The fact that the two ears are stimulated to different extents enables animals to estimate the direction from which the sound comes. Most mammals can also move their ear pinnae to a favourable position for receiving the sound and so obtain a more accurate bearing.

A source of sound which is equidistant from both ears is difficult to locate because it can be below eye-level, directly in front, directly above or behind the head and still stimulate both ears equally. In such a situation dogs will cock their heads on one side, resulting in one ear being stimulated more than the other.

A dog can locate the position of a sound in one out of thirty-two positions all round it, while humans are accurate in the perception of one out of only eight possible sources. Cats have been shown capable of distinguishing the position of two sounds only half a metre apart and eighteen metres away.

Semicircular canals and sense of balance (Fig. 26.13)

The *utriculus*, *sacculus* and *semicircular* canals are organs of balance and posture. In the fluid-filled cavities of the utriculus and sacculus are gelatinous plates containing chalky granules, called *otoliths*, which lie above sensitive patches of the lining. Sensory fibres from these patches are embedded in the otoliths

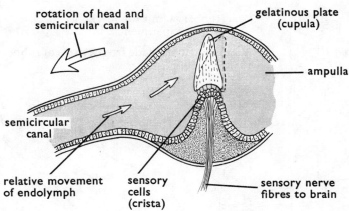

Fig. 26.14(*a*) Diagrammatic section through ampulla

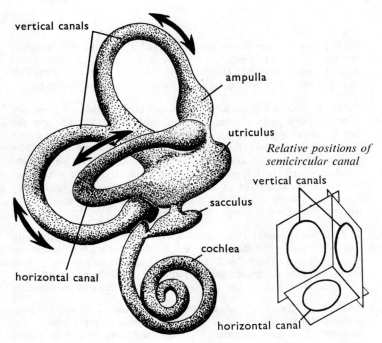

Fig. 26.13 Semicircular canals
Arrows show the direction of rotation which stimulates each canal.

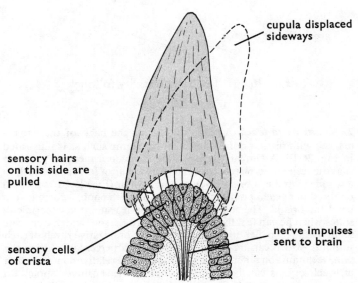

Fig. 26.14(*b*) Detail of crista and cupula

and when the head is tilted, the otoliths pull on the fibres. The nervous impulses fired off from these organs as a result of such stimulation reach the brain and set off a reflex tending to return the body to its normal posture.

The semicircular canals contain a fluid and in the *ampullae* are sense organs which respond to movements of the fluid. When the head rotates, the fluid tends to remain stationary for a time and is thought to displace the *cupula* in the ampulla (Fig. 26.14). The three canals are in planes at right angles to each other and are stimulated by rotation in their respective planes.

The utricles are regarded as responding to the tilting movements of the head or body while the semicircular canals are stimulated by accelerations of a rotary kind in their particular plane. If the utricles and semicircular canals did not function properly, animals would keep falling over unless they relied on their eyes. Without the semicircular canals one could probably stand upright if quite stationary, but it would be very difficult to maintain balance while moving or changing direction.

Experiences of vertigo during night flying show that we normally rely on information from the eyes and other sense organs, as well as from the utriculus and semicircular canals, for accurate sensations of position and balance.

PRACTICAL WORK

SKIN SENSE. 1. *To find the sensitivity of the skin to touch.* A simple apparatus such as is shown in Fig. 26.15 can be used. The distance apart of the points is measured, and starting at about 5 cm apart, the experimenter touches the two points simultaneously on parts of the skin of a "volunteer" whose eyes are closed. In the more sensitive regions of the skin, the two points are felt as separate stimuli. Elsewhere, e.g. the back of the hand or the neck, only a single stimulus is felt, perhaps because there are fewer touch endings. By reducing the distance between the points from time to time, the degree of sensitivity of different regions of the skin can be mapped out.

It is best to vary the stimulus, using sometimes one point and sometimes two so that the subject does not know in advance which is to be used.

wire "hair pin"

Fig. 26.15

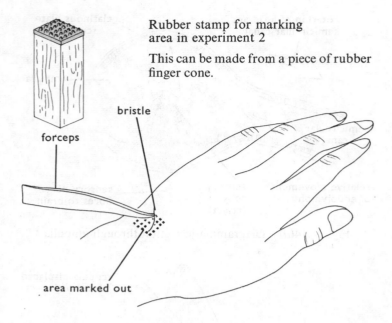

Rubber stamp for marking area in experiment 2

This can be made from a piece of rubber finger cone.

Fig. 26.16   Testing sensitivity to touch

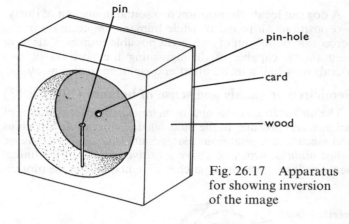

Fig. 26.17   Apparatus for showing inversion of the image

2. *Sensitivity to touch.* A patch of skin on the back of the wrist is marked with regular dots, using a rubber stamp such as is illustrated in Fig. 26.16. A similar pattern is stamped on a piece of paper so that the results can be recorded. Held by a pair of forceps or stuck to a wooden handle, a bristle such as a horse hair (Fig. 26.16) is pressed on the skin at each point marked by a dot, with enough force just to bend the bristle. The subject, who must not watch the experiment, states when he can feel the stimulus. A third person, with a duplicate set of marks on paper, indicates the positive or negative result of each stimulus, which can finally be expressed as a percentage. Using the same technique on different parts of the skin the relative concentration of touch organs can be estimated though the sensitive "spots" do not correspond to single nerve-endings.

3. *Location of the source of stimulation.* If a dried pea or marble is rolled about on a table between the crossed tips of the first and second fingers, an impression of two solid objects is received in the brain. The eyes should be closed while doing this so that only sensations of touch are received. Normally these regions of the fingers are stimulated simultaneously only by two separate objects, and it is this impression that has been learned by the brain.

EYES 4. *Inversion of the image.* If the apparatus shown in Fig. 26.17 is held close to the eye and the pin observed by looking through the pin-hole, an upright silhouette of the pin's head is seen. The apparatus is now reversed so that the pin is nearer to the eye, and moved until the pin-head can be seen against the outline of the pin-hole. In this case, an upright and enlarged shadow is cast on to the retina, and the brain makes the usual correction so that the impression gained is of the pin-head upside down.

5. *The double image.* If a nearby object is observed, and a finger pressed against the lower lid of one eye so as to displace the eye-ball slightly, an impression of two separate images will result, one formed in each eye.

6. *The blind spot.* Hold the book about 60 cm away. Close the left eye and concentrate on the cross with the right eye. Slowly bring the book closer to the face. When the image of the dot falls on the blind spot it will seem to disappear.

7. *To find which eye is used more.* A pencil is held at arm's length in line with a distant object. First one eye is closed and opened and then the other. With one the pencil will seem to jump sideways. This shows which eye was used to line up the pencil in the first place.

EARS. 8. *Location of sound.* Two large funnels, held in clamps, are connected to lengths of rubber tubing which can be inserted into the ears of the subject. The subject holds the tubes in his ears so that only sounds entering the funnels will reach the ear-drums. The subject is blindfolded and the two funnels crossed over so that the one leading to the left ear is pointing to the right and vice versa. When sounds are made on the right of the subject, he thinks they are coming from the left. By altering the position of the funnels, without the subject's knowledge, interesting results can be obtained.

9. *Location of sound.* A ticking clock is held, in turn, above, behind, and in front of a blindfolded subject, so that it is always equidistant from both ears. The subject is asked to indicate its position. These results are then compared with the number of successes scored when the clock is held in similar positions but to the sides of the subject.

TASTE. 10. *Sensitivity of the tongue.* Solutions of sucrose (sweet), sodium chloride (salt), citric acid (sour), and quinine (bitter) are prepared. The subject puts out his tongue and the experimenter places a drop of one of the solutions at one point of the tongue using a pipette. Without withdrawing the tongue, the subject tries to identify the taste. The solutions are applied in turn to all parts of the tongue, washing the pipette between each application. In this way it may be possible to determine which regions of the tongue are most sensitive to particular groups of chemicals.

See *Experimental Work in Biology 8: Human Senses* (p. 1) for other experiments.

QUESTIONS

1. Most animals have a distinct head end and tail end. Why do you think the main sensory organs are confined to the head end?
2. Chemicals such as sugar and saccharine both taste sweet and yet they are chemically quite different. Middle C on the piano has a frequency of 264 vibrations per second whereas D has 297 vibrations per second. The difference is small and yet the two notes are easily distinguished.
   What are the properties of the sense organs concerned which make for poor discrimination of chemicals and precise discrimination of sounds?
3. In what functional way does a sensory cell in the retina differ from a sensory cell in the cochlea?
4. An eye defect known as "cataract" results in the lens becoming opaque. To relieve the condition, the lens can be removed completely. Make a diagram to show how an eye without a lens could, with the aid of spectacles, form an image on the retina. What disadvantages would result from such an operation?
5. In poor light, an object can be seen more clearly in silhouette by looking to one side of it than by looking at it directly. Explain this phenomenon.
6. A person whose ear ossicles are ineffective can often hear a ticking watch pressed against his head better than he can if it is held close to his ear. Explain this effect.

27 Co-ordination

THE various physiological processes in living animals have been described so far as if they were quite separate functions of the body, the total result of which constitutes a living organism. Although these processes can be usefully considered individually, they are in fact all very closely linked and dependent on each other. The digestion of food, for example, would be of little value without a blood stream to absorb and distribute the products.

The working together of these systems is no haphazard process. The timing and location of one set of activities is closely related to the others. For example, in walking, the legs are moved alternately, not both at the same time, without the walker's having to think consciously about it. During exercise, when the body needs more food and oxygen, the breathing rate is automatically increased and the heart beats faster, so sending a greater volume of oxygenated blood to the muscles. When one is eating a meal the position of the food is recorded by the eyes, and as a result of this information the arms are moved to the right place to take it up, not by trial and error, but with precision and accuracy. As the food is raised to the mouth the latter opens to receive it at just the right moment, chewing movements begin, and saliva is secreted. At the moment of swallowing, many things happen simultaneously as described on p. 86. In the stomach, the gastric glands begin to secrete enzymes which will digest the food when it arrives.

In the sequences described above, many bodily functions come into action at just the right moment, with the result that no unnecessary movements are made and no enzymes are wasted by their being secreted when no food is present.

The linking together in time and space of these and other activities is called co-ordination. Without co-ordination the bodily activities would be thrown into chaos and disorder: food might pass undigested through the alimentary canal for lack of enzyme secretion, even assuming it negotiated the hazard of the windpipe; both legs might be bent simultaneously in an attempt at walking; a runner would collapse after a few metres from lack of an increased oxygen supply, and so on.

Co-ordination is effected by the nervous system and the *endocrine system*. The nervous system is a series of conducting tissues running to all parts of the body (Fig. 27.1), while the endocrine system is a number of glands in the body which produce chemicals that circulate in the blood stream and stimulate certain organs (p. 144).

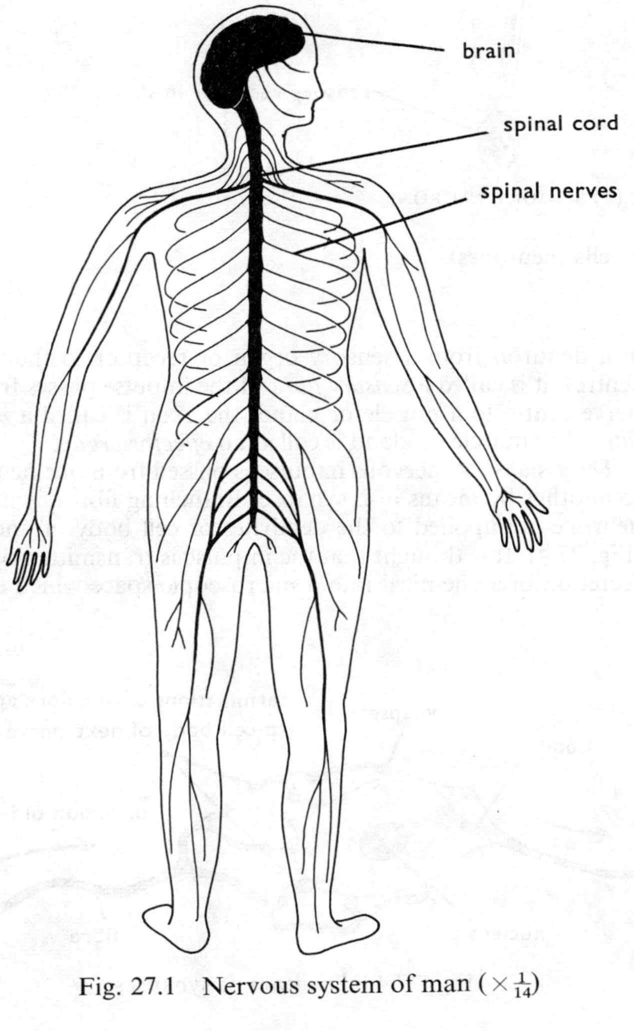

Fig. 27.1 Nervous system of man ($\times \frac{1}{14}$)

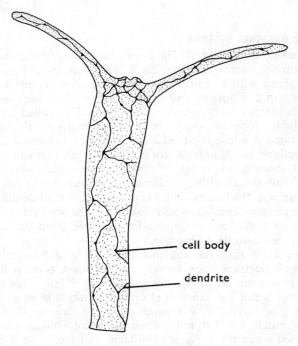

Fig. 27.2 Generalized diagram of nerve network of a Hydra ($\times 10$)

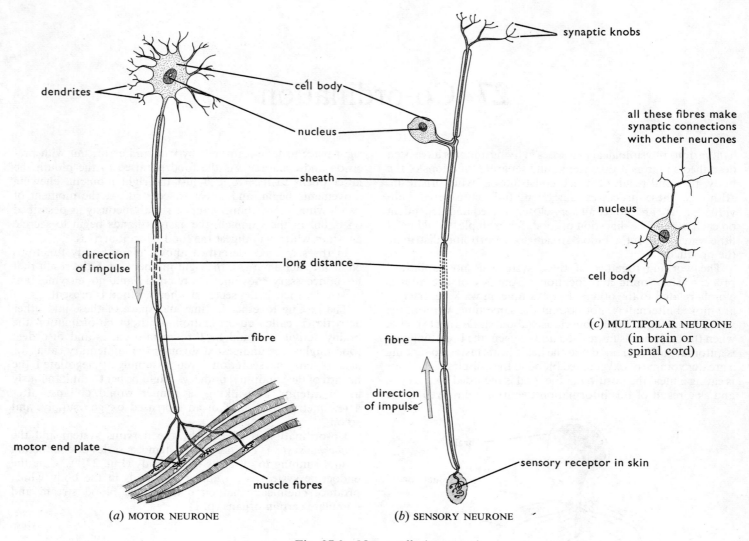

Fig. 27.3   Nerve cells (neurones)

The nervous system

Nerve cells (*neurones*). The units which make up the nervous system are nerve cells (Fig. 27.3). These are small masses of cytoplasm with a central nucleus. One or more branching, cytoplasmic filaments called *dendrites* conduct impulses towards the cell body, while a single, long fibre called an *axon* conducts impulses away. In sensory neurones the single, elongated dendrite is called a *dendron*. Dendrons and axons, collectively called nerve fibres in this account, consist of fluid-filled, cytoplasmic tubes, in certain cases surrounded by an insulating sheath of fatty material. In mammals, the cell body is usually in the brain (Plate 41) or spinal cord, while the axon or dendron extends the whole distance to the organ concerned, often for considerable lengths, for example, from the base of the spine down to the big toe.

The nerve fibre has the special property of being able to transmit electrical impulses very rapidly down its entire length and pass them on to the next nerve cell in line. This transmission is not the same as electrical conduction in a metal, where the current flow depends on the voltage applied. The axon builds up within itself an electrical charge which is released when the nerve is stimulated and has to be built up again before the next impulse can pass. Nerve cells usually transmit impulses in one direction only. If the impulse travels

in a dendron from a sensory organ or receptor to the nerve centres, it is called a *sensory fibre*. If the impulse passes from a nerve centre to a muscle or gland, the axon is called a *motor fibre*. The muscle or gland is called an *effector organ*.

The synapse. A nervous impulse is passed from one neurone to another by means of a synapse. Branching fibres from one neurone are applied to the dendrites or cell body of another (Fig. 27.4). It is thought that the impulse is transmitted by the secretion of a chemical into a microscopic space which exists

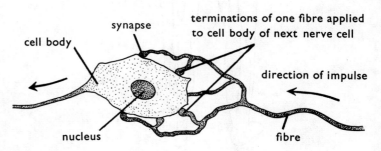

Fig. 27.4   Diagram of synapses

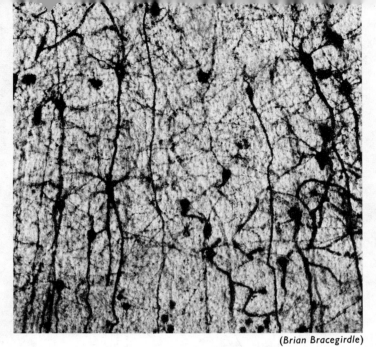

(Brian Bracegirdle)

Plate 41. MULTIPOLAR NEURONES IN BRAIN CORTEX (×350)

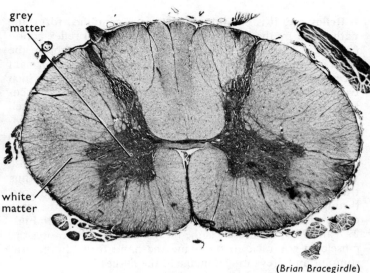

(Brian Bracegirdle)

Plate 42. SECTION THROUGH SPINAL CORD (×7)

between the termination of the fibre and the membrane of the cell body. A single impulse does not necessarily get across the synapse. It may take two or three impulses arriving in rapid succession or perhaps arriving simultaneously from two or more fibres, to start an impulse in the next neurone.

An individual cell body may have synapses with many incoming fibres and it is via the synapses that the different parts of the body and brain are kept in communication. Because of the enormous possibilities of inter-connexion and since the simultaneous arrival of impulses at a cell body may stimulate or inhibit the relaying of a subsequent impulse, the synapse is probably the basic "computer" unit of the central nervous system, making possible effective co-ordination and learning.

The grey matter of the brain and spinal cord consists of cell bodies and their synapses. The white matter consists of large tracts of fibres.

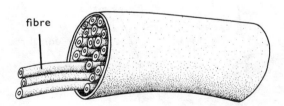

Fig. 27.5   Diagram to show nerve fibres grouped into a nerve

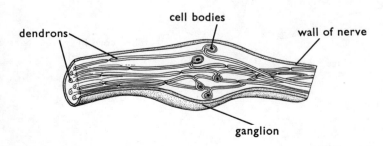

Fig. 27.6   Diagram of cell bodies forming a ganglion

Nervous systems. In relatively simple animals like the sea-anemone and Hydra, the nerve cells spread fairly evenly in all directions (Fig. 27.2), so that an impulse started at one point spreads out slowly in all directions encountering many synapses. In higher animals the nerve cells are bundled together into nerves which run in distinct paths from the nerve centres to important organs. Nerves are bundles of fibres (Fig. 27.5), the cell bodies of sensory fibres sometimes forming a bulge, or *ganglion*, in their length (Fig. 27.6). Most nerves contain both motor and sensory fibres, though one or other type may predominate.

Central nervous system. In vertebrate animals the central nervous system consists of the brain and spinal cord (Fig. 27.1). Most of the cell bodies lie in the central nervous system. Since all impulses from or to the body pass through it, it is possible to have a vast number of cross-connexions and linkages that could not arise if the nerves simply ran from one organ to another. The following analogy of the telephone exchange illustrates the advantages of a centralized nervous system—though in their actual mechanism the nervous system and the telephones are completely different.

If you were to connect your telephone directly to all the people you were likely to want to call up, hundred of wires would be needed, and if everybody did this the numbers and confusion of wires would be overwhelming. Even so, you would be limited to only those few hundred possible calls. A telephone exchange makes it possible for you to be placed in communication with anyone in the country, rapidly and efficiently. This gives some idea of the increased efficiency of co-ordination as a result of having a central nervous system.

Reflex action is a rapid, automatic response to a stimulus, by an organ or system of organs, which does not involve the brain for its initiation. For example, the iris of the eye can contract or dilate the pupil in response to changing light intensity without our being aware that it is happening. More commonly we are aware of a reflex occurring but are unable to control it. Blinking when a foreign particle touches the cornea is a reflex action which protects the eyes. We know it is happening but can do nothing to prevent it or modify it. Sneezing is a reflex response to a stimulus in the nose. The knee-jerk is another example. If the right leg is crossed over the left and struck sharply just above or below the knee cap, the lower leg jerks outwards by reflex action.

Reflex arc. It is possible to trace, in a simplified form, the path taken by the impulses involved in a spinal reflex action (*see* Fig. 27.7). If one unexpectedly touches a hot object, the hand is rapidly removed from the source of heat. Heat or pain receptors in the skin are stimulated and fire off impulses which travel along the sensory fibres in a nerve of the arm. The sensory fibres enter the spinal cord via the *dorsal root*, their cell bodies producing the swelling known as the *dorsal root ganglion*. In the grey matter of the spinal cord, the impulses pass from the sensory neurone to a relay or association neurone across a synapse. The relay neurone, in turn, makes a synapse with one or more motor neurones. The impulses are thus transmitted to the motor fibres which leave the spinal cord through the *ventral root* and pass in a nerve, probably the same one in which the sensory impulse travelled, to a muscle. In Figs. 27.8 and 27.9, the muscle is the biceps. The impulse causes the muscle to contract, so removing the hand from the painful stimulus and preventing damage to the tissues.

A similar reflex arc produces the knee jerk, but in this case the sensory neurone makes a synapse directly with the motor neurone and there is no relay neurone. Striking the tendon below the knee cap stimulates stretch receptors in the leg extensor muscle. The impulse travels round the reflex arc and causes the same muscle to contract. Reflex actions are not usually so simple as described here. Several receptors of different kinds may be stimulated at once and many sets of muscles or glands may be brought into action, involving many more than the three nerve cells mentioned in the reflex arc described above. The relay nerve cell usually allows connexions to many other motor neurones. In addition, since we are usually aware that a reflex is taking place, nerve fibres must conduct the impulses passing in the reflex arc up the spinal cord to the brain.

The term spinal reflex refers to reflex actions in regions below the head, and in an animal such as the frog spinal reflex actions will occur even if the brain is destroyed. Reflex actions concerning organs of the head take place in the brain (cranial reflex).

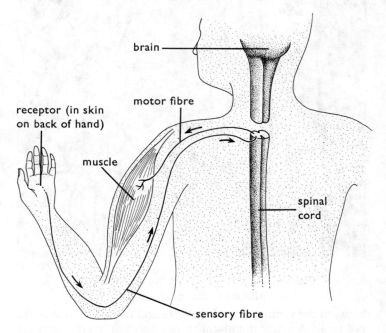

Fig. 27.7   A reflex pathway

Conditioned reflexes. In most simple reflexes, the stimulus and response are related. For example, the chemical stimulus of food in the mouth produces the reflex of salivation. After a period of learning or training, however, it is possible for a different and often irrelevant stimulus to produce the same response. In such a case, a "conditioned reflex" has been established, and the animal is said to be conditioned to this stimulus. Pavlov, a Russian biologist, carried out with dogs a great many experiments on conditioned reflexes, one of which is now something of a classic.

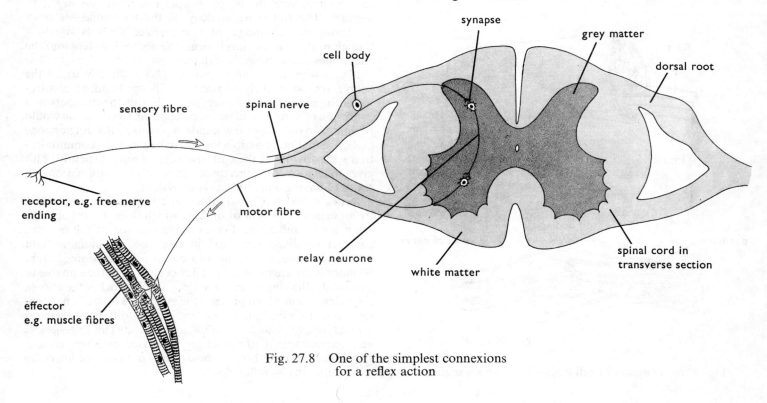

Fig. 27.8   One of the simplest connexions for a reflex action

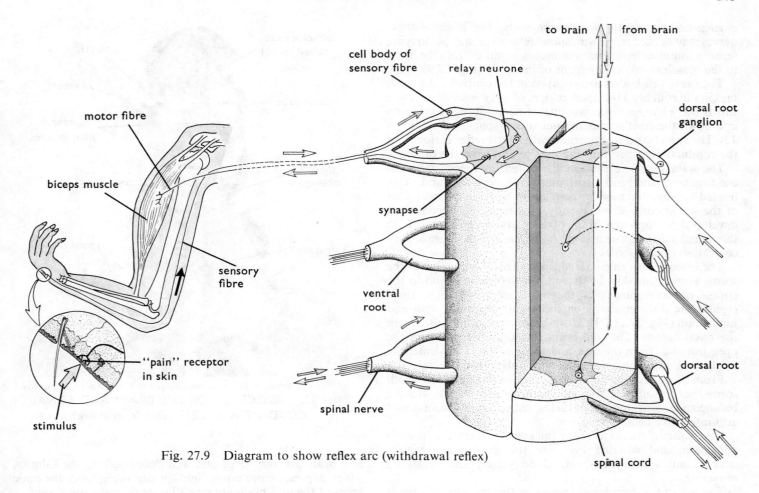

Fig. 27.9   Diagram to show reflex arc (withdrawal reflex)

The smell and taste of food is a stimulus that activates a dog's salivary glands, making its mouth water. For several days, Pavlov rang a bell at the time the food was given to the dogs. Later, the sound of the bell alone was a sufficient stimulus to cause a dog's mouth to water, without sight or smell of the food. The original chemical stimulus of the food had been replaced by an unrelated stimulus through the ears (Fig. 27.10).

The training of animals is done largely by conditioning them to respond to new stimuli. Many of our own actions, such as walking and riding a bicycle, are complicated sets of conditioned reflexes which we acquired in the first place by concentration and practice.

The **spinal cord** consists of a great number of nerve cells, both fibres and cell bodies, grouped into a cylindrical mass, running from the brain to the tail and protected by the bone of the spinal column. From between the vertebrae, spinal nerves

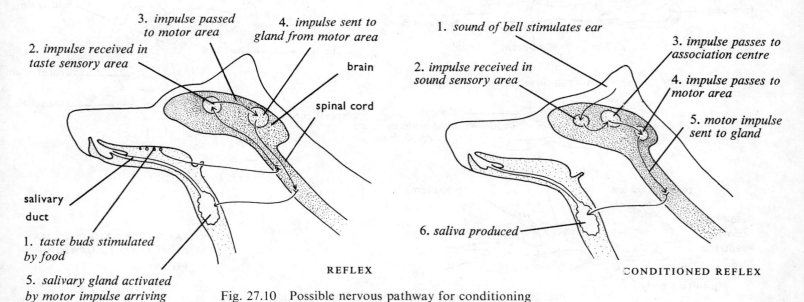

Fig. 27.10   Possible nervous pathway for conditioning

emerge and run to all parts of the body. The fibres of these nerves may be concerned with spinal reflexes or may be carrying sensory impulses to the brain or motor impulses from the brain to the muscles and other organs of the body (Fig. 27.9).

The nerve cell bodies are grouped in the centre of the cord, making a roughly H-shaped region of *grey matter*. Outside this is the *white matter* consisting of nerve fibres running up and down the cord or passing out to the spinal nerves (Plate 42). The spinal cord is concerned with spinal reflex actions and the conduction of nervous impulses to and from the brain.

The brain. During evolution, the increasing specialization of the organs of the head, particularly the eyes, ears and nose, has led to more and more sensory fibres entering the front part of the spinal cord. Consequently, this region has grown and developed to form the brain of the vertebrate animal. Like the spinal cord, it consists of nerve cells with a great concentration of cell bodies.

The brain is thus an enlarged, specialized front region of the spinal cord (Plate 43). In the simpler vertebrates, and in the course of development of the more advanced ones, three regions are distinguishable in the brain: the fore-, mid- and hind-brain (Fig. 27.11). The fore-brain receives impulses from the nasal organs. The mid-brain receives impulses from the eyes. Impulses from the ears, and semicircular canals enter the hind-brain, as do also the sensory impulses from the skin.

From the roof of the hind-brain a thickening develops which forms the *cerebellum*. This region controls and co-ordinates the balancing organs and the muscles, thus making precise and accurate movements possible.

The floor of the hind-brain thickens to form the *medulla oblongata*, and situated here are the involuntary centres which control the heart-beat, blood vessels and breathing movements.

The size of these principal regions of the brain usually bears a relation to the most important senses of the animal. In the dogfish, which hunts its prey by smell, the front lobes of the

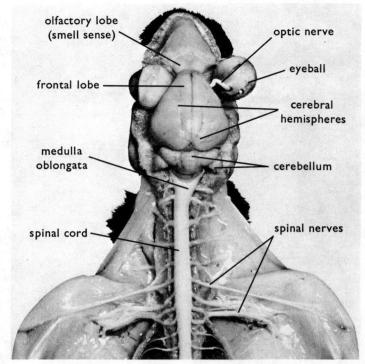

(Dissection by Griffin and George Ltd. Gerrard Biological Centre.)

Plate 43.   DISSECTION OF THE BRAIN AND SPINAL CORD OF A RABBIT (seen from above)

fore-brain are very large and well developed. In the salmon, which depends more on its sight for capturing food, the optic lobes of the mid-brain are much larger than the fore-brain.

MOTOR AREAS. As well as receiving impulses from the sense organs, the brain can send off from certain motor areas

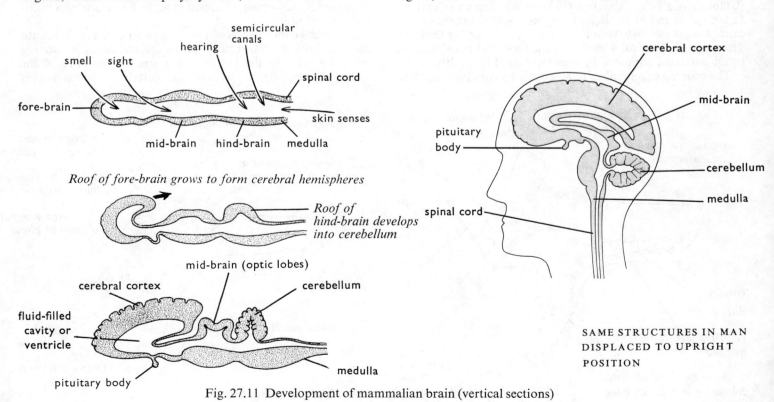

Fig. 27.11 Development of mammalian brain (vertical sections)

Roof of fore-brain grows to form cerebral hemispheres

Roof of hind-brain develops into cerebellum

SAME STRUCTURES IN MAN DISPLACED TO UPRIGHT POSITION

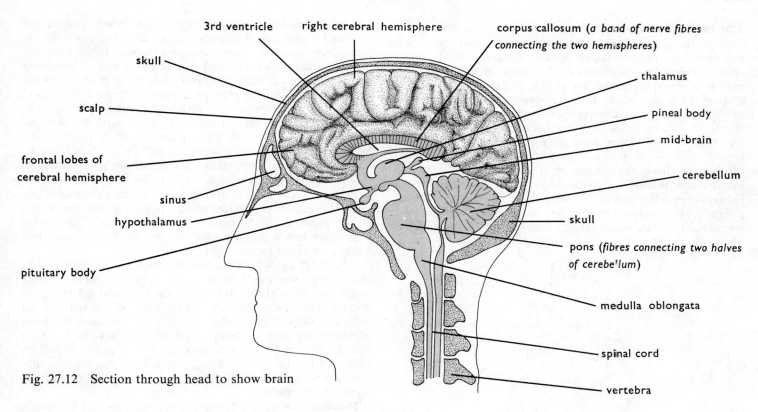

Fig. 27.12 Section through head to show brain

impulses which initiate activity in the body. Sometimes these are simple reflexes in response to external stimuli or responses to internal stimuli such as sensations of hunger from the stomach.

ASSOCIATION CENTRES. Certain areas of the brain are supplied with fibres from the principal sense centres of the fore-, mid- and hind-brain, so that several impulses from different sense organs may be correlated. For example, the smell of meat may stimulate the nose of a dog and cause sensory impulses to be sent to the fore-brain. These would normally be relayed to the motor areas and produce co-ordinated movement towards the food, but the sight of another dog in possession and the sound of its ferocious growling will stimulate other sense organs and other brain centres. All these impulses will be relayed to the association centres and, according to the strength of the stimulation and the past experience "stored" in the brain, the motor areas will receive impulses from the association centre resulting in fight or flight.

Without association centres, conditioned reflexes and learning would not be possible. In Pavlov's experiment, sensations of hearing a bell would be most unlikely to produce salivation if only taste and smell centres were connected to the salivary glands.

CEREBRAL HEMISPHERES (*cerebrum*). In mammals, large outgrowths from the fore-brain spread backwards over the rest of the brain and form two lobes called the cerebral hemispheres (Figs. 27.11 and 27.12). These are important association centres where linkages between thousands of nerve cells allow intelligent behaviour, memory and, in ourselves at least, consciousness of our own activities. In animals without cerebral hemispheres the behaviour is a matter of simple and conditioned reflexes and inborn behaviour patterns called instinct.

Certain regions of the cerebral hemispheres have been shown to affect particular regions of the body or to be concerned with impulses from a particular sense organ (Fig. 27.13).

Functions of the brain. To sum up:

1. The brain receives impulses from all the sensory organs of the body.

2. As a result of these sensory impulses, it sends off motor impulses to the glands and muscles, causing them to function accordingly.

3. In its association centres it correlates the various stimuli from the different sense organs.

4. The association centres and motor areas co-ordinate bodily activities so that the mechanisms and chemical reactions of the body work efficiently together.

5. It "stores" information so that behaviour can be modified according to past experience.

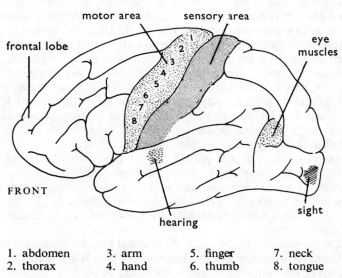

| 1. abdomen | 3. arm | 5. finger | 7. neck |
| 2. thorax | 4. hand | 6. thumb | 8. tongue |

Fig. 27.13 Localization of areas in the left cerebral hemisphere

The endocrine system (Fig. 27.14)

Co-ordination is also effected by chemicals called hormones, secreted from the endocrine glands. These glands have no ducts or openings. The chemicals they produce enter the blood stream as it passes through the glands and they are circulated all over the body. When the hormones reach particular parts of the body they cause certain changes to take place. Their effects are much slower and more general than nerve action and they control rather long-term changes such as rate of growth, rate of activity and sexual maturity. When they pass through the liver, the hormones are converted to relatively inactive compounds which are excreted, in due course, by the kidneys (hence the tests on urine for the hormonal products of pregnancy). The liver in this way limits the duration of a hormonal response which might otherwise persist indefinitely.

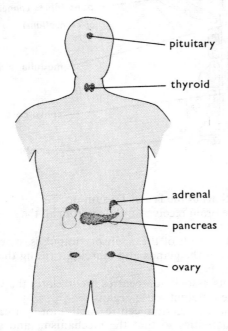

Fig. 27.14   Position of endocrine glands in the body

Thyroid. The thyroid gland is in the neck, in front of the wind-pipe. It produces a hormone, *thyroxine*, which in young animals controls the rate of growth and development. In tadpoles, for example, thyroxine brings about metamorphosis. Feeding tadpoles on thyroxine induces early metamorphosis. In adult humans thyroxine controls the rate of chemical activity, particularly respiration: too little tends to lead to over-weight and sluggish activity; too much can cause thinness and over-activity. Deficiency of the thyroid in infancy causes a certain type of mental deficiency called *cretinism*, which can be cured in the early stages by administering thyroxine.

Adrenal. The adrenal glands are situated just above the kidneys. The outer layer of the adrenal body, the cortex, produces several hormones, including *cortisone*, one of whose functions is to accelerate the conversion of proteins to glucose (p. 90). Secretion by the adrenal cortex is stimulated by certain pituitary hormones.

The inner zone, medulla, of the adrenal gland is stimulated by the nervous system and produces *adrenaline*. When the sense organs of the animal transmit to the brain, impulses which are associated with danger or other situations needing vigorous action, motor impulses are relayed to the adrenal medulla which releases adrenaline into the blood. When this reaches the heart it quickens the heart-beat. In other regions it diverts blood from the alimentary canal and the skin to the muscles; it makes the pupils dilate and speeds up the rate of breathing and oxidation of carbohydrates. All these changes would increase the animal's efficiency in a situation that might demand vigorous activity in running away or putting up a fight. In ourselves, they do the same but, together with the nervous system, they produce also the sensation of fear: thumping heart, hollow feeling in the stomach, pale face, etc. In humans, adrenaline may be secreted in many situations which promote anxiety or excitement, and not only in the face of danger.

Pancreas. As well as containing cells which secrete digestive juices, the pancreas contains endocrine cells which control the use of sugar in the body. The hormone is called *insulin*; it determines how much sugar is converted to glycogen and how much is oxidized for energy.

Insulin (*a*) accelerates the rate at which blood sugar is converted to glycogen in the liver (p. 91), (*b*) promotes the uptake of glucose from the blood by muscle cells and (*c*) increases protein synthesis in some cells. The failure of the pancreas to produce sufficient insulin leads to *diabetes*. The diabetic cannot effectively regulate the blood sugar level. It may rise to above 160 mg/100 cm^3 and so be excreted in the urine, or fall to below 40 mg/100 cm^3 leading eventually to convulsions and coma. The diabetic condition can be corrected by regular injections of insulin.

Reproductive organs. The ovary produces several hormones called *oestrogens* of which oestradiol and oestrone are the most potent. These oestrogens (i) control the development of the secondary sexual characters at puberty (*see* p. 117), (ii) cause the lining of the uterus to thicken just before an ovum is released, and (iii), in some mammals at least, oestradiol brings the animal "on heat", i.e. prepares it to accept the male. Progesterone, the hormone produced from the corpus luteum (p. 115) after ovulation, promotes the further thickening and vascularization of the uterus. Progesterone also prevents the uterus from contracting until the baby is due to be born.

Testosterone is the male sex hormone, produced by the testis. It promotes the development of the masculine secondary sexual characters.

Duodenum. The presence of food stimulates the lining of the duodenum to produce a hormone, *secretin*, which on reaching the pancreas in the blood stream, initiates the production of pancreatic enzymes. In this way, the enzymes are secreted only when food is present.

Pituitary. The pituitary gland is an outgrowth from the base of the fore-brain (Figs. 27.11 and 27.12). It releases into the blood several different hormones. Some of them appear to have a direct effect on the organ systems of the body. For example, *anti-diuretic hormone* (ADH) controls the amount of water reabsorbed into the blood by the kidneys (*see* p. 108). *Growth hormone* influences the growth of bone and other tissues. Injection of growth hormone in experimental animals causes them to grow larger and to continue growing for longer than usual. Growth, however, is affected by other endocrine glands as well, in particular the thyroid and pancreas, and the growth hormone may exert its influence through these glands rather than directly on the tissues.

In fact, the majority of the pituitary hormones do act upon and regulate the activity of the other endocrine glands to such an extent that the pituitary is sometimes called the "master gland". It is a pituitary hormone FSH which, acting on the ovary, causes the Graafian follicle to develop and secrete its

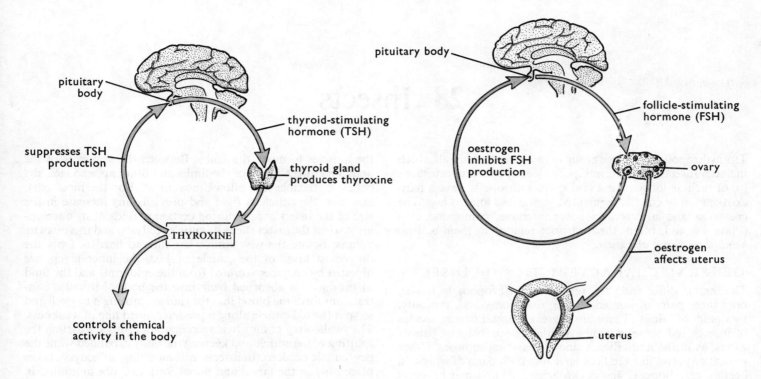

Fig. 27.15 "Feedback"

own hormone, oestrogen. Another pituitary hormone stimulates the thyroid gland to grow and to produce thyroxine and a third acts on the cortex of the adrenal gland and promotes the production of cortisone.

Homeostasis

The foregoing account shows how the hormones (e.g. adrenaline) co-ordinate the organs of the body to meet various contingencies, to produce rhythmic patterns of activity (e.g. the sex hormones), and to maintain control over long-term processes such as the rate of growth (thyroid and pituitary). It can also be seen that they fulfil a homeostatic function in regulating the composition of the internal environment (*see* p. 108). If the blood sugar level rises, the pancreas is stimulated to secrete insulin which increases the amount of glucose removed from the blood and stored as glycogen in the muscles and liver. A fall in the blood sugar level suppresses the production of insulin from the pancreas. A fall in the osmotic potential of the blood results in the release of ADH from the pituitary gland and the consequent reabsorption of water from the kidney tubules.

Interaction and "feed-back"

For effective control, two opposing systems are needed. The car needs an accelerator and brakes, a muscle must have its antagonistic partner (p. 122). The hormones too have antagonistic effects. A pancreatic hormone called *glucagon* promotes the release of sugar into the blood while insulin has the opposite effect. A fine adjustment of the balance of such antagonistic hormones helps to maintain the controlled growth, development and activity in constantly changing conditions.

The balance is maintained partly by the "feed-back" effect of hormones, i.e. a system whereby "information" is "fed

back" to a source "telling it" about events in the body and so enabling it to adjust its output accordingly. A pituitary hormone stimulates the thyroid to produce thyroxine but thyroxine production is kept in check by the fact that when thyroxine reaches the pituitary via the circulation, production of thyroid-stimulating hormone is suppressed. The "feed-back" of thyroxine to the pituitary regulates the output of the latter. The ovarian follicles are stimulated to produce oestrogen by the pituitary hormone FSH (Fig. 27.15) but when the oestrogen in the blood reaches a certain level it suppresses the secretion of follicle-stimulating hormone by the pituitary. A delay in the feed-back effect leads to rhythmic changes. For example, it may take two weeks for the level of oestrogen in the blood to affect the pituitary, by which time the uterus lining has thickened and the ovum has been released from the follicle. The output of follicle-stimulating hormone is diminished as a result of increasing oestrogen and this in turn reduces the output of oestrogen from the ovary which, in the absence of fertilization and the development of the corpus luteum, leads to the breakdown of the uterine lining, characteristic of menstruation.

QUESTIONS

1. List the differences between control by hormones and control by the nervous system.
2. Trace by diagram or description, the possible reflex arc involved in (*a*) sneezing, (*b*) blinking. Do not attempt to describe the effector systems in detail and treat the brain as simply an enlarged region of the spinal cord.
3. All nervous impulses, whether from the eyes, ears, tongue or skin are basically the same. This implies that the information reaching the brain is little more than a rapid series of electrical pulses of identical strength. How then is it possible for us to distinguish between light and sound, heat and touch?
4. It is possible to train a dog to seek food concealed behind one of several identical doors by flashing a light over the appropriate door. Suggest a nervous pathway by which this behaviour is established.

28 Insects

THE arthropods are a large group of invertebrate animals which includes insects, spiders, millipedes, centipedes and crustacea (p. 6) such as lobsters and crabs. All arthropods have a hard exoskeleton or cuticle, segmented bodies and jointed legs. The crustacea and insects also have antennae, compound eyes (Plate 48) and, often, three distinct regions to their bodies: *head*, *thorax* and *abdomen*.

GENERAL CHARACTERISTICS OF INSECTS

The insects differ from the rest of the arthropods in having only three pairs of jointed legs on the thorax and, typically, two pairs of wings. There are a great many different species of insects and some, during evolution, have lost one pair of wings, as in the houseflies, crane-flies and mosquitoes. Other parasitic species like the fleas have lost both pairs of wings. In beetles, grasshoppers and cockroaches, the first pair of wings has become modified to form a hard outer covering over the second pair.

Cuticle and ecdysis. The value of the external cuticle is thought to lie mainly in reducing the loss from the body of water vapour through evaporation, but it also protects the animal from damage and bacterial invasion, maintains its shape and allows rapid locomotion. The cuticle imposes certain limitations in size, however, for if arthropods were to exceed the size of some of the larger crabs, the cuticle would become too heavy for the muscles to move the limbs. Between the segments of the body and at the joints of the limbs and other appendages, the cuticle is flexible and allows movement. For the most part, however, the cuticle is rigid and prevents any increase in the size of the insect except during certain periods of its development when the insect sheds its cuticle (ecdysis) and increases its volume before the new cuticle has time to harden. Only the outermost layer of the cuticle is shed, the inner layers are digested by enzymes secreted from the epidermis and the fluid so produced is absorbed back into the body. Muscular contractions force the blood into the thorax, causing it to swell and so split the old cuticle along a predetermined line of weakness. The swallowing of air often accompanies ecdysis, assisting the splitting of the cuticle and keeping the body expanded while the new cuticle hardens. In insects, this moulting, or ecdysis, takes place only in the larval and pupal form and not in adults. In other words, mature insects do not grow.

Breathing. Running through the bodies of all insects is a branching system of tubes, *tracheae* (Plate 44), which contain air. They open to the outside by pores called *spiracles* (Figs. 28.1 and 28.24a) and they conduct air from the atmosphere to all living regions of the body (Fig. 28.1). The tracheae are lined with cuticle which is thickened in spiral bands (Fig. 28.2). This thickening keeps the tracheae open against the internal pressure of body fluids. The spiracles, typically, open on the flanks of each segment of the body, but in some insects there are only one or two openings. The entrance to the spiracle is usually supplied with muscles which control its opening or closure.

Since the spiracles are one of the few areas of the body from which evaporation of water can occur, the closure of the spiracles when the insect is not active and therefore needs less oxygen, helps to conserve moisture.

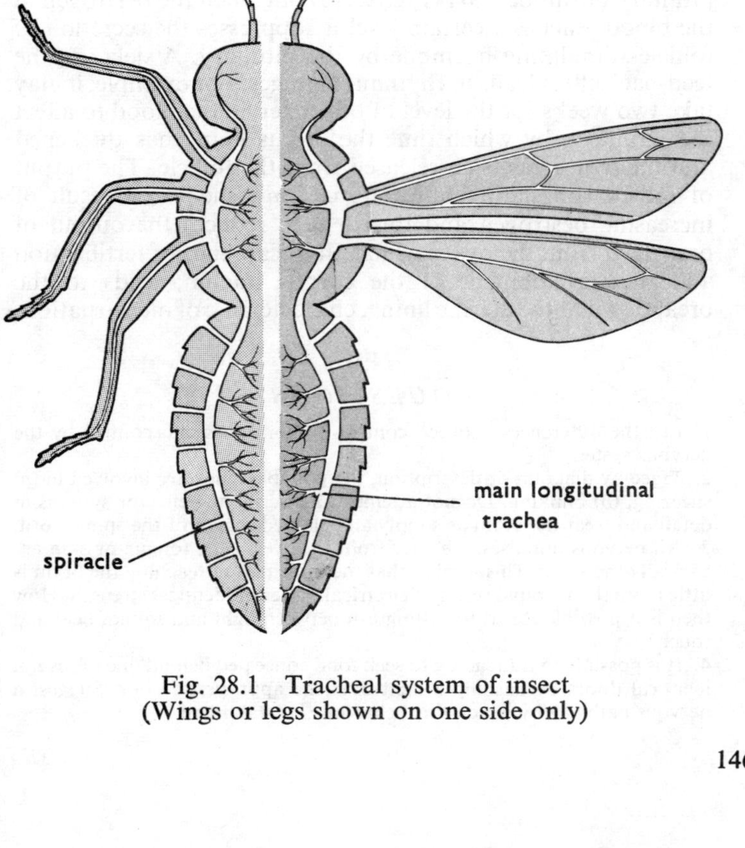

spiracle

main longitudinal trachea

Fig. 28.1 Tracheal system of insect
(Wings or legs shown on one side only)

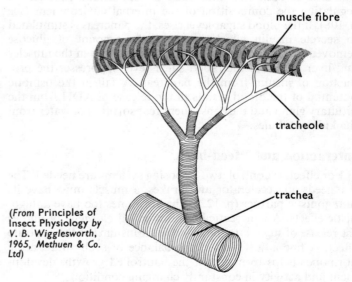

muscle fibre

tracheole

trachea

(From Principles of Insect Physiology by V. B. Wigglesworth, 1965, Methuen & Co. Ltd)

Fig. 28.2 To show how the tracheae supply oxygen to the muscles

146

tracheoles is most dense in the region of very active muscle, e.g. the flight muscles in the thorax.

The movement of oxygen from the atmosphere, through the spiracles, up the tracheae and tracheoles to the tissues, and the passage of carbon dioxide in the opposite direction, can be accounted for by simple diffusion but in active adult insects there is often a ventilation process which exchanges up to 60 per cent of the air in the tracheal system. In many beetles, locusts, grasshoppers and cockroaches, the abdomen is slightly compressed vertically (dorso-ventrally) by contraction of internal muscles. In bees and wasps the abdomen is compressed rhythmically along its length, slightly telescoping the segments. In both cases, the consequent rise of blood pressure in the body cavity compresses the tracheae along their length (like a concertina) and expels air from them. When the muscles relax, the abdomen springs back into shape, the tracheae expand and draw in air. Thus, unlike mammals, the positive muscular action in breathing is that which results in expiration.

This tracheal respiratory system is very different from the respiratory systems of the vertebrates, in which oxygen is absorbed by gills or lungs and conveyed in the blood stream to the tissues. In the insects, the oxygen diffuses through the trachea and tracheoles directly to the organ concerned. The carbon dioxide escapes through the same path although a proportion may diffuse from the body surface.

Blood system. The tracheal supply carrying oxygen to the organs gives the circulatory system a rather different role in insects from that in vertebrates. Except where the tracheoles terminate at some distance from a cell, the blood has little need to carry dissolved oxygen and, with a few exceptions, it contains no haemoglobin or cells corresponding to erythrocytes. There is a single dorsal vessel (Fig. 28.3) which propels blood forward and releases it into the body cavity, thus maintaining a sluggish circulation. Apart from this vessel, the blood is not confined in blood vessels but occupies the free space between the cuticle and the organs in the body cavity. The blood therefore serves mainly to distribute digested food, collect excretory products and, in addition, has important hydraulic functions in expanding certain regions of the body to split the old cuticle and in pumping up the crumpled wings of the newly emerged adult insect (Plate 45).

(Shell International Petroleum Co. Ltd)

Plate 45. THE FINAL ECDYSIS OF THE LOCUST

The tracheae branch repeatedly until they terminate in very fine *tracheoles* which invest or penetrate the tissues and organs inside the body. The walls of the tracheae and tracheoles are permeable to gases and oxygen is able to diffuse through them to reach the living cells. As might be expected, the supply of

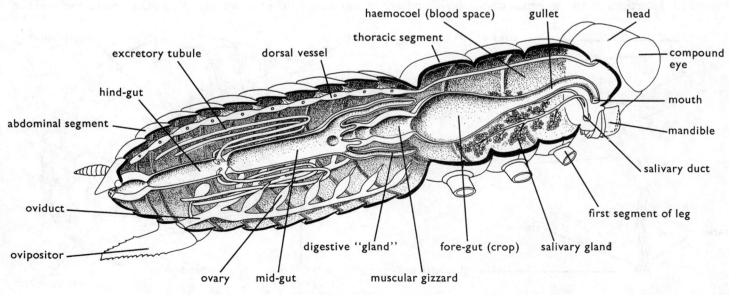

Fig. 28.3   Insect anatomy

(After Nicholas Jago)

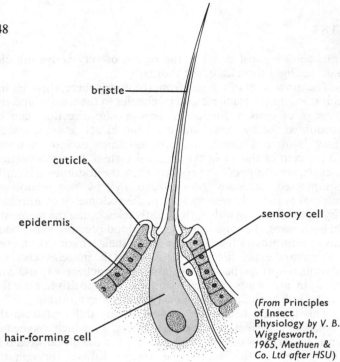

Fig. 28.4 Touch-sensitive bristle

(From Principles of Insect Physiology by V. B. Wigglesworth, 1965, Methuen & Co. Ltd after HSU)

(Rothamsted Experimental Station)

Plate 46. THE OAK BEAUTY MOTH (male) showing feathery antennae

Sensory system

Touch. From the body surface of the insect there arises a profusion of fine bristles most of which have a sensory function, responding principally to touch, vibration, or chemicals. The tactile (touch-sensitive) bristles are jointed at their bases (Fig. 28.4) and when a bristle is displaced to one side, it stimulates a sensory cell which fires impulses to the central nervous system. The tactile bristles are numerous on the tarsal segments (Fig. 28.5), the head, wing margins, or antennae according to the species and as well as informing the insect about contact stimuli, they probably respond to air currents and vibrations in the ground or in the air.

Proprioceptors. Small oval or circular areas of cuticle are differentially thickened and supplied with sensory fibres. They probably respond to distortions in the cuticle resulting from pressure, and so feed back information to the central nervous system about the position of the limbs. Organs of this kind respond to deflections of the antennae during flight and are

thought to "measure" the air speed and help to adjust the wing movements accordingly. In some insects there are stretch receptors associated with muscle fibres, apparently similar to those in vertebrates (p. 128).

Sound. The tactile bristles on the cuticle and on the antennae respond to low-frequency vibrations but many insects have more specialized sound detectors in the form of a thin area of cuticle overlying a distended trachea or air sac and invested with sensory fibres. Such *tympanal organs* appear on the thorax or abdomen or tibia according to species and are sensitive to sounds of high frequency. They can be used to locate the source of sounds as in the case of the male cricket "homing" on the sound of the female's "chirp", and in some cases can distinguish between sounds of different frequency.

Smell and taste. Experiments show that different insects can distinguish between chemicals which we describe as sweet, sour, salt and bitter, and in some cases more specific substances. The organs of taste are most abundant on the mouth parts, in the mouth, and on the tarsal segments but the nature of the sense organs concerned is not always clear.

Smell is principally the function of the antennae. Here there are bristles, pegs or plates with a very thin cuticle and fine perforations through which project nerve endings sensitive to chemicals. Sometimes these sense organs are grouped together and sunk into *olfactory pits*. In certain moths the sense of

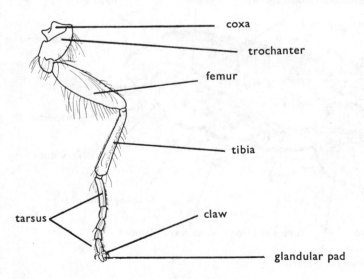

Fig. 28.5 Leg of house-fly

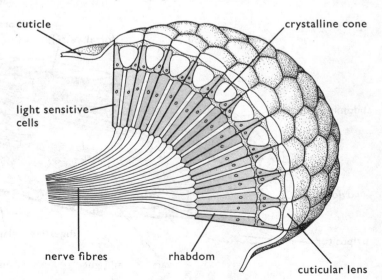

Fig. 28.6 Compound eye

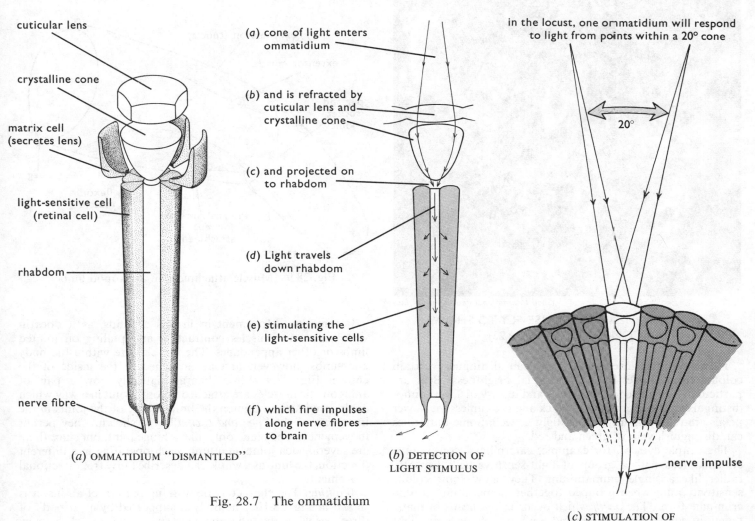

cuticular lens

crystalline cone

matrix cell
(secretes lens)

light-sensitive cell
(retinal cell)

rhabdom

nerve fibre

(*a*) OMMATIDIUM "DISMANTLED"

(*a*) cone of light enters
ommatidium

(*b*) and is refracted by
cuticular lens and
crystalline cone

(*c*) and projected on
to rhabdom

(*d*) Light travels
down rhabdom

(*e*) stimulating the
light-sensitive cells

(*f*) which fire impulses
along nerve fibres
to brain

(*b*) DETECTION OF
LIGHT STIMULUS

in the locust, one ommatidium will respond
to light from points within a 20° cone

20°

nerve impulse

(*c*) STIMULATION OF
AN OMMATIDIUM

Fig. 28.7   The ommatidium

Plate 47.   SECTION THROUGH LOCUST'S
COMPOUND EYE (×70)

(Harris Biological Supplies Ltd)

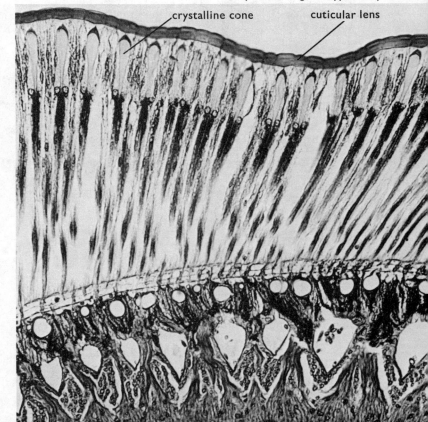

crystalline cone   cuticular lens

smell is very highly developed. The male Emperor moth will fly to an unmated female from a distance of a mile, attracted by the "scent" which she exudes. Plate 46 shows a male moth's antennae which may carry many thousand chemo-receptors.

Sight. The compound eyes of insects consist of thousands of identical units called *ommatidia* (Fig. 28.7), packed closely together on each side of the head (Plate 48). Each ommatidium consists of a lens system formed partly from a thickening of the transparent cuticle and partly from a special *crystalline cone* (Fig. 28.6 and Plate 47). This lens system concentrates light from within a cone of 20°, on to a transparent rod, the *rhabdom*. The light, passing down this rhabdom, stimulates the eight or so *retinal cells* grouped round it to fire nervous impulses to the brain. Each ommatidium can therefore record the presence or absence of light, its intensity, in some cases its colour and, according to the position of the ommatidium in the compound eye, its direction. Although there may be from 2000 to 10,000 or more ommatidia in the compound eye of an actively flying insect, this number cannot reconstruct a very accurate picture of the outside world (Fig. 28.8). Nevertheless, the *"mosaic image"* so formed, probably produces a crude impression of the form of well-defined objects enabling bees, for example, to seek out flowers and to use landmarks for finding their way to and from the hive. It is likely that the construction of compound eyes makes them particularly sensitive to moving objects, e.g. bees are more readily attracted to flowers which are being blown by the wind.

Plate 48.   HEAD OF HOUSE-FLY TO SHOW
COMPOUND EYES (×25)

Flower-visiting insects, at least, can distinguish certain colours from shades of grey of equal brightness. Bees are particularly sensitive to blue, violet and ultra-violet but cannot distinguish red and green from black and grey unless the flower petals are reflecting ultra-violet light as well. Some butterflies can distinguish yellow, green and red.

The simple eyes of, for example, caterpillars, consist of a cuticular lens with a group of light-sensitive cells beneath, rather like a single ommatidium. They show some colour sensitivity and, when grouped together, some ability to discriminate form. The *ocelli* which occur in the heads of many flying insects probably respond only to changes in light intensity.

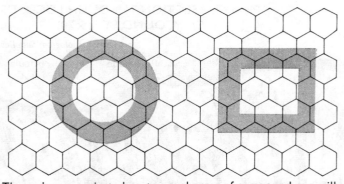

These shapes projected on to equal areas of compound eye will

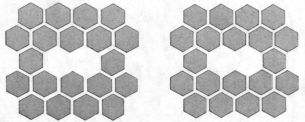

stimulate the same pattern of ommatidia and so produce indistinguishable mosaic images

Fig. 28.8   Limitation of interpretation by compound eye

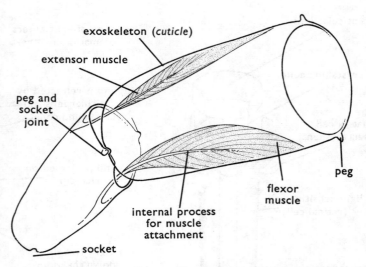

Fig. 28.9   Muscle attachment in arthropod limb

Locomotion. Movement in insects depends, as it does in vertebrates, on muscles contracting and pulling on jointed limbs or other appendages. The muscles are within the body and limbs, however, and are attached to the inside of the cuticle. Fig. 28.9 shows diagrammatically how a pair of antagonistic muscles are attached across a joint in a way which could bend and straighten the limb. Many of the joints in the insect are of the "*peg and socket*" type shown. They permit movement in one plane only, like a hinge joint, but since there are several such joints in a limb, each operating in a different direction, the limb as a whole can describe fairly free directional movement.

Walking. The characteristic walking pattern of an insect is shown in Fig. 28.10. The body is supported by a "tripod" of three legs while the other three are swinging forward to a new position. On the tarsi are claws and, depending on the species,

the shaded legs are moving more or less at the same time, the others are in contact with the ground

(From Principles of Insect Physiology by V. B. Wigglesworth, 1965, Methuen & Co. Ltd after LENGERKEN)

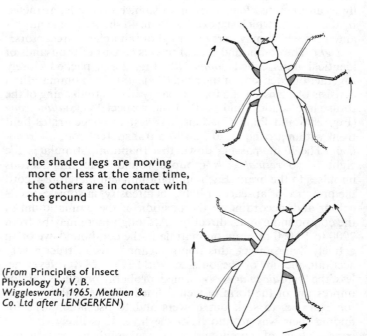

Fig. 28.10   Walking pattern in a beetle

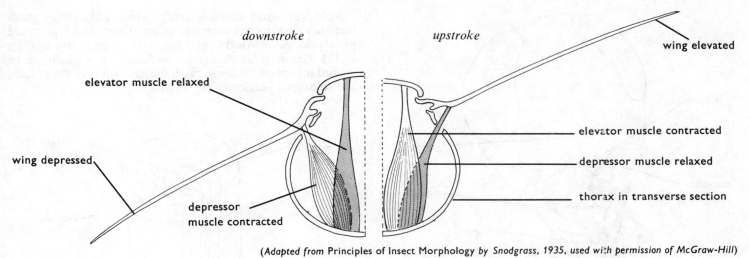

(Adapted from Principles of Insect Morphology *by* Snodgrass, 1935, used with permission of McGraw-Hill*)*

Fig. 28.11   Action of direct flight muscles (e.g. dragonflies)

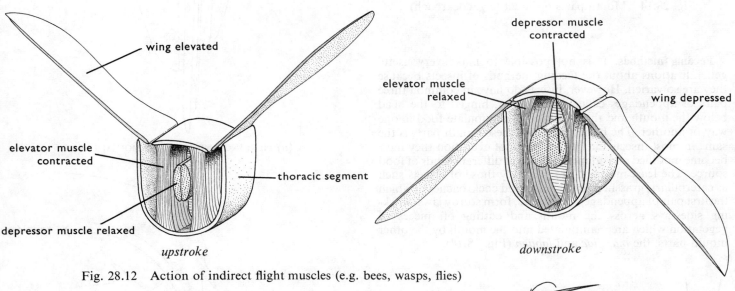

Fig. 28.12   Action of indirect flight muscles (e.g. bees, wasps, flies)

adhesive pads which enable the insect to climb very smooth surfaces. The precise mechanism of adhesion is uncertain.

Modification of the limbs and their musculature enables insects to leap, e.g. grasshopper, or swim, e.g. water beetles.

Flying. Figs. 28.11 and 12 represent sections through the thorax of an insect. In insects such as dragonflies, the muscles which pull the wings down are attached directly to the wing base. The elevator muscles, which raise the wings, are attached to the roof of the thorax. When they contract, they pull the thorax roof down and pivot the wings upwards. The depressor muscles of insects such as bees, wasps and flies are attached to the walls of the thorax and not to the wing base. This is characteristic of compact insects with small wings and a rapid wing beat. In both cases there are direct flight muscles which, by acting on the wing insertion, can alter its angle in the air. Fig. 28.13 shows that during the downstroke (1 and 2) the wing is held horizontally, so thrusting downwards on the air and producing a lifting force. During the upstroke the wing is rotated vertically and offers little resistance during its upward movement through the air.

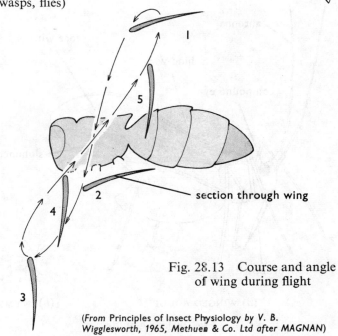

Fig. 28.13   Course and angle of wing during flight

(From Principles of Insect Physiology *by V. B. Wigglesworth, 1965, Methuen & Co. Ltd after MAGNAN)*

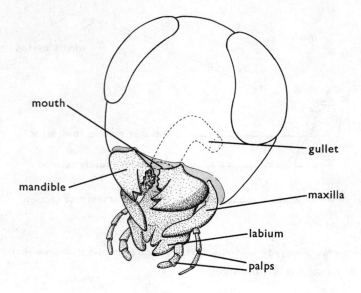

Fig. 28.14   Mouth parts of insect (e.g. cockroach)

Aphids are small insects (greenfly) which feed on plant juices that they suck from leaves and stems. Their mouth parts are greatly elongated to form a piercing and sucking proboscis (Fig. 28.15). The maxillae fit together to form a tube which can be pushed into plant tissues to reach the sieve tubes of the phloem and so extract nutrients (Fig. 28.16).

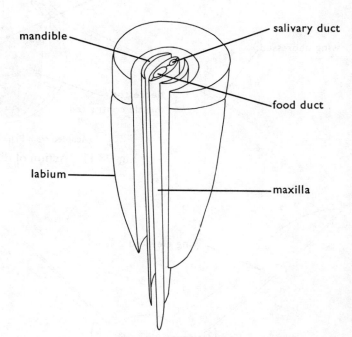

(a) PIERCING AND SUCKING MOUTH PARTS

Feeding methods. It is not possible to make very useful generalizations about the feeding methods of insects because they are so varied. However, insects do have in common three pairs of appendages called *mouth parts*, hinged to the head below the mouth and these extract or manipulate food in one way or another. The basic pattern of these mouth parts is the same in most insects but in the course of evolution they have become modified and adapted to exploit different kinds of food source. The least modified are probably those of insects such as caterpillars, grasshoppers, locusts and cockroaches in which the first pair of appendages, *mandibles*, form sturdy jaws, working sideways across the mouth and cutting off pieces of vegetation which are manipulated into the mouth by the other mouth parts, the *maxillae* and *labium* (Fig. 28.14).

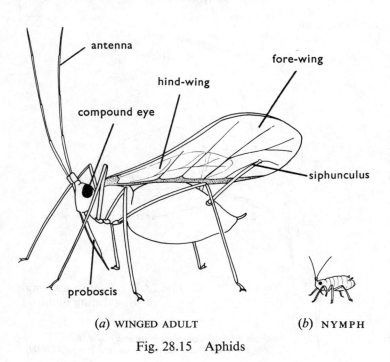

(a) WINGED ADULT   (b) NYMPH

Fig. 28.15   Aphids

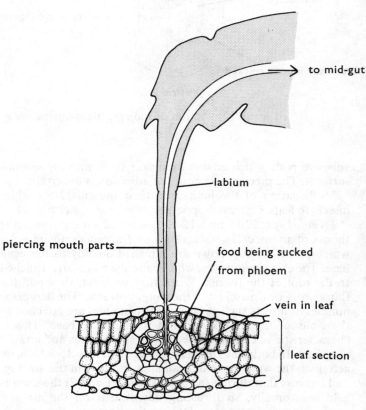

(b) SECTION THROUGH HEAD DURING FEEDING

Fig. 28.16   Feeding method of aphids

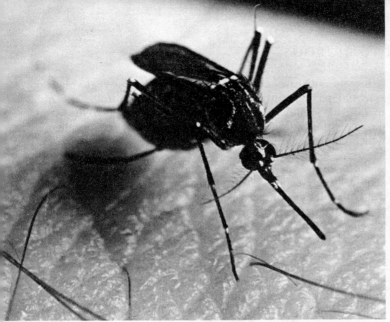

(Shell International Petroleum Co. Ltd)

Plate 49.   MOSQUITO ABOUT TO FEED ON HUMAN ARM (×5)

proboscis

(Heather Angel)

Plate 50.   RED ADMIRAL SUCKING NECTAR FROM FLOWER

The mosquito has mandibles and maxillae in the form of slender, sharp stylets which can cut through the skin of a mammal as well as penetrating plant tissues. To obtain a blood meal the mosquito inserts its mouth parts through the skin to reach a capillary and then sucks blood through a tube formed from the *labrum* or "front lip" which precedes the mouth parts. Another tubular structure, the *hypopharynx*, serves to inject into the wound a substance which prevents the blood from clotting and so blocking the tubular labrum. In both aphid and mosquito the labium is rolled round the other mouth parts, enclosing them in a sheath when they are not being used (Figs. 28.17, 28.21 and Plate 49).

In the butterfly, only the maxillae contribute to the feeding apparatus. The maxillae are greatly elongated and in the form of half tubes, i.e. like a drinking straw split down its length. They can be fitted together to form a tube through which nectar is sucked from the flowers (Fig. 28.18 and Plate 50).

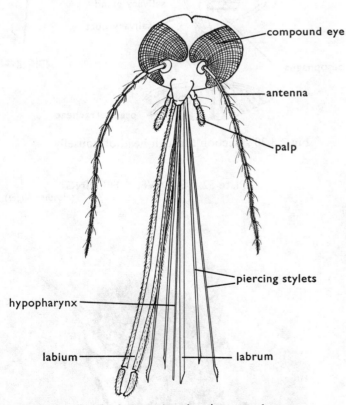

Fig. 28.17   Head of mosquito showing mouth parts

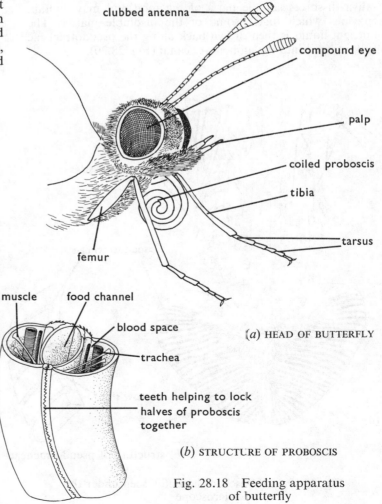

clubbed antenna

compound eye

palp

coiled proboscis

tibia

tarsus

femur

muscle   food channel

blood space

trachea

teeth helping to lock halves of proboscis together

(a) HEAD OF BUTTERFLY

(b) STRUCTURE OF PROBOSCIS

Fig. 28.18   Feeding apparatus of butterfly

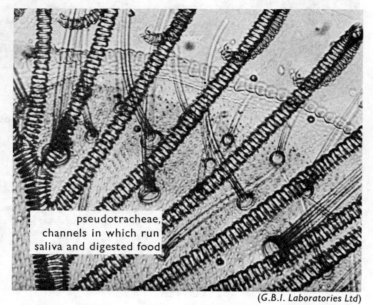

Plate 51.   PART OF PROBOSCIS OF HOUSEFLY (× 200)

(G.B.I. Laboratories Ltd)

Insects as disease carriers. Many insects are known to play an essential part in transmitting diseases. Mosquitoes transmit malaria and yellow fever, tsetse flies carry sleeping sickness, and fleas harbour bubonic plague. Animals which carry organisms which can cause disease in other animals or plants are called *vectors*.

Malaria and the mosquito. Malaria is caused by a microscopic, single-celled parasite which enters and eventually destroys a large number of red blood cells. The parasites are transmitted from man to man by female mosquitoes of the genus *Anopheles*, which pierce the skin with their sharp mouth parts and feed on the blood which they suck from the superficial skin capillaries (*see* Fig. 28.21). If the blood so taken contains the malarial parasites, these undergo a complicated series of changes within the mosquito, including extensive reproduction, and eventually accumulate in large numbers in the salivary glands. If this mosquito now bites a healthy person, saliva containing hundreds of parasites is injected into his blood stream and he may develop malaria.

The housefly also sucks liquid but its mouth parts cannot penetrate tissue. Instead, the labium is enlarged to form a *proboscis* which terminates in two pads whose surface is channelled by grooves called *pseudotracheae* (Fig. 28.19 and Plate 51). The fly applies its proboscis to the food (Plate 52) and pumps saliva along the channels and over the food. The saliva dissolves soluble parts of the food and may contain enzymes which digest some of the insoluble matter. The nutrient liquid is then drawn back along the pseudotracheae and pumped into the alimentary canal (Fig. 28.20).

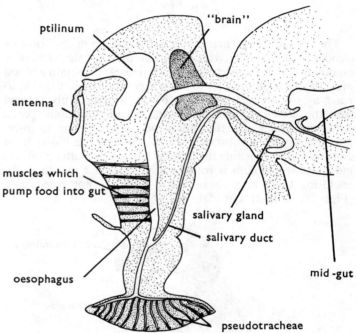

Fig. 28.20   Section through head of housefly

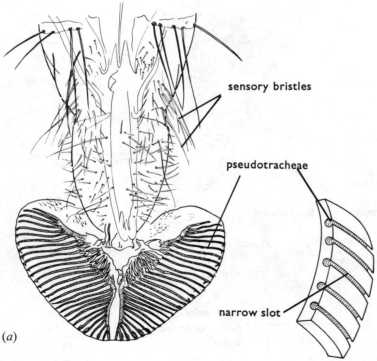

(a)

(b) structure of pseudotracheae

Fig. 28.19. Proboscis of housefly, seen under the microscope

Plate 52.   BLOWFLY FEEDING

(Heather Angel)

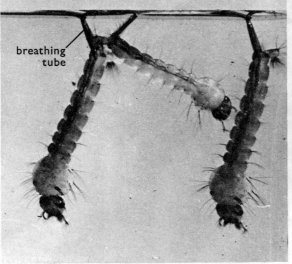

breathing tube

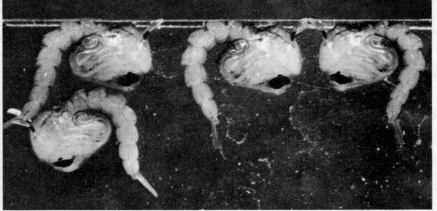

Plate 54.   PUPAE OF MOSQUITO (*Culex molestus*) (×8)

Plate 53.   LARVAE OF MOSQUITO
(*Culex molestus*) (×8)

If mosquitoes could be prevented from biting humans, the disease could not be transmitted. Thus methods of controlling the disease, apart from drugs which kill the malarial parasite in the blood, concentrate largely on eliminating the mosquito. The species of mosquito which normally rest in dwellings can be attacked by sprays containing DDT or BHC. The spray remains effective on the walls of dwellings for several months and will kill any insects which settle on the sprayed surface. It is known that, although the adult mosquito spends its life on land, the larvae and pupae live in water. The female mosquito lays her eggs in the static water of lakes, ponds, ditches or even water collected in puddles, drinking troughs or tin cans. The eggs soon hatch to larvae which breathe air at the surface through a tracheal tube (Plate 53) and feed on microscopic algae in the water. The larva eventually pupates and although the pupa does not feed, it still breathes air (Plate 54). Finally, the pupal skin splits open, the imago emerges and flies away. Knowledge of this life cycle leads to methods of mosquito eradication directed at the larval and pupal stages. By draining swamps and turning sluggish rivers into swifter streams, the breeding grounds of the mosquito are destroyed. In towns and villages, water must not be allowed to collect in any container, e.g. tanks, pots or tins, accessible to the mosquito. Spraying stagnant water with oil and insecticides suffocates or poisons the larvae and pupae. Such spraying must include not only lakes and ponds but any accumulation of fresh water which mosquitoes can reach, e.g. drains, gutters and the receptacles mentioned above.

In 1956, The World Health Organization started a programme for eradication of malaria, using DDT to kill mosquitoes. Although about fifteen countries have virtually eliminated malaria, mosquitoes have developed resistance to DDT. Eradication of malaria now seems unlikely but a combination of control measures may reduce the incidence of the disease.

Houseflies and disease. Although there is little direct evidence, the housefly is thought to help in the spread of sixty or more diseases. A great many harmful bacteria and viruses are found in or on houseflies but, unlike the mosquito and malaria, flies are incidental rather than essential to the spread of disease. The indiscriminate feeding habits of flies constitute some of the circumstantial evidence against them. They will settle on decaying organic matter in refuse tips where bacteria are abundant, or alight on human faeces containing, possibly, the germs of typhoid, cholera, poliomyelitis or dysentery. The

germs adhere to their legs or their proboscis; they lodge in the pseudotracheae or pass through the alimentary canal to be released with the faeces. If a contaminated fly alights and feeds on food intended for human consumption, it seems very likely that the germs from the feet, proboscis or faeces will arrive on the food and eventually be ingested by people.

The principal methods of preventing transmission of disease by houseflies try to (*a*) prevent flies having access to food for human consumption, (*b*) stop flies getting to human faeces and (*c*) reduce the population of flies by means of insecticides directed at the adults and the breeding grounds of the larvae.

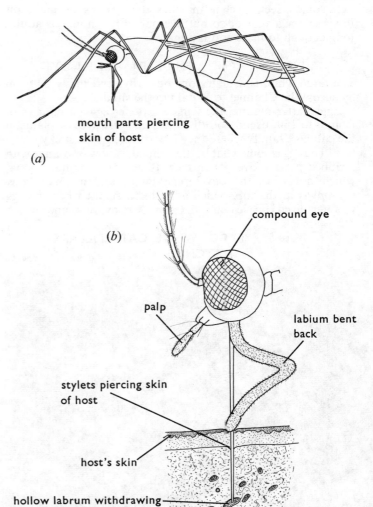

mouth parts piercing skin of host

(*a*)

compound eye

(*b*)

palp

labium bent back

stylets piercing skin of host

host's skin

hollow labrum withdrawing blood from capillary

Fig. 28.21   Mosquito feeding

Life history and metamorphosis

(*a*) *Complete metamorphosis*. Insects lay eggs which hatch into *larvae*. These larvae are usually quite unlike the adult and are called grubs, maggots or caterpillars according to the species of insect. Generally the larva is the feeding and growing stage, eating voraciously, shedding its cuticle several times and growing rapidly. When it has reached full size, the larva becomes inactive, neither moving nor feeding, and extensive breakdown and reorganization takes place within its body, giving rise eventually to the adult or imago form. The stage in the insect's life when these changes take place is called the *pupa* and the changes are called *metamorphosis*. The adults then mate and lay eggs.

(*b*) *Incomplete metamorphosis*. With insects such as the mayfly or dragonfly the egg does not hatch into a larva but a *nymph* which, though still very different from the imago, more closely resembles it than does a larva. The nymph has three pairs of jointed legs, compound eyes and rudimentary wings. At each moult, changes occur which bring it nearer to the adult form. There is no prolonged "resting" stage as there is in a pupa, though the final ecdysis usually reveals drastic changes that have occurred in the final weeks of the nymph's development. In both types of metamorphosis the habitat, behaviour, locomotion and feeding habits of the adult are quite different from the larvae or nymphs.

In some insects, such as mayflies, the nymphs live and grow in water for a year or so but live only a few hours as adults, long enough to mate and lay eggs.

THE LARGE WHITE BUTTERFLY

The large white is a migratory butterfly, and many of them fly across the Channel to Britain in the summer.

Eggs. After mating, the female lays her eggs in batches of from 6 to 100, usually on the underside of leaves of plants in the cabbage family where they are attached by a sticky secretion. The egg is somewhat bottle-shaped, yellow or orange with a ribbed pattern over its surface (Plate 55). It contains yolk, which nourishes the developing larva, and a small hole, *micropyle*, at the top, which admitted a sperm when the egg was fertilized on leaving the female's oviduct and which

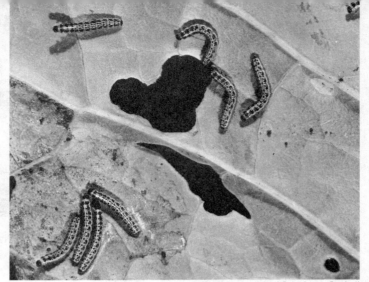

(Shell International Petroleum Co. Ltd)

Plate 56.   LARVAE OF LARGE WHITE BUTTERFLY (× 1)

now allows air to reach the embryo. In warm weather, after about a week the eggs hatch and small larvae called caterpillars eat their way out of the egg shells and devour the remains of them. Their egg shells, indeed, seem to be quite vital to their continued development. Next, they feed on the cabbage leaf, biting off pieces with their jaws (Fig. 28.22).

Caterpillars. The body of the caterpillar is cylindrical and yellow with black markings (Fig. 28.24*a* and Plate 56). It consists of a head and thirteen segments. On the head are biting jaws which move transversely, and a pair of small antennae. Inside the head a pair of salivary glands form silk-producing organs, and their ducts project in a tube, the *spinneret*, behind the jaws. Liquid secreted by these glands passes through the spinneret and hardens in the air to form a thread of silk. The caterpillar uses this to attach itself when walking on a slippery surface, secreting a zig-zag trail of thread which sticks to the surface and can be gripped by the legs. There are six simple eyes on each side of the head (Fig. 28.22). They consist of a single lens with light-sensitive cells beneath.

The first three body segments correspond to the thorax of the adult and the remaining ten become its abdomen. There is a spiracle on both sides of the first thoracic segment and the first eight abdominal segments. The thoracic segments each bear a pair of jointed "true" legs (Figs. 28.23*b* and 28.24*a*), with claws

Plate 55.   LARGE WHITE: CATERPILLARS
HATCHING (× 6)

(Stephen Dalton NHPA)

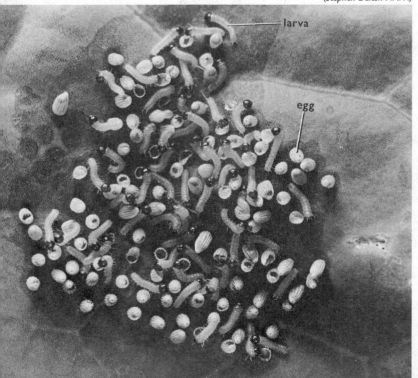

larva

egg

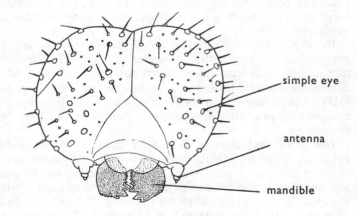

simple eye

antenna

mandible

FROM IN FRONT

Fig. 28.22   Caterpillar's head

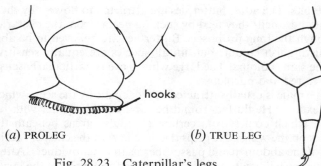

(*a*) PROLEG   (*b*) TRUE LEG

Fig. 28.23   Caterpillar's legs

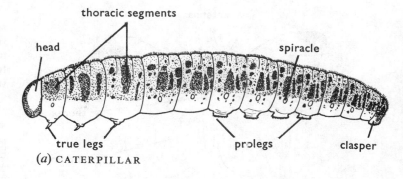

(*a*) CATERPILLAR

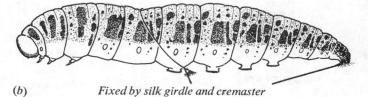

(*b*)   *Fixed by silk girdle and cremaster*

(*c*)   *Rhythmic contractions; cuticle splits in thoracic region*

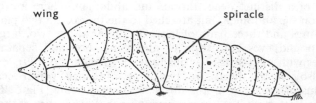

(*d*)   *Last larval cuticle finally cast. Pupa revealed, pale in colour*

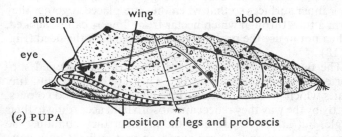

(*e*) PUPA

Fig. 28.24   Large white butterfly: stages in metamorphosis

at the end. These correspond to the legs of the adult. The third to sixth abdominal segments bear a pair of unjointed, fleshy projections with rows of tiny hooks at the end. These are called prolegs (Figs. 28.23*a* and 28.24*a*) and are not present in the imago. The last segment bears a pair of claspers similar to the prolegs. As it grows, the caterpillar sheds its cuticle four times.

Pupation. When it has reached full size the caterpillar leaves the cabbage plant and migrates to a dry, sheltered place such as a wall or a tree. It settles vertically on this, head uppermost, and attaches its thorax to it by spinning a girdle of silk (Fig. 28.24*b*). Its body shortens and swells in the thoracic region, splitting its last larval cuticle down the back after about two days (Fig. 28.24*c*). By rhythmic contractions of the body it pushes off the old cuticle. On the last segment has formed a group of hooks, the *cremaster*, and when the cuticle is shed the hooks anchor into a pad of silk that the caterpillar has previously spun.

What emerges from this last larval ecdysis is called the pupa, or *chrysalis* (Plate 57). It is pale and soft at first (Fig. 28.24*d*) but hardens and darkens, often approximately matching the colour of its background. From the appearance of the pupa it is clear that before the final moult extensive changes have taken

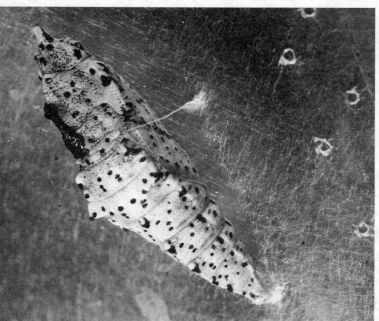

(*Shell International Petroleum Co. Ltd*)

Plate 57.   CHRYSALIS OF LARGE WHITE BUTTERFLY
(×5)
Note girdle of silk around anterior and pad of silk at posterior end

place, since the outline of the adult's legs, proboscis, eyes and wings can be seen in its cuticle (Fig. 28.24*e*). During the next two or three weeks the pupa remains more or less motionless, but inside it larval organs are being digested away and cells which have remained dormant begin to multiply and give rise to the organs of the adult. At the end of this period the pupal skin splits down the back and the adult insect pulls itself out of the skin. The wings are crumpled and folded at this stage but blood is forced into them down the "veins" and they expand and finally harden and dry.

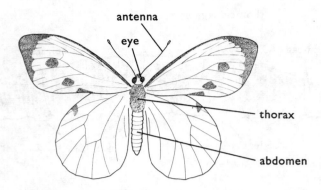

Fig. 28.25   Large white butterfly female

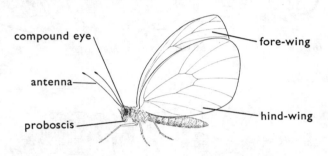

Fig. 28.26   Large white butterfly from side

Imago (Figs. 28.25 and 28.26). The body of the imago consists of a distinct head, thorax and abdomen. There are no legs on the abdomen, but attached to the thorax are two pairs of wings and three pairs of legs. On the head are two, large, compound eyes consisting of about six thousand separate lenses with light-sensitive cells beneath them (*see* p. 149).

Also on the head are two long antennae which bear organs of smell. In moths this sense is very acute (Plate 46), but in butterflies it is less important, probably because of their day-flying habits.

The *proboscis* consists of two long processes, grooved along their inner surfaces so that when they are placed together they form a tube through which nectar from flowers can be sucked. When not in use the proboscis is coiled beneath the head (Fig. 28.18).

The two pairs of wings on the thorax are broad and supported by a network of "veins". They are covered with tiny scales which give the characteristic wing patterns and colours, partly by the way they reflect and absorb light according to their angle, but largely because of the pigment they contain. The fore-wings overlap the hind-wings, so that in flight both pairs move together. Butterflies at rest often hòld their wings vertically above their bodies, while most moths rest with outspread wings.

The legs consist of nine segments: the first two, *coxa* and *trochanter* are very short; the next two, *femur* and *tibia* are long, and there are five short joints in the foot or *tarsus* (Fig. 28.18). In some species the last pair of legs bear taste organs in the tarsi, and the butterfly will uncoil its proboscis if these legs are dipped into sugar solution or fruit juice.

The thorax is covered with "hairs" or *setae*. These are present on the bodies of all insects. In structure they are quite unlike hairs of mammals and many have sensitive organs of touch at their bases (*see* p. 148).

Habits. The adult butterflies are attracted to flowers by their sight and smell; they feed by sucking nectar from the nectaries through the long proboscis. Butterflies are believed to be able to distinguish colour, but not all species are equally sensitive to the same colours. The large white's eye is particularly sensitive to red and purple.

The male is usually attracted to the female by the scent which she exudes. He flutters round her and stimulates her with his scent which comes from certain glandular areas beneath the scales on the wings. He grips the female with the claspers at the end of the abdomen and passes sperms into her oviducts. Afterwards she will lay fertilized eggs on the food plant of the larva. There may be two or more broods during the season. The adults do not survive the winter but the pupae of the last brood hatching from eggs laid in the autumn can do so, in which case pupation lasts several months until the following May.

The caterpillars of the large white are often attacked by an *ichneumon fly* called *Apanteles glomeratus*. The female ichneumon lays eggs inside the body of the caterpillar through her tubular *ovipositor*. The eggs hatch to larvae inside the caterpillar and these feed on certain food reserves in the body but do not harm the vital organs, with the result that the caterpillar continues to live and grow. By the time the caterpillar is in position for pupation it dies and the *Apanteles* grubs emerge through its skin and spin yellow cocoons on the surface before pupating.

Butterflies in general. The account of structure and life-history given here is of only one species though, in general, it applies to most British butterflies and moths. Some of the main differences from other species are worth mentioning.

Frequently, the eggs are not laid in batches but singly, though nearly always on the food plant of the caterpillars. Some butterflies scatter their eggs at random over grassland while in flight. Many caterpillars feed on only one species or family of plants while others can exist on a great variety of food. About half the British species pupate as described in the above account of the large white, but most of the others pupate hanging head downwards from the cremaster, and a few spin cocoons of silk round themselves before the last larval ecdysis.

In winter the majority of butterflies hibernate as caterpillars, a number as pupae, a few as eggs, and a few (such as the small tortoise-shell) as adults.

QUESTIONS

1. What structural features do nearly all insects have in common?
2. The mandibles, maxillae and labium represent the basic external feeding apparatus of most insects. Say how these head appendages are modified to cope with the feeding methods of the butterfly, the aphid, the mosquito and the housefly.
3. Outline the stages in metamorphosis of (*a*) a butterfly, (*b*) a mosquito and (*c*) a mayfly.
4. State briefly how the larva and imago of a butterfly are adapted in their feeding and locomotion to exploit differing aspects of their environment.
5. What features of a vertebrate eye enable it to convey a more detailed impression of the environment to the brain than does an insect's compound eye?
6. Mention the sense organs involved and the part they play in the event of a butterfly approaching a flower, alighting and seeking nectar with its proboscis.
7. In what ways does a knowledge of the life history and habits of a harmful insect (e.g. a vector of disease organisms) enable man to work out methods designed to control it?

29 Fish

FISH are vertebrate animals living in fresh- or sea-water. They are *poikilothermic* and reproduce by laying eggs. Fish are streamlined in their shape, their bodies are covered with scales, they possess fins and breathe by means of gills.

Poikilothermic. An animal whose body temperature varies with that of its surroundings is said to be poikilothermic. The temperature of such an animal is usually a few degrees above that of its environment, but a rise or fall in the temperature of the air or water in which it lives will produce a corresponding change in the animal's body temperature.

Since the speed of most chemical changes that take place in living organisms is increased by a rise in temperature, it follows that the rate of activity of a poikilothermic animal will depend to a large extent on the surrounding temperature. Some animals may, at a low temperature, be reduced to a state of torpor, while high temperatures may lead to vigorous activity.

The popular term "cold-blooded" is unsatisfactory since it conveys the impression of a constant temperature; for example, fish living in "warm" tropical seas would, according to popular definition, be correspondingly "warm-blooded"!

External features (Fig. 29.1)

Nostrils. The nostrils of fish do not open into the back of the mouth as do those of mammals, and are not, therefore, used for breathing. They lead into organs of smell which are, as a rule, very sensitive, so that a fish can detect the presence of food in the water at considerable distances. The nostrils are double, so allowing water to pass through the organ of smell.

Eyes. The eyes of a fish have large round pupils which do not vary in size.

Hearing. Although fish have no ears visible externally they can hear by transmission of vibrations through the body to sensitive regions of the sacculus or utriculus in the inner ear. There are no ossicles or cochlea.

Mouth. The mouth serves for taking in food; also for the breathing current of water.

Scales are bony plates made in the skin. In sharks, rays and dogfish the scales grow out through the skin but in other fish they are covered by skin. They overlap each other and give a protective covering. Under the microscope, rings can be seen in the scales, and from these rings the age of the fish can be estimated. One ring does not correspond to one year, but the groups of rings appear spaced close together or far apart according to the rate of feeding of the fish, since its growth rate affects the increase in size of the scales.

The **operculum** is a bony structure covering and protecting the gills; it plays an important part in the breathing mechanism.

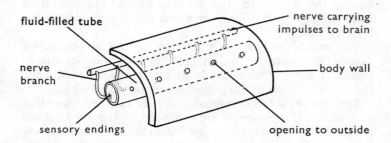

Fig. 29.2   Diagram of lateral line

The **lateral line** (Fig. 29.2) is a fluid-filled tube or canal just below the skin. It opens to the water outside by a series of tiny pores. Its function is to detect movements in the water. A disturbance set up, for example, by a person's hand moving in the water, will cause the fluid in the tube to vibrate. The canal is lined with nerve endings which are stimulated by vibrations and send impulses to the brain. The fish can thus detect the direction and intensity of water movements and this sense helps it to navigate round obstacles or to avoid enemies even if its vision is impaired, e.g. in muddy water.

Fins give stability, and control direction of movement during swimming, as explained later.

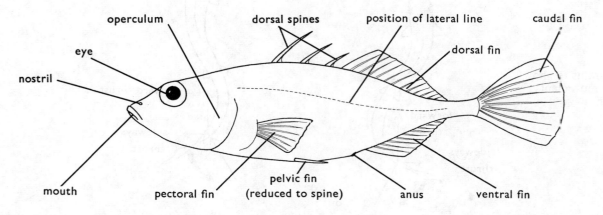

Fig. 29.1   Three-spined stickleback

Swimming

The vertebral column consists of a series of *vertebrae* held together by *ligaments*, but not so tightly as to prevent slight sideways movement between each pair of vertebrae. The whole spine is, therefore, flexible.

The muscles on each side of the spine contract in a series from head to tail and down each side alternately, causing a wave-like movement to pass down the body (Fig. 29.3). Such a movement may be very pronounced in fish such as eels, and hardly perceptible in others, e.g. mackerel. The frequency of the waves varies from about 50/min in the dogfish to 170/min in the mackerel. The greater weight and limited flexibility of the head leads to a far greater movement at the tail as a result of these waves of contraction.

The sideways and backwards thrust of the tail and body against the water results in the resistance of the water pushing the fish sideways and forwards in a direction opposed to the thrust. When the corresponding set of muscles on the other side contracts, the fish experiences a similar force from the water on that side. The two sideways forces are equal and opposite, unless the fish is making a turn, so they cancel out, leaving the sum of the two forward forces. The tail, in its final lash, may contribute as much as 40 per cent of the forward thrust.

The swimming speed of fish is not so fast as one would expect from watching their rapid movements in aquaria or ponds. A 10 kg salmon may be able to reach 16 km/h, while the majority of small fish probably do not exceed 4 km/h.

Function of the fins in swimming. It must be emphasized that the swimming movements are produced by the whole of the muscular body, and in only a few fish do the fins contribute any propulsive force. Their main function is to control the stability and direction of the fish.

The *median fins*, that is, the *dorsal*, *anal* and *ventral* fins, control the rolling and yawing movements of the fish by increasing the vertical surface area presented to the water (Fig. 29.4).

The *paired fins*, *pectoral* and *pelvic*, act as hydroplanes and control the pitch of the fish, causing it to swim downwards or upwards according to the angle to the water at which they are held by the muscles. The pectoral fins lie in front of the centre of gravity and, being readily mobile, are chiefly responsible for sending the fish up or down. The paired fins are also the means by which the fish slows down and stops.

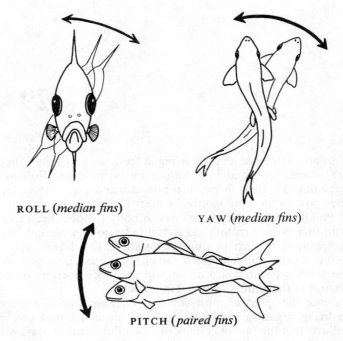

ROLL (*median fins*)   YAW (*median fins*)

PITCH (*paired fins*)

Fig. 29.4   Movements controlled by fins

Swim bladder. Fish other than those of the shark family have in their body-cavity a long air-filled bladder running just beneath the spinal column (Fig. 29.5).

In some fish the bladder opens into the gut and the air pressure in it may be increased or decreased by gulping or releasing air through the mouth. In others, the bladder has no such opening, and the blood vessels surrounding it secrete or absorb air and so control the pressure in it.

The swim bladder makes fish buoyant so that, unlike the shark or dogfish, they do not sink when they stop swimming. When the fish swims to a different depth the pressure needs to be regulated. Some lung-fish which live in poorly oxygenated water in swamps use their swim bladders for breathing air.

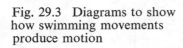

Fig. 29.3   Diagrams to show how swimming movements produce motion

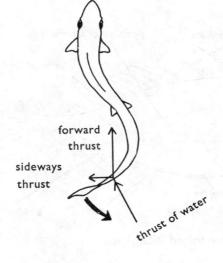

forward thrust

sideways thrust

thrust of water

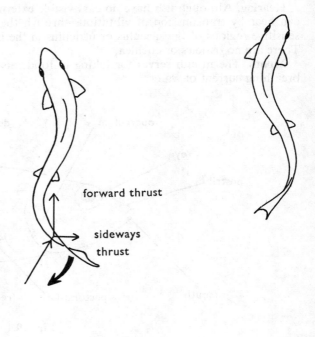

forward thrust

sideways thrust

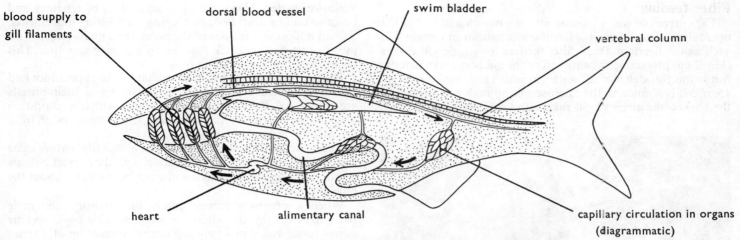

Fig. 29.5   Diagram showing position of swim bladder

Breathing

Oxygen dissolved in the water is absorbed by the gills. The movements of the mouth and operculum are co-ordinated to produce a stream of water, in through the mouth, over the gills and out of the operculum.

There are usually four gills on each side consisting of a curved bony gill-bar bearing many fine filaments (Fig. 29.6). Through the gill-bar run blood vessels which send branches into the gill filaments. The filaments bear smaller filaments down their length which, in turn, divide into smaller branches (Fig. 29.7). So great a number of minute branches provides a very large surface area when the gills are immersed in water. The walls of the gill filaments are very thin, enabling the oxygen to diffuse rapidly into the blood. A convenient way of visualizing the gills is as an orderly system of blood capillaries exposed to the water in such a way as to absorb oxygen.

Although there is more oxygen in air than in water, a fish will suffocate in air. This is probably because the muscular system of mouth and operculum which can work in water will not function in air. In other words, the valve system which is water-tight is not air-tight. Another important reason is that when a fish is out of water, the surface tension of the water-film covering the gill filaments sticks them together so that the total surface exposed is very much reduced.

The mechanism for pumping water over the gills varies in detail with the type of fish, but in general the pressure in the mouth cavity is reduced by the floor of the mouth being lowered so that water enters through the mouth. The free edge of the operculum is pressed against the body wall by the higher pressure outside, so preventing the entry of water by this route. Next, the volume of the mouth cavity is decreased by raising the floor of the mouth. The escape of water from the mouth is prevented by the closure of two inturned folds of skin along the upper and lower jaws.

The pressure thus forces water between the gill filaments, assisted by an outward movement of the operculum which "sucks" the water from the front to the back of the mouth cavity and over the gills. Finally, as the mouth and operculum close, the fold of skin along the free edge of the operculum is forced outwards and water escapes between the operculum and body wall (Fig. 29.8).

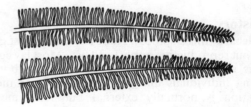

Fig. 29.7   Tips of filaments seen under the microscope

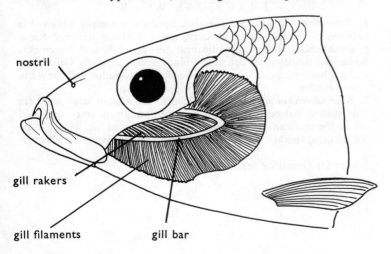

Fig. 29.6   Herring (operculum cut away to show gills)

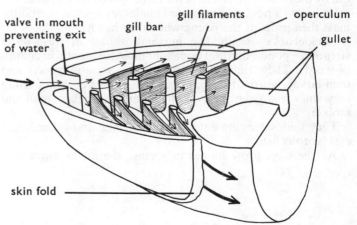

Fig. 29.8   Diagram to show respiratory currents
(gill rakers not shown)

Filter feeding

The current of water passing into the mouth and out of the operculum also serves as a feeding mechanism in certain fish such as the herring. These filter feeders have long gill rakers (Fig. 29.6) projecting forwards from the gill bars and when the fish swims through surface waters in which zooplankton (Plate 15, p. 57) is abundant, the microscopic animals are trapped in the basket-like array of gill rakers and eventually swallowed.

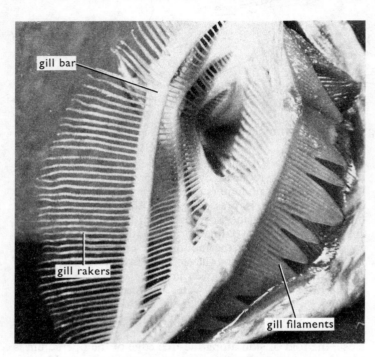

Plate 58.   GILL RAKERS OF HERRING (×3)

Life history

Stickleback (three-spined). None but the vaguest generalizations about reproduction apply to all fish. Most lay eggs though some are *viviparous*, that is, the young fish are born as free-swimming individuals and not in egg-cases or membranes. Fertilization is normally external but sometimes internal fertilization occurs. In many species of fish, once the eggs are laid, there is no parental care, but if there is, it is usually carried out by the male. This is true of the three-spined stickleback. The male builds a nest, one or more females lay eggs in it, and the male then protects the young when they hatch.

Sticklebacks usually live in small shoals in fresh-water streams or ponds, and will be caught most frequently in clumps of weeds. They are carnivorous, eating aquatic insects and their larvae, worms, fish spawn and fry. If kept in an aquarium they will rapidly devour every living animal that is small and mobile.

They themselves are eaten by perch, pike and other larger carnivorous fish.

As the days grow longer in spring, the males move into shallower water. They take up a small area of territory and begin to build a nest; first, by digging, with biting movements, a small hollow in the floor of the pond, then by roofing it with pieces of vegetation stuck together by a kidney secretion. This results in a tunnel-like shelter.

During this time the males have changed in appearance and behaviour. Their eyes become bright blue and their breasts are orange-red. If another male stickleback with this coloration invades the territory, the occupant drives him away by a threatening posture or a fighting attack.

Experiments with models show that a sexually mature male of this kind will attack a crude model with the correct colours much more vigorously than it will a perfect model without the blue and red colours.

When a pregnant female enters his territory, the male recognizes her by the abdomen swollen with eggs, and he swims towards her in a "zig-zag dance" consisting of a series of swoops. The female responds to the male's dance and to his colouring by swimming towards him, and he turns and leads her to the nest, pointing out the entrance with his head. The female enters and is stimulated to lay eggs by rapid thrustings of his snout against her body (Fig. 29.9). Two or three females may be induced to spawn in this way. Immediately the female leaves the nest the male enters and fertilizes the eggs by shedding sperms on them.

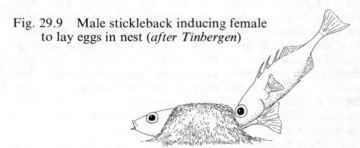

Fig. 29.9   Male stickleback inducing female to lay eggs in nest (*after Tinbergen*)

For the next 7–10 days he aerates the developing eggs by fanning water through the nest with his pectoral fins, after which the eggs hatch and the young fry are kept in the nest by the male for a period of up to a month. During this time he will protect them from intruders, but gradually his bright colours fade and he allows the young fish to leave the nest and disperse.

PRACTICAL WORK

1. *Dissection and external features.* Sprats are usually obtainable between November and February, and are cheap material for a practical class. Herring are cheapest between July and December. Scales and gill filaments can be examined microscopically. Gill rakers and gill bars can be seen easily, and a simple dissection will show the swim bladder.
2. *Sticklebacks in aquaria.* If sticklebacks are kept in large numbers they should not be placed in a balanced aquarium since they will eat all the small animals and disturb the balance of life. They can be fed on dried food or water fleas and insect larvae specially collected.

[*Note*: for Questions *see* p. 173.]

30 Frogs and Tadpoles

THE frogs are some of the few remaining members of the amphibia, a group which flourished 250 million years ago. Other present-day members of the group are the toads, newts (Fig. 30.2) and salamanders. The amphibia are so adapted that they can, in general, move, feed and breathe as well on land as they do in fresh water. At certain times of their life history or at particular seasons, however, they show a dependence on or preference for one or the other.

External features (Fig. 30.1)

The frog is *poikilothermic* (*see* p. 159) with a loose-fitting moist skin. Its eyes protrude in such a way that they are above water when the rest of the body is immersed (Fig. 30.3). The eyes have movable lids but, in addition, the whole eyeball can be withdrawn farther into the head by muscles. This can be seen to happen sometimes when the frog is swallowing. Its nostrils are so situated that air can be breathed while the frog is swimming at the surface: they can also be closed. Behind the eyes are circular ear-drums. Sounds in the air or water set these thin membranes vibrating, the vibration being transmitted by a small bone to a sensory region which sends nervous impulses to the brain.

Locomotion

The frog's powerful hind legs are adapted for both swimming and leaping. The strong *extensor muscles* of the thigh contract, extending the limb and thrusting the foot against the ground or against the water. The thrust is transmitted through the body of the frog by the *pelvic girdle* and the spine so that the whole animal is pushed forward (Fig. 30.4). (*See* also p. 124.)

In the water the webbed hind feet provide a greater surface area for pushing backwards on the water. The smaller fore-limbs help to steer when the frog is swimming and absorb the shock of landing after a jump on land. On moving from water to land or over rough ground the frog will crawl rather than leap.

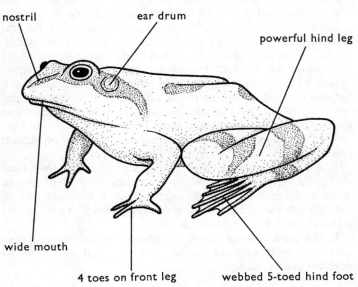

Fig. 30.1   External appearance of frog

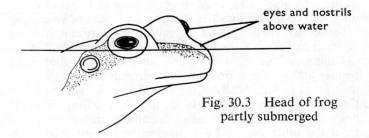

Fig. 30.2   Smooth newt (female)

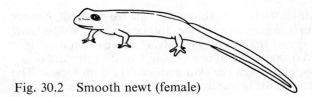

Fig. 30.3   Head of frog partly submerged

Fig. 30.4   Frog leaping
(From James Gray, How Animals Move, Cambridge University Press)

1. At rest: hind legs flexed

2. Hind legs extended pushing frog forward and upwards

3. Front legs being extended: beginning to draw up the hind legs

4. Front legs extended to take first shock of landing: hind legs drawn up

Breathing

The frog's skin is smooth and moist, fairly thin, and well supplied with blood vessels which branch into a fine network of thin-walled capillaries (Fig. 30.5). Oxygen from the air or water, dissolves in the film of moisture over the skin, diffuses through the skin, through the walls of the blood capillaries and

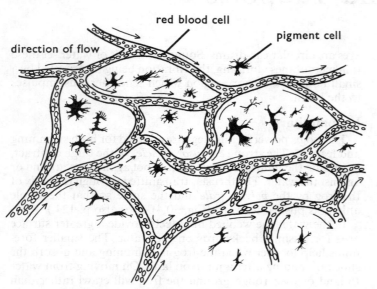

Fig. 30.5   Capillary circulation in tadpole's tail

into the blood. Here it combines with the red pigment, *haemoglobin*, and is carried away in the circulation, back to the heart and then round the rest of the body. Excess carbon dioxide is eliminated from the blood in a similar way, diffusing out of the blood vessels, through the skin and into the atmosphere. The skin is effective for breathing on land or in water and is in use continuously.

When the frog is inactive the skin absorbs enough oxygen to meet its needs. During and after activity it may breathe air into its lungs by gulping movements of the floor of its mouth. The lungs can be used only when the frog is on land or swimming at the surface, the nostrils having valves which prevent the entry of water and control the flow of air movement into the lungs.

Unlike the mammals, amphibia and reptiles do not make regular and rhythmic breathing movements but gulp air into their lungs spasmodically as the need arises. Air is forced into the lungs by raising the floor of the mouth. The lungs lie in the body cavity and, unlike those of mammals, are not separated from the other organs by a diaphragm (p. 86). The lungs can be inflated to many times their relaxed size, so apparently inflating the entire frog. The moist lining of the large mouth is also a respiratory surface. Like the skin, it is in constant use, except when submerged, but the movements of the mouth-floor can be used to exchange the air in it.

Feeding

Adult frogs are carnivorous, feeding on worms, beetles, flies and other insects. Worms and beetles may simply be picked up by the mouth but flying insects can be caught on the wing. On occasions the frog will leap towards the insect and trap it in its wide, gaping mouth; at others its tongue is used. The tongue is attached to the front of the mouth and can be rapidly extended by muscles. It is shot out in a half circle, and the

insect is trapped by the sticky saliva covering the surface (Fig. 30.6). Insects can be picked off the ground or vegetation in a similar way. The prey is swallowed whole but there are rows of tiny, closely set teeth in the upper jaw and in the roof of the mouth (which prevent its escape). In swallowing, the eyes are often pulled farther into the head and press down on the prey.

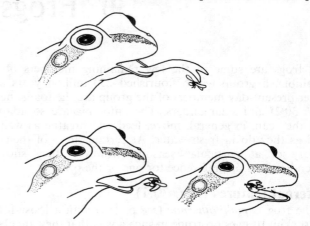

Fig. 30.6   Diagrams to show frog catching a fly with its tongue

Skin and colour

The predominant colours in the common frog are green, yellow, brown and black, with whitish areas on the underside. The mottled patterning of these colours is thought to be of protective value in camouflaging the animal in its natural surroundings. The frog's colour can change to some extent, expansion or contraction of pigment cells in the skin making the frog darker or lighter. Such tones may correspond more closely to the frog's background in different circumstances and help to conceal it, but in many amphibia, temperature and humidity play a part in causing colour changes. The colour change operates through the sense organs and brain. A hormone from the pituitary gland circulates in the blood stream and has an effect on the pigment in certain cells.

In the skin are mucous glands which make the slimy fluid that covers the body. The sliminess makes the frog difficult to catch and keeps the skin moist. Without a constant film of moisture on the surface, oxygen could not go into solution and so reach the blood beneath the skin.

In the toad, a special group of glands behind the eyes produces an unpleasant poisonous substance that may serve as protection against enemies.

Life-cycle

(*a*) **Mating.** Nearly all amphibians must return to water to breed although some species make a "pond" in a rolled up leaf or hollow tree, the "pond" being derived from liquefaction of the jelly round the eggs. In one species the female places the fertilized eggs in little pouches on her back where they develop through all the tadpole stages into tiny but fully formed frogs.

In this country in spring, usually during March, male and female frogs move out of their winter quarters to the nearest pond. The females are usually larger, their bodies swollen with mature eggs, while the males have developed black, horny pads on their thumbs. These pads enable the male to grip the female behind the fore-limbs, in which way the male may be carried about on the female's back, mostly in the water, for several days.

(Heather Angel)

Plate 59. TOADS MATING

6. The cells continue to divide again and again until the "egg" becomes a hollow ball of tiny cells. The cells are very numerous and too small to be seen even with a hand lens. There is little increase in size during this division or *cleavage* but a great increase in the number of cells and nuclei. The egg has become an embryo, but to a casual observer it is still a spherical black ball and there is little evidence of the vigorous activity that has been going on (Fig. 30.8 *f* and *g*).

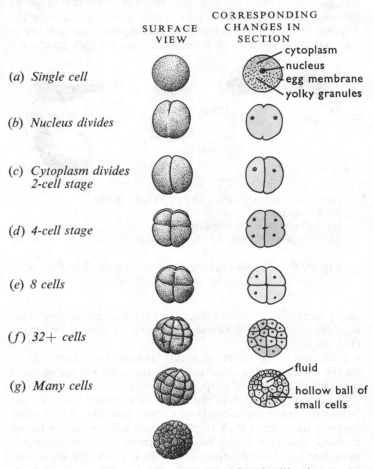

Fig. 30.8   Changes in the ovum after fertilization

When the female lays the eggs, the males produce a *seminal fluid* containing *sperms*. This pours on to the eggs as they leave the female's body, and the sperms swim through the jelly and fertilize the eggs. Fertilization occurs when the nuclei of the sperm and egg meet and join together or fuse.

The eggs or sperms leave the body through an opening, the *cloaca*, just above the region where the hind legs join the body.

Since the jelly, or albumen, round the eggs swells on contact with the water, fertilization would be impossible unless carried out at the moment the eggs leave the female's body. Thus, although fertilization is external, the pairing of frogs ensures that this happens.

(*b*) **Development.** 1. The jelly round the eggs in the familiar frog-spawn (Fig. 30.7) has several advantages. It sticks them together and prevents their being swept away or eaten. It protects them from mechanical injury, from drying up and, probably, from attacks by fungi and bacteria.

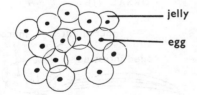

Fig. 30.7 Appearance of frog's spawn

2. The egg itself is a small sphere, a semi-liquid cytoplasm in a tough, black egg membrane. There is a nucleus, and the lower cytoplasm contains yolky granules that are the only food supply for the first weeks of development. Sufficient oxygen must be able to diffuse through the jelly and the egg to allow the vital processes to go on (Fig. 30.8*a*).

3. Shortly after fertilization, the time depending partly on the temperature, the nucleus of the egg divides into two smaller nuclei which separate. The cytoplasm then divides to include each nucleus in a separate unit of cytoplasm so that there now appear two smaller cells, each with a nucleus (Fig. 30.8 *b* and *c*).

4. A similar division takes place again in each cell but at right angles to the first division, making four smaller, roughly equal cells (Fig. 30.8*d*).

5. A third division takes place in the four cells, this time at right angles to the other two, round the "equator", forming eight cells of which the lower four are slightly larger than the upper four (Fig. 30.8*e*).

Plate 60.   FROG TADPOLE (× 3)   (Heather Angel)

7. (Fig. 30.9.) All this happens in the first day or two. Later the sphere begins to elongate and develop a distinct head and tail. Meanwhile the cells are being organized internally to form the structures and organs of the tadpole. The energy and raw materials for this process come from the yolk.

8. After about ten days the jelly immediately round the tadpole liquefies and the tadpole can be seen moving about inside. The liquefaction also makes it easier for the tadpole to wriggle out of the jelly and into the water. At this stage its mouth has not yet opened and it is still digesting and using the remains of the yolk in its intestine. It clings to water weed or to the surface of its jelly by its *mucous glands* which produce a

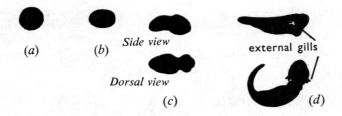

(*a*) Small sphere consisting of 2 or 3 layers of cells
(*b*) Elongates
(*c*) Tail and head distinguishable
(*d*) Wriggling in jelly. Ready to emerge

Fig. 30.9   Changes taking place during the 4 to 5 days after laying (×3)

sticky secretion. It is quite black, and the rudimentary external gills are visible, but it breathes through its skin at this stage (Fig. 30.9*d*).

Although the tadpoles are merely attached to and not feeding on the water weed, a good deal of spasmodic wriggling takes place in the clusters of tadpoles.

9. In two or three days the mouth has opened and the tadpoles can scrape the coatings of microscopic plants and other deposits from the surface of pond weeds using a pair of horny toothed lips (*see* Plate 60). The three pairs of external, branched gills have developed (Fig. 30.10*a*) and the blood can be seen circulating in them under the low power of a microscope (Fig. 30.11). These gill filaments are thin-walled and present a fairly large surface area to the water. Oxygen dissolved in the water passes through the filament walls and into the blood close to the surface.

10. After about three weeks the mucous glands "disappear" and a distinct division into body and tail occurs, together with a rapid increase in size. Internal gills are formed, opening by slits from the mouth cavity, or pharynx, to the outside. A fold of skin from the front of the head grows back (Fig. 30.10 *d* and *e*) and covers the slits on both sides, fuses with the skin behind the slits and so forms a continuous chamber, the *atrium*, round the slits, opening to the outside by a single hole, the *spiracle*, on the left side. The fold of skin enclosing the space outside the gills is called the *operculum*.

By now the external gills have shrivelled and been reabsorbed into the body. In breathing, water is taken in through the mouth, passed over the gills, through the gill slits into the gill chamber formed by the operculum and, finally, out through the spiracle (Fig. 30.12*a*). As the water passes over the gill filaments, dissolved oxygen diffuses into the blood.

11. The tail elongates and develops a broad, transparent web along its dorsal and ventral surfaces. Vigorous wriggling

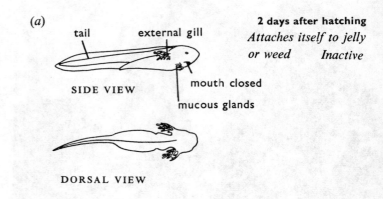

(*a*)   2 days after hatching
Attaches itself to jelly or weed   Inactive

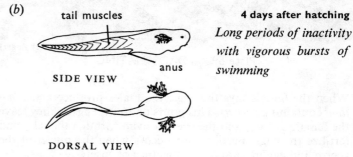

(*b*)   4 days after hatching
Long periods of inactivity with vigorous bursts of swimming

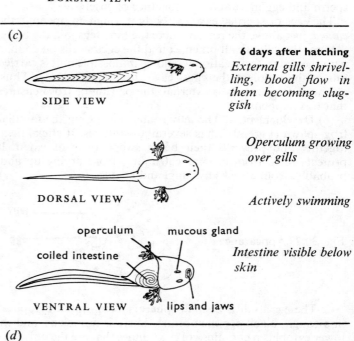

(*c*)   6 days after hatching
External gills shrivelling, blood flow in them becoming sluggish

Operculum growing over gills

Actively swimming

Intestine visible below skin

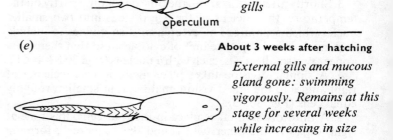

(*d*)   *Diagram showing how operculum grows over gills*

(*e*)   About 3 weeks after hatching
External gills and mucous gland gone: swimming vigorously. Remains at this stage for several weeks while increasing in size

Fig. 30.10   Changes in the first days after hatching (×3)

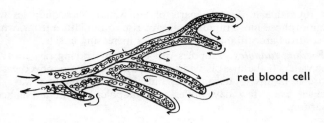

Fig. 30.11   Diagram of external gill showing blood circulation

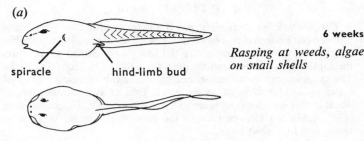

movements of the body and tail propel the tadpole through the water in a similar way to a fish but with less speed and precision.

A long, coiled intestine has developed and can be seen through the skin of the abdomen. The long intestine is adapted to the digestion of an exclusively vegetable diet. Eyes and nostrils are easily seen at this time. In this stage the tadpole grows considerably in size with little pronounced change in form for two or three weeks.

12. At two months from hatching, the tadpole comes to the surface frequently to gulp air into its lungs, which have begun to form. The hind-limb buds near the junction of the body and tail begin to grow and develop into perfect legs (Fig. 30.12 *a* and *b*). The front legs also grow but do not yet appear because they are covered by the operculum; nevertheless they can be seen bulging beneath the skin in this region. The hind limbs are not yet used for locomotion but hang limply by the side of the body while the fish-like wriggling movements take place. The diet changes from vegetation, the tadpoles nibbling preferentially at dead animals or raw meat, at least in the aquarium, and associated with this is the shortening of the intestine and, later, the narrowing of the abdominal region.

13. **Metamorphosis.** At about three months, the front legs break through the operculum, the left leg appearing first by pushing through the spiracle (Fig. 30.12*c*) while the right has to rupture the operculum (Fig. 30.12*d*). The tail shortens (Fig. 30.12*e*), being internally digested and absorbed, so providing a source of nutriment for the tadpole which has stopped feeding. The skin is shed, taking with it the larval lips and horny jaws, leaving a much wider mouth. Finally, the young frog climbs out of the pond on to the land, still with a tail stump, but using its legs for jumping and swimming. These changes take place within about four weeks. The young frogs remain in the damp vegetation and long grass in the vicinity of the pond, catching and eating small insects. In four years the frog will be old enough to breed. The times given for development are only approximate since the temperature of the pond-water can alter the rate of metamorphosis from days to weeks and vice versa. It is more reliable to refer to the phase of development, e.g. hind-leg stage or external-gill stage, rather than age in days.

Habitat

In early spring, during the breeding season, frogs spend their time in ponds and lakes with a steady flow of water. They are not usually found in swiftly running water. After egg-laying they are more likely to be encountered in damp vegetation than in water. They are unlikely to be found in any dry situation where their skins could lose water and dry up and so seriously impair their breathing. In winter they hibernate in the sense that they are dormant and do not feed. They lie up in the mud at the bottom of ponds, in damp moss or holes in the ground and their eyes, mouth and nostrils are closed.

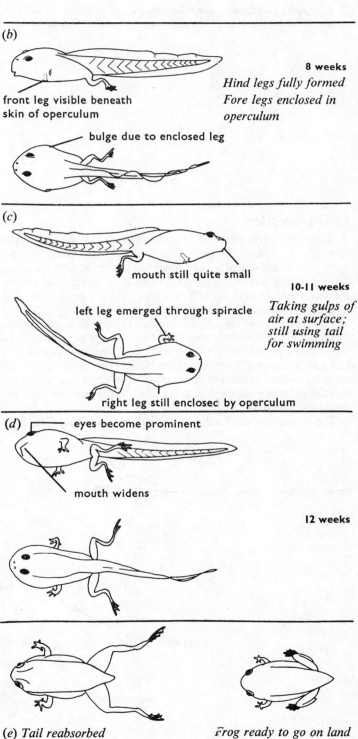

Fig. 30.12   Changes leading up to metamorphosis ($\times 1\frac{1}{2}$)

PRACTICAL WORK

1. *Circulation in a tadpole's tail.* If a tadpole with a well-developed tail web but no legs is placed on a slide with as little water as possible, and the tail web observed under the low power of the microscope, the capillary circulation and the pigment cells can be seen (Fig. 30.5). The tadpole may be anaesthetized beforehand by placing a drop of 1 per cent chlorobutanol in its water. One or two minutes out of water in this way does not seem to harm the tadpole. This circulation (Fig. 30.11) can also be seen in the gills of a younger tadpole or better still in a newt larva.

2. *Control of metamorphosis by thyroxine.* A group of tadpoles is separated from others of the same age, and placed in a solution of 1 part thyroxine to 20×10^6 parts of tap water. These tadpoles will metamorphose more rapidly and give rise to miniature frogs, while the controls are still immature tadpoles increasing in size.

3. *Feeding tadpoles.* Tadpoles can be fed by leaving raw meat in their water. If the meat is tied to a thread it can be removed easily before it begins to decompose, but the water should always be changed every few days unless the tadpoles are in a balanced aquarium.

[*Note*: for Questions *see* p. 173.]

31 Birds

Characteristics

Birds are warm-blooded (*homoiothermic, see* p. 110) vertebrates, with fore-limbs modified to wings, and skins covered with feathers. Typically they have the power of flight, and all reproduce by laying eggs. The skull and lower jaw are extended forward into *mandibles* which make a beak. (*See* Fig. 31.1 for external features.)

The feathers are the single external feature that distinguish birds from other vertebrates. The feathers are produced from the skin which is loose and dry, without sweat glands, and they form an insulating layer round the bird's body, helping to keep its temperature constant, and repelling water. The wings are specially developed for flight, having a large surface area and very little weight.

The *barbules* of the feathers interlock in such a way (Fig. 31.2)

that should a feather be damaged in flight, for example, preening with the beak will re-form it perfectly.

The feather quills have attached to them muscles which can alter the angles of the feathers; for example, when a bird fluffs its feathers out in cold weather. They also have a nerve supply which, when the feathers are touched, is stimulated in a similar way to a cat's whiskers.

The down feathers are fluffy (Fig. 31.2), trapping a layer of air close to the body. The *flight feathers* and *coverts* are broad and flat and offer resistance to the passage of air.

The shape of the bird and the lay of its feathers make it streamlined in flight.

The bird's legs and toes are covered with overlapping scales.

Birds possess a third, transparent eyelid, the *nictitating membrane*, which can move across the eye.

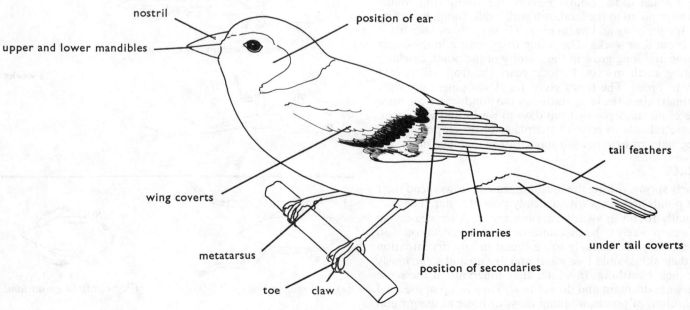

Fig. 31.1 External features of a chaffinch

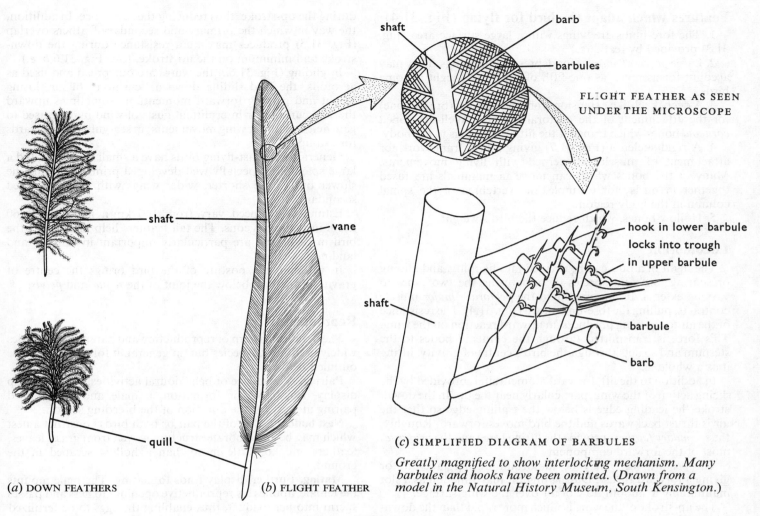

shaft

barb

barbules

**FLIGHT FEATHER AS SEEN
UNDER THE MICROSCOPE**

shaft

vane

shaft

(a) DOWN FEATHERS

quill

(b) FLIGHT FEATHER

hook in lower barbule
locks into trough
in upper barbule

barbule

barb

(c) SIMPLIFIED DIAGRAM OF BARBULES

*Greatly magnified to show interlocking mechanism. Many
barbules and hooks have been omitted. (Drawn from a
model in the Natural History Museum, South Kensington.)*

Fig. 31.2   Feather structure

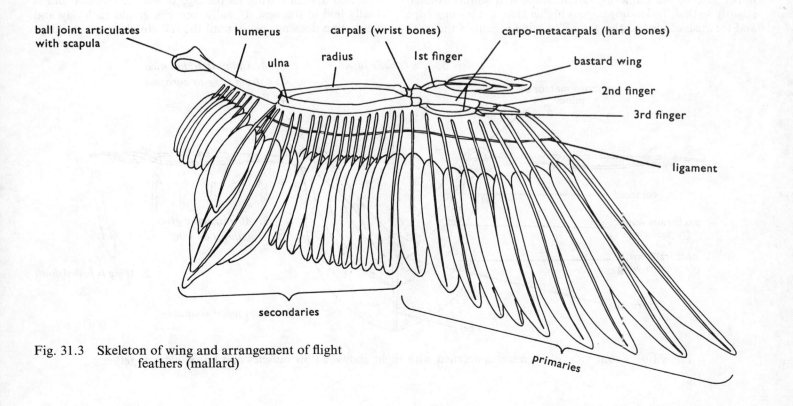

ball joint articulates
with scapula

humerus

ulna

radius

carpals (wrist bones)

1st finger

carpo-metacarpals (hard bones)

bastard wing

2nd finger

3rd finger

ligament

secondaries

primaries

Fig. 31.3   Skeleton of wing and arrangement of flight
feathers (mallard)

Features which adapt the bird for flying (Fig. 31.4)

1. The fore-limbs are wings with a large surface area (Fig. 31.3) provided by feathers.

2. Large *pectoral muscles* for depressing the wings. They may account for as much as one-fifth of the body weight in some birds.

3. A deep, keel-like extension from the *sternum* (breast bone) for the attachment of the pectoral muscles. Well-developed *coracoid* bones which transmit the lift of the wings to the body.

4. A rigid skeleton (Fig. 31.7) giving a firm framework for attachment of muscles concerned with flying movements. Many of the bones which can move in mammals are fused together in birds; for example, the vertebrae of the spinal column in the body region.

5. Hollow bones, which reduce the bird's weight.

Locomotion

The flight of a bird can be divided into flapping, and gliding or soaring, different species of birds using the two types to varying extents. In flapping flight the *pectoralis major* muscle contracts, pulling the fore-limb down (Fig. 31.4). The resistance of the air to the wing produces an upward reaction on the wing. This force is transmitted through the coracoid bones to the sternum and so acts through the bird's centre of gravity, lifting it as a whole.

In addition to the lift, forward momentum is provided by the slicing action of the wing, particularly near the tip. In the down-stroke the leading edge is below the trailing edge so that the air is thrust backwards and the bird moves forward. Roughly, the *secondary feathers* provide the lifting force and the *primaries* most of the forward component.

The *bastard wing* may be important during take-off for giving a forward thrust. During flight it may function as a slot maintaining a smooth flow of air over the wing surface.

The up-stroke of the wing is much more rapid than the down-stroke. The *pectoralis minor* contracts and raises the wing, since its tendon passes over a groove in the coracoid to the upper side of the humerus. Often the arm is simply rotated slightly so that the leading edge is higher than the trailing edge and the rush of air lifts the wing. The wing is bent at the wrist

during the up-stroke, thus reducing the resistance. In addition, the way in which the primary and secondary feathers overlap (Fig. 31.5) produces maximum resistance during the down-stroke and minimum on the up-stroke. (*See* Fig. 31.6 *b–e*.)

In gliding (Fig. 31.6*a*) the wings are outspread and used as aerofoils, the bird sliding down a "cushion" of air, losing height and gaining forward momentum. Sometimes upward thermal currents or intermittent gusts of wind may be used to gain height without wing movements; in sea-gulls and buzzards for example.

Generally, the fast-flying birds have a small wing area and a large span, with specially well-developed primaries, while the slower birds have shorter, wider wings with well-developed secondaries.

Estimates of speed vary from 160 km/h in swifts to 60 km/h in racing pigeons. The tail feathers help to stabilize the bird in flight and are particularly important in braking and landing.

In walking, the posture of the bird brings the centre of gravity of the bird below the joint of the *femur* and *pelvis*.

Reproduction

The detailed pattern of reproduction and parental care varies widely in different species but, in general, it follows the course outlined below.

Pairing. A sequence of behavioural activities, e.g. courtship display, leads to pair formation, a male and female bird pairing at least for the duration of the breeding season.

Nest building. One of the pair or both birds construct a nest which may be an elaborate structure woven from grass, leaves, feathers, etc., or little more than a hollow scraped in the ground.

Mating. Further display leads to mating. The male mounts the female, applies his reproductive openings to hers and passes sperm into her oviduct, thus enabling the eggs to be fertilized internally.

Egg laying. The fertilized egg is enclosed in a layer of albumen and a shell during its passage down the oviduct and is finally laid in the nest. Usually, one egg is laid each day and incubation does not begin until the full clutch has been laid.

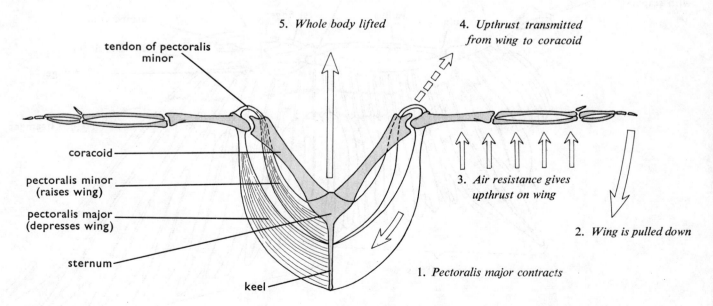

Fig. 31.4   Front view of skeleton concerned with flight showing how muscles and bones work together

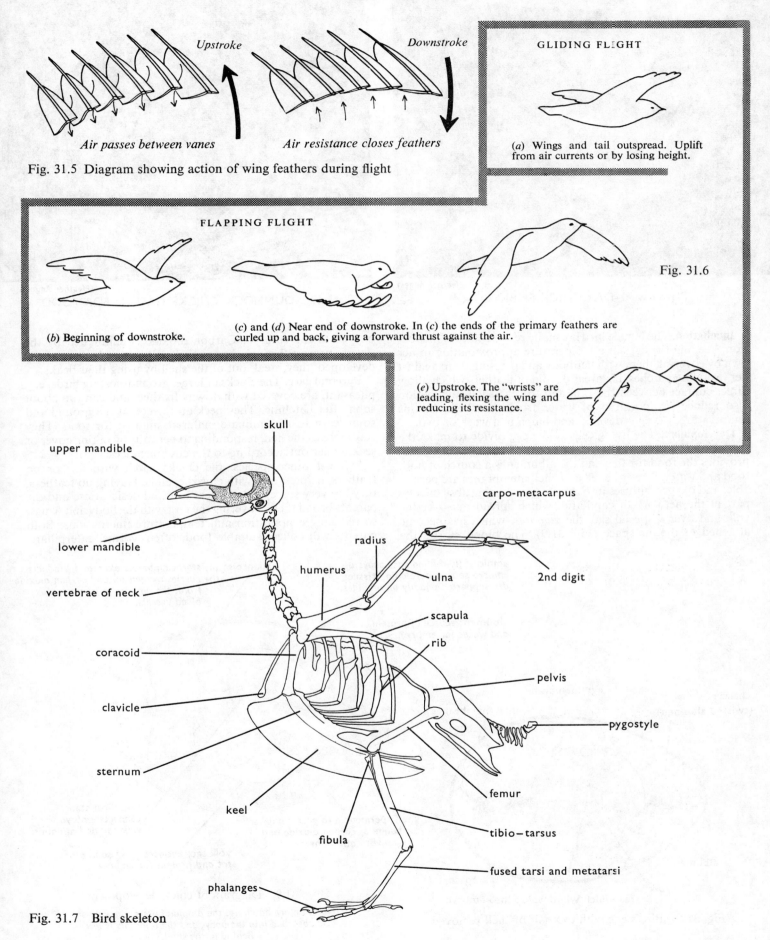

Upstroke

Air passes between vanes

Downstroke

Air resistance closes feathers

Fig. 31.5 Diagram showing action of wing feathers during flight

GLIDING FLIGHT

(*a*) Wings and tail outspread. Uplift from air currents or by losing height.

FLAPPING FLIGHT

Fig. 31.6

(*b*) Beginning of downstroke.

(*c*) and (*d*) Near end of downstroke. In (*c*) the ends of the primary feathers are curled up and back, giving a forward thrust against the air.

(*e*) Upstroke. The "wrists" are leading, flexing the wing and reducing its resistance.

upper mandible

skull

lower mandible

vertebrae of neck

humerus

radius

carpo-metacarpus

ulna

2nd digit

coracoid

scapula

rib

clavicle

pelvis

sternum

pygostyle

keel

femur

fibula

tibio—tarsus

fused tarsi and metatarsi

phalanges

Fig. 31.7 Bird skeleton

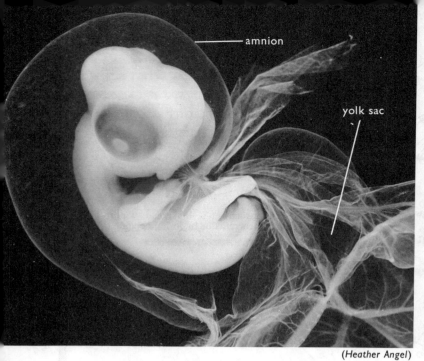
amnion

yolk sac

(Heather Angel)
Plate 61. 7-DAY CHICK EMBRYO (× 4)

(Heather Angel)
Plate 62. DUNNOCK CHICKS GAPING FOR FOOD

Incubation. The female bird is usually responsible for incubation, keeping the eggs at a temperature approximating to her own by covering them with her body and pressing them against her brooding patches, i.e. areas devoid of feathers which allow direct contact between the skin and the egg shell. Incubation also reduces evaporation of water from the shell. At this temperature, the eggs develop and hatch in a week or two.

Development. The living cells in the egg divide to make the tissues and organs of the young birds. The yolk (Fig. 31.8) provides the food for this and the albumen is a source of both food and water. The egg shell and shell membranes are permeable, and oxygen diffuses into the air space, being absorbed by part of the network of capillaries which spread out over the yolk and over a special sac, the *allantois*, which has become attached to the air space (Fig. 31.9). The blood carries the

oxygen to the embryo. Carbon dioxide is eliminated by the reverse process through the egg shell. When the chicks are fully developed, they break out of the shell by using their beaks.

Parental care. The chicks of large, ground-nesting birds, e.g. pheasant, are covered with downy feathers and can run about soon after hatching. They peck at objects on the ground and soon learn to discriminate material suitable for food. They stay close to the hen, responding to her calls by taking cover or seeking her out according to the circumstances.

In most other species, the chicks hatch with few or no feathers, helpless and with closed eyelids. Having no feathers, they are very susceptible to heat loss and desiccation, and the parents brood them, covering the nest with the body and wings, so reducing evaporation and temperature fluctuations. Both parents will collect suitable food, often worms, caterpillars,

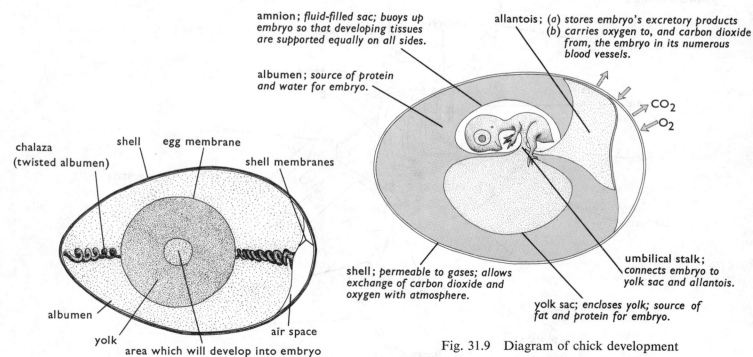

amnion; *fluid-filled sac; buoys up embryo so that developing tissues are supported equally on all sides.*

albumen; *source of protein and water for embryo.*

allantois; (*a*) *stores embryo's excretory products* (*b*) *carries oxygen to, and carbon dioxide from, the embryo in its numerous blood vessels.*

CO_2

O_2

chalaza (twisted albumen)

shell

egg membrane

shell membranes

albumen

yolk

area which will develop into embryo

air space

Fig. 31.8 Bird's egg with top half of shell removed

shell; *permeable to gases; allows exchange of carbon dioxide and oxygen with atmosphere.*

umbilical stalk; *connects embryo to yolk sac and allantois.*

yolk sac; *encloses yolk; source of fat and protein for embryo.*

Fig. 31.9 Diagram of chick development
(*Note. Before hatching, the amnion bursts, the yolk sac is absorbed into the body, and the allantois is left behind in the shell.*)

Plate 63. YELLOW HAMMER FEEDING YOUNG

insects and other materials equally rich in protein. The sound or sight of the parents approaching the nest causes the nestlings to stretch their necks and gape their beaks. The bright orange colour inside the beaks induces the parent to thrust the food it is carrying into the open beaks (Plates 62 and 63).

Adaptations of beaks and feet

Many birds show interesting variations in the structure of their beaks and feet. These differences are thought to be adaptations to the mode of life and methods of feeding and locomotion. Some are illustrated in Fig. 30.10.

CURLEW

Long, narrow beak for probing into mud and sand on the shore and in estuaries to reach burrowing worms and molluscs. Characteristic of most waders, e.g. Sandpipers, Redshanks.

BUZZARD

Powerful, sharp, hooked beak, for tearing the flesh from small birds and mammals. This type of beak is characteristic of most birds of prey, including Hawks, Falcons, Eagles and Owls.

LITTLE OWL

Three toes directed forward, and one back, but they can be bent to meet. They are powerful, with sharp, curved talons for catching and killing prey. Characteristic of many predatory birds such as Falcons and Hawks.

HERRING GULL

Hind toe very small. The web between the three front toes is for swimming, and for walking on soft surfaces. Characteristic of other gulls, sea birds and ducks.

Fig. 31.10
Beaks and feet

After a week or two, the young birds begin to climb out of the nest and sit in the bush or tree but the parents still find and feed them. When the primary and secondary feathers have developed, the fledglings begin short practice flights. This is one of the most dangerous periods of their lives since they can feed themselves to only a limited extent and cannot escape from predators such as cats and hawks. Some estimates suggest that only 25 per cent of the eggs laid in open nests of this kind reach the stage of fully independent birds.

QUESTIONS (Chapters 29–31)

Apart from simple recall of facts, the most likely types of essay question in public examinations are those demanding a comparison of these organisms from the point of view of their methods of locomotion, breathing or reproduction, e.g. the contrast between the lack of parental care in the frog and the highly evolved parental behaviour of birds; the use of trachea in insects, and gills in fish for gaseous exchange. The best practice for answering questions of this type is to select pairs of animals and draw up lists of features common to both in connexion with breathing, locomotion or reproduction and then tabulate the differences relevant to those activities. The similarities will usually be concerned with fundamental principles, e.g. oxygen and carbon dioxide are exchanged in both lungs and gills but the methods of ventilation are quite different.

To make the answers relevant, only points of similarity or difference should be mentioned. Try to start all sentences with either "Both . . ." or "Whereas . . .", at least in practice answers, e.g. "Both fish and frog can exchange oxygen and carbon dioxide with the water surrounding them", "Whereas the frog uses only its skin for this process, the fish has specialized structures, gills, with a greatly increased surface area for gaseous exchange."

Bear in mind that a "compare" question requires both similarities and differences while a "contrast" question demands only differences. For either type of question, two separate accounts will not do.

32 Simple Organisms and Unicells

As examples of living organisms, the animals and plants described so far in this book are extremely complicated in their structure and physiology. They all consist of thousands of cells which are specialized in certain ways and have specific functions. Some cells are specialized for conduction of fluids, some for contraction and some for the manufacture of chemicals.

There are in existence many organisms in which all these processes are carried out, not by specialized groups of cells, but in the single cell which makes up the entire organism. These unicellular creatures are microscopic. Some examples will be described and their biological significance discussed.

Spirogyra

Spirogyra is not a unicellular organism, but consists of thin filaments made up of identical cylindrical cells. The filaments are visible to the naked eye and, in their thousands, produce a green, slimy growth in ponds and lakes.

All the cells in a filament are identical and there is no specialization of any cell for a particular function. Any one of the cells could lead an independent existence and divide to form a new filament. In this respect, Spirogyra is similar to the unicells. Each cell consists of a cellulose wall surrounding a large central vacuole lined with cytoplasm. The nucleus is supported by strands of cytoplasm. One or two ribbon-like chloroplasts form a helix (cylindrical spiral) in the cytoplasm (Fig. 32.1 and 32.2).

Spirogyra belongs to a group of plants called *Algae* which includes such diverse organisms as the single-celled desmids in Fig. 32.5, and the much larger multi-cellular seaweeds.

Nutrition. Spirogyra makes its food in a similar way to green plants, but without the elaborate system of roots, stem and leaves of the higher plants. It is surrounded by water containing dissolved carbon dioxide and salts so that in the light, with the aid of its chloroplast, it can build up starch by photosynthesis. From this carbohydrate, with additional elements, it can synthesize all the other materials necessary for its existence.

Growth. Any of the cells can divide transversely, the daughter cells remaining together, so forming the filament and increasing its length. The retention of daughter cells after division represents the way in which many-celled organisms could have arisen from single-celled creatures.

Reproduction. Asexual reproduction is an outcome of the cell division and growth described above. Pieces of the growing filament simply break away and continue to grow independently of the "parent" filament.

Sexual reproduction takes place in Spirogyra under certain conditions when two filaments lie side by side (Fig. 32.3). In the cells of each filament the cytoplasm shrinks away from the cell wall, the detailed structure of chloroplast, nucleus and cytoplasm seems to disappear, and a rounded protoplasmic mass is formed. Meanwhile outgrowths from the walls of opposite cells in the two filaments have met and joined to form a continuous tube through which the cell contents of one filament pass into the other, leaving one filament quite empty. In the cells of the other filament the protoplasmic masses fuse and a thick wall is secreted round the resulting zygote, forming a zygospore which, when released by the breakdown of the cell wall of the filament, falls to the bottom of the pond. The zygospore can remain dormant and withstand adverse conditions. In this form it may also be transported to fresh ponds in the dried dust or mud from its original habitat.

When the zygospore germinates, its cytoplasm extends from the split wall of the spore and grows into a new filament (Fig. 32.4).

(*a*) FILAMENT

chloroplast

Fig. 32.1   Spirogyra

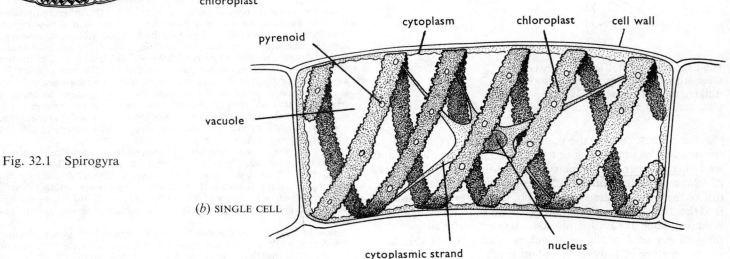

pyrenoid

cytoplasm   chloroplast   cell wall

vacuole

(*b*) SINGLE CELL

cytoplasmic strand   nucleus

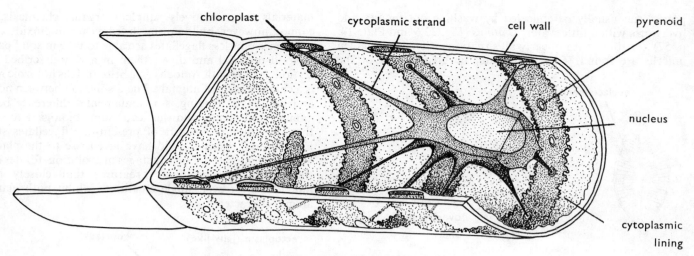

Fig. 32.2 Stereogram of Spirogyra cell in section

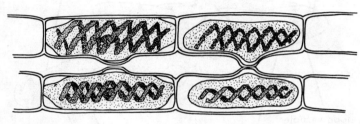

(a) Adjacent cells develop protuberances. Cell contents shrink and round off.

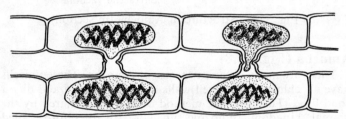

(b) Protuberances meet and form tube. Cell contents from one filament pass into the other.

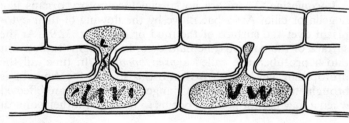

(c) Cell contents fuse.

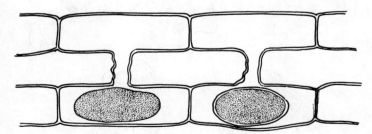

(d) Zygote secretes thick wall.

Fig. 32.3 Sexual reproduction in Spirogyra

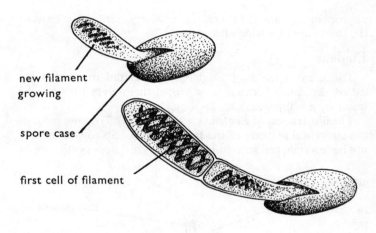

new filament growing

spore case

first cell of filament

Fig. 32.4 Zygospore of Spirogyra germinating

The process of reproduction described above is called *conjugation*. The entire cell-contents form the gametes and, since the cells of a particular filament all discharge their gametes or receive the gametes of the other filament, the filaments can be said to exhibit a difference of sex, although no structural differences are visible before conjugation.

Other plant-like unicells

Fairly familiar unicellular algae are those which make up the powdery green dust on tree trunks. These belong to a genus called *Pleurococcus*. Related to Spirogyra are some unicellular plants called *Desmids* (Fig. 32.5) which, apart from inhabiting

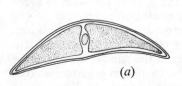
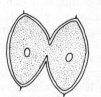

(a) (b)

Fig. 32.5 Examples of common desmids

ponds, can usually be collected by washing out mosses and liverworts with a little water. *Diatoms* (Fig. 32.6 and Plate 14, p. 57) are abundant in nearly all situations where other green unicells are found. They constitute the greater part of the

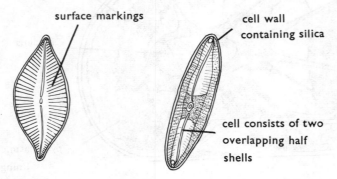

Fig. 32.6   Two kinds of diatom (greatly magnified)

phytoplankton and are therefore very important as the basis of the food chain on which fish depend.

Euglena

The many species of Euglena and related organisms are called *flagellates* because they propel themselves through the water by a lashing action of a long *flagellum*.

The illustration of *Euglena gracilis* (Fig. 32.7) shows most of the structural features of the family. Unlike Spirogyra, it does not have a rigid cell wall, but like Spirogyra it does synthesize its

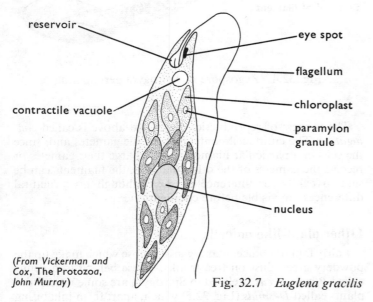

(From Vickerman and Cox, The Protozoa, John Murray)

Fig. 32.7   *Euglena gracilis*

food with the aid of its choloroplasts. Unicells, which feed like green plants, using chlorophyll, are called *protophyta*.

If Euglena gracilis is kept in the dark it will lose its green colour and become unable to photosynthesize, but it will continue to live if suitable organic matter is present in the water. In this respect it is very similar (some biologists think it is identical) to some of its relatives in the genus *Astasia*. These colourless flagellates live only in water rich in organic materials such as the fluid in the puddles round cowsheds. They absorb the organic substances from their environment and use them as food, although they may still be able to synthesize complex

materials from relatively simple inorganic chemicals. They cannot, however, build up starch from carbon dioxide.

Other colourless flagellates are able to take in solid particles of food material and digest them in a way described in the section dealing with Amoeba (*see* below). This holozoic feeding is a characteristic of animals. The distinction between holozoic and holophytic feeding is a fundamental difference between animals and plants, and this leads some biologists to suggest that the remote ancestors of present-day flagellates such as Euglena and Astasia could have given rise to the plant and animal kingdoms in the early stages of evolution; for it is only in this group of unicellular creatures that closely related individuals, or even single creatures, exhibit both plant and animal characteristics.

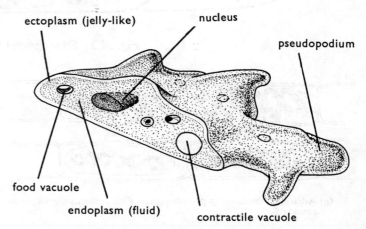

Fig. 32.8   Structure of Amoeba (in section)

Amoeba (Fig. 32.8)

There are many different species of Amoeba. These unicells have no chlorophyll or cell walls and they take in and digest solid food. This holozoic method of feeding is shared by the animal kingdom; holozoic unicells are therefore called *protozoa*. Amoebae live in ponds, ditches and other moist places and in the soil.

Locomotion. There are no special locomotory organs like flagella or cilia. Amoebae move by the flowing of their cytoplasm over the surface of the mud or soil (Fig. 32.9). At the surface of the Amoeba the fluid cytoplasm begins to flow out into a protuberance called a *pseudopodium*. In time, all the protoplasm will have flowed into this so that the Amoeba is brought to a new position. Changes of direction are effected when a new pseudopodium begins to form at another point of

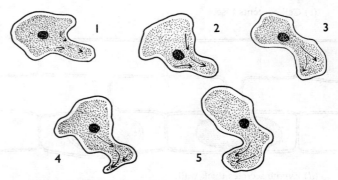

Fig. 32.9   Diagrams to show flowing movements of Amoeba

the Amoeba's surface. The direction of movement is probably determined by local differences in the water. Slight acidity or alkalinity may cause the protoplasm to start flowing or prevent its doing so altogether. The chemicals diffusing from suitable food material may cause the protoplasm to flow in that direction.

Feeding. When an Amoeba encounters a microscopic alga or flagellate, pseudopodia flow out rapidly and surround it (Fig. 32.10) so that the prey is ingested with a drop

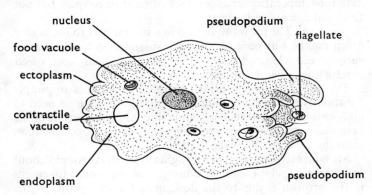

Fig. 32.10   Amoeba ingesting a flagellate

of water into the cytoplasm of the Amoeba. This forms a *food vacuole*. The surrounding cytoplasm secretes into the food vacuole enzymes which digest parts of the prey. The soluble, digested materials are then absorbed into the surrounding cytoplasm, and the undigested residue left behind or egested, as Amoeba flows on its way.

Food vacuoles with material in various stages of digestion can be seen in the cytoplasm of the Amoeba. Ingestion and egestion can take place at any point on the surface; there is no "mouth" or "anus".

Osmo-regulation. The cell boundary of Amoeba is selectively permeable, with the result that the low osmotic potential of the solutions in the *endoplasm* causes water to enter. This excess water is collected up in a spherical *contractile vacuole* which gradually swells, and then seems to contract or burst, liberating the accumulated water to the exterior.

Reproduction. Amoeba stops moving and its nucleus

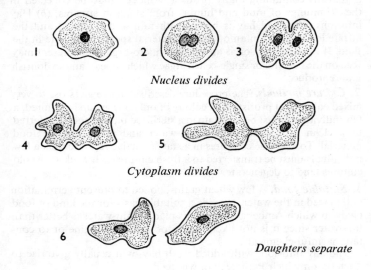

Fig. 32.11   Amoeba reproducing by binary fission

divides. The cytoplasm then divides to make two daughter individuals (Fig. 32.11). This is *binary fission*.

Some species of soil Amoebae are able to survive dry conditions by *encystment*. They round off and secrete a protective wall round the cell. This prevents desiccation. When conditions are favourable, the cysts break open and the amoebae emerge. *Amoeba proteus*, the species most frequently studied in schools, does not encyst.

There is no evidence of any form of sexual reproduction.

Related protozoa

Ciliates. The protozoans most frequently encountered, belong to the ciliates. These move by means of rows of cytoplasmic filaments, *cilia*, extending from their surface. These cilia flick in rhythmic waves and propel the microscopic organisms smoothly forwards or backwards.

There is a more complicated organization in the body than appears in Amoeba. For example, in the *Paramecium* illustrated in Fig. 32.12, a row of special cilia waft food particles into a shallow gullet, and ingestion can take place only at the end of this gullet. The food vacuoles move in a very definite path through the body of Paramecium, and egestion takes place at only one point near the region of ingestion.

Enterozoic protozoa. There are many protozoa which live only in the gut or other regions within the bodies of higher

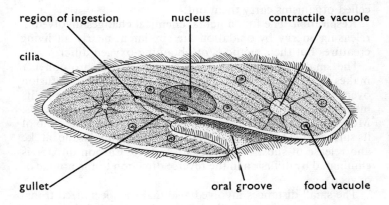

(a) PARAMECIUM

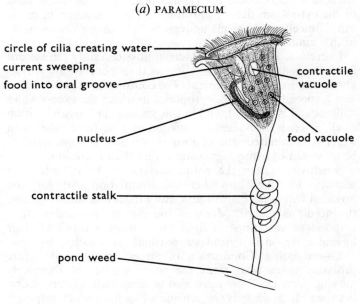

Fig. 32.12   Two ciliates   *(b)* VORTICELLA

animals. They are often quite harmless or even beneficial, and some play an important part in the digestion of cellulose in the paunch of the cow and the caecum of the rabbit, to mention only two examples. If those present in the alimentary canals of certain termites are killed, the termite can no longer digest the woody material on which it feeds and it will die of starvation.

There are some protozoa, however, like the malarial parasite, which lives in red blood cells, and others like the dysentery amoeba in the intestine, which cause severe symptoms of disease.

Biological significance of single-celled organisms

Single-celled organisms are of particular interest because similar creatures were probably some of the earliest living organisms to inhabit the earth, and many of the present-day animals and plants may have evolved from such a stock.

The other point of interest is the way in which these minute creatures carry out all the vital processes associated with living, without the complicated cellular organization seen in the higher organisms. In many instances the greater size of the higher organisms would not be possible without an organization into cells and the development of special organs. It is worthwhile to consider some of the vital processes and to see how the single-celled organisms carry them out.

Respiration. The fundamental chemical changes involved in releasing energy by oxidation are similar in nearly all living creatures but their methods of obtaining oxygen differ.

The small size of the *Protista* (protozoa and protophyta) makes special respiratory organs unnecessary since diffusion over small distances, from the outside of the protistan to its innermost protoplasm, is rapid enough to meet its needs. Moreover, the smaller an organism is, the greater is the ratio of its surface area to its volume, and so the more rapid is the intake and release of substances from its surface. Carbon dioxide is eliminated by diffusion in the reverse direction to the uptake of oxygen.

The small distances involved also make a specialized transport system unnecessary. The digested materials from a food vacuole in an Amoeba are rapidly distributed by the flowing of the cytoplasm during movement. The cytoplasm in many other unicellar organisms undergoes "streaming" movements of this kind.

Excretion. Excretory products almost certainly diffuse out from the general surface as soon as their concentration rises above that in the environment. The contractile vacuoles represent a specialized area of cytoplasm in which the excess water collects and is expelled, and in this sense is an "organ" within a cell. There is no evidence to suggest that anything other than water is eliminated: the contractile vacuole is considered to be an organ of osmo-regulation rather than excretion.

Sensitivity. Often the entire surface of the cytoplasm is sensitive to touch and chemical stimulation. An Amoeba prodded with a needle will flow away from it no matter where the needle is placed. Many of the mobile unicellular algae respond to variations in light intensity as a result of their having a region of cytoplasm particularly sensitive to light.

Co-ordination. Conduction of an electrochemical nature probably takes place over the entire surface of the freely moving protista. There must also be some path of conduction between a light-sensitive region and a flagellum which responds to the stimulus.

Reproduction. In binary and multiple division of the single-celled creatures the entire protoplasm becomes incorporated in the offspring and nothing is left behind to die of old age. In the higher animals only the small fraction of their protoplasm, represented by the sperm and ovum of a fertilized egg continues the existence of the adult. The greater proportion of the protoplasm and other materials in the body is bound to die sooner or later. The protistan may die as a result of desiccation, extreme temperature changes, lack of food or oxygen, but not from old age or senile decay.

Nutrition. The protophyta feed in a similar way to the higher green plants, but because they are surrounded by the water, carbon dioxide and salts that they use, no specialized conducting systems are needed.

The protozoa break down complex substances to simple substances just as do the higher animals, but have no need of elaborate digestive tracts or glands since almost any part of the cytoplasm is capable of secreting enzymes in food vacuoles once the food has been ingested.

Movement. Movement in the higher animals is brought about by a complex relationship of muscles and skeleton. Movement in the protista is due to the flowing of the cytoplasm or the lashing of cilia or flagella. Although the flagellum has no organization comparable to that of a limb, it is not structureless. The organization depends on the arrangements of the tiny cytoplasmic filaments, just as in other cell "organs" the chemical nature of the local cytoplasm or the arrangement of its molecules may confer special properties upon it.

In conclusion, the organization, specialization and division of labour which occur in these simple organisms result from the different nature of the cytoplasm in certain regions and are not due (as in higher organisms) to the existence of various groups of special cells.

PRACTICAL WORK

1. *Sources.* (a) Stagnant, green-looking water in puddles, ditches, drinking troughs and roof gutters usually has abundant microscopic organisms containing many protista. Amoeba may be collected in the skimmings of mud and humus from sluggish streams. (b) Hay infusions. Chopped hay is boiled with rain-water, filtered and the filtrate left to stand in an open jar. Bacterial spores germinate in the fluid. If fresh hay or soil is added, any micro-organisms present may feed on the bacteria or single-celled algae which appear, and so flourish and reproduce.

2. *Culture methods.* The procedure described above gives rise to very mixed colonies of protista. If a colony of only one group is required, a few individuals must be taken up in a sterilized pipette and transferred to a clean dish with filtered rain-water and given suitable food material. To keep the creatures in a state of active reproduction a few individuals must be transferred to a fresh culture each week as the old cultures tend to degenerate.

3. *Suitable food.* A few wheat grains, boiled to prevent germination and placed in the water, provide a suitable basis for the kind of food chain in which Amoeba can take its place. Rain-water is better than tap-water since it is not likely to be too acid or alkaline or to contain harmful chemicals.

The hay infusion, with added fresh hay, will usually give rise to bacteria on which Paramecium will feed.

Water boiled with guano provides a suitable medium for Euglena and other flagellates.

33 Chromosomes and Heredity

MOST living organisms start their existence as a single cell, a *zygote* (fertilized egg, p. 112). This single cell divides into two cells, four, eight and so on to produce eventually the thousands of cells which make up the new organism.

If the new cells were all the same they would produce only a mass of structureless tissue. The cells, however, become different from each other in structure and function. Therefore, in the cell divisions which turn a zygote into an organism there must be forces directing the process which determine that some cells become muscle, some skin, some bone or blood and these cells must be directed into groups in the right order and in the right place to produce tissues, organs and ultimately the complete, integrated, co-ordinated organism.

Moreover, a zygote does not produce just any organism. It will produce one which resembles the parents from whom the zygote was derived. The zygotes of a human and a rabbit may look identical, a nucleus surrounded by a little cytoplasm, but they will develop in quite different ways. The rabbit zygote will produce a rabbit and not a man. A chaffinch zygote will produce a chaffinch and not a robin. The study of the mechanism by which the characteristics of the parents are handed on to the offspring is known as genetics.

The zygote of a chaffinch develops into a chick inside the egg without any outside interference other than incubation. It follows therefore, that the "instructions" for building a chaffinch chick from a single-celled zygote must reside somewhere inside the zygote. What is more, the "instructions" must be present in the two gametes which fuse to form the zygote. They could be in the cytoplasm, the nucleus or both.

When one examines the gametes of most animals, the egg usually has a relatively large volume of cytoplasm associated with its nucleus. The male gamete, on the other hand, consists of little more than a nucleus with a very thin layer of cytoplasm round it, and a tail. However, there is nothing to suggest that the male's contribution to the "instructions" or *genotype* of the zygote is any less than the female's so it looks as if the bulk of the genotype, the "blue-print" for building a rabbit or a man, resides in the nucleus.

The next question is, can the "blue-print" be seen or studied in some way? One would not expect to see printed directions but there might be structures to be seen which would give some idea about the nature of the "instructions". Thus, a study of the nucleus would seem to be the most profitable course, particularly at a time when the nucleus is dividing because the genetic information, the genotype, must be handed on intact and undiminished to each cell. For example, if the two cells resulting from the first division of a frog's zygote (p. 165) are separated, each cell can develop into a complete frog. This is true of cells even at the 4- and 8-celled stage. Thus the "frog-building blue-print" is intact and complete in each of these cells after cell division.

Thus, if one could observe a structure or structures, reproduced exactly in the nucleus and shared equally between the two nuclei at cell division, this might give a clue to the site of the genetic information. The next section, consequently, examines the events in the nucleus at cell division in some detail.

Cell division

In the early stages of growth and development of an organism, all the cells are actively dividing to produce new tissues and organs. Later, particularly when the cells become specialized, this power of division is lost and only a limited number of unspecialized cells retain the power of division, e.g. cambium cells in plants and cells of the Malpighian layer in the skin which produce new epidermis. In those cells that continue to divide, the sequence of events leading to cell division is basically the same. Firstly, the nucleus divides into two and then the whole cell divides, separating each nucleus in a unit of cytoplasm, so that two cells now exist where previously there was only one. Both cells may then enlarge to the size of the parent cell. Such cell division and enlargement gives rise to growth.

The detailed sequence of events which takes place when the nucleus of a cell divides has been worked out over the last eighty years and is called *mitosis*.

Mitosis (Fig. 33.2 and Plate 64)

Prior to division, the nucleus of the cell enlarges and in the nucleus there appear a definite number of fine, coiled, thread-like structures called *chromosomes* (Fig. 33.1). The behaviour of these chromosomes during cell division is usually described as a series of stages, prophase, metaphase, etc., though, in fact,

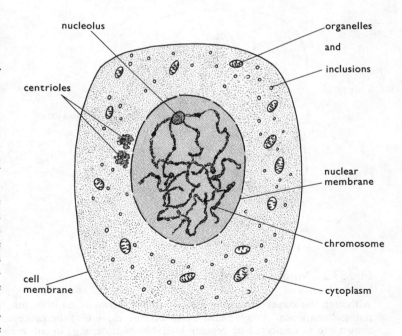

Fig. 33.1   Animal cell at early prophase

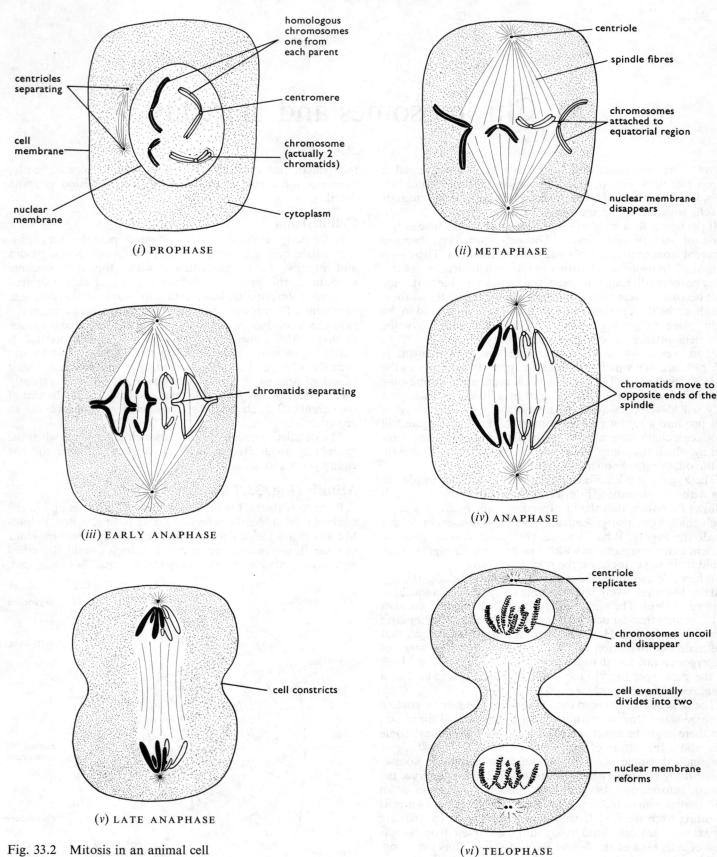

Fig. 33.2   Mitosis in an animal cell

Although the stages of mitosis are necessarily shown as static events, it must be emphasized that the process is a continuous one and the names "anaphase", "metaphase", etc., do not imply that the process of mitosis comes to a halt at this juncture. Moreover, the stages shown are not selected at regular intervals of time, e.g. in the embryonic cells of a particular grasshopper the timing at 38°C is as follows: **prophase** 100 min, **metaphase** 15 min, **anaphase** 10 min, **telophase** 60 min.

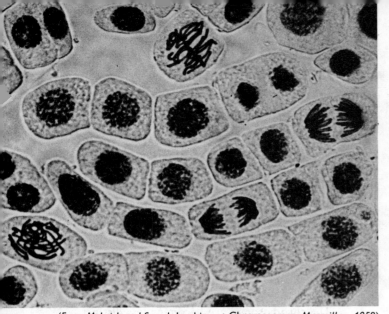

(From McLeish and Snoad, Looking at Chromosomes, Macmillan, 1958)

**Plate 64.   CELLS FROM A ROOT TIP SHOWING
MITOSIS (×500)**

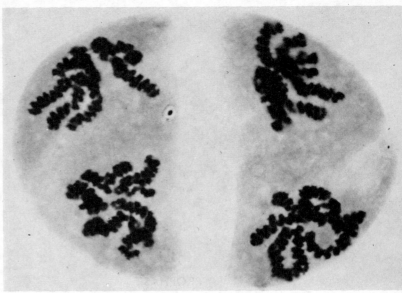

(A. H. Sparrow, Ph.D., Brookhaven National Laboratory, U.S.A.)

**Plate 65.   CHROMOSOMES OF *TRILLIUM ERECTUM* AT
TELOPHASE OF MEIOSIS (p. 188), SHOWING COILING
(×2000)**

the events occur in a smoothly continuous pattern and do not occupy equal periods of time (*see* p. 180).

1. **Prophase.** The chromosomes become more pronounced (that is, they react more readily to stains and chemical fixatives). They shorten and thicken (Plate 66*a*), probably by coiling like a helical spring, but with the coils so close to each other that they are not visible at low magnifications (Fig. 33.3 and Plate 65). The nuclear membrane dissolves, leaving the chromosomes suspended in the cytoplasm, and at the same time the one or more *nucleoli* disappear.

2. **Metaphase.** In the cells of animals and some of the simpler plants there is a pair of minute bodies called the *centrioles* which lie just outside the nucleus. At this stage they move away from each other and migrate to opposite ends of the cell. From a region near the centriole there radiate cytoplasmic fibres which meet and join near the centre of the cell. This system of "fibres" makes a web-like structure called the *spindle* (Fig. 33.2*ii*) and the chromosomes become attached by their centromeres (Fig. 33.3) to the equatorial region of the spindle. (Most plant cells do not have centrioles but a spindle is formed nevertheless.) By this time it is apparent that each chromosome consists of two parallel strands, called *chromatids* (Plate 66*b*), joined in one particular region, the *centromere*. In forming two chromatids, the chromosome *replicates*; that is, it produces an exact copy of itself but the two chromatids

remain in contact along their length. This replication has occurred before prophase but is more evident during metaphase.

3. **Anaphase.** The two chromatids now separate at the centromere and begin to migrate in opposite directions towards either end of the spindle (Plate 66*c*). Experiments show that the spindle fibres play some part in separating the chromatids. The appearance is that of the chromatids first repelling each other at the centromere and then being pulled entirely apart by the shortening spindle fibres, although such a mechanism has not yet been verified.

4. **Telophase.** The chromatids, now chromosomes, collect together at the opposite ends of the spindle (Plate 66*d*) and become less distinct, probably by becoming uncoiled and therefore thinner. The one or more nucleoli reappear, and a nuclear membrane forms round each group of daughter chromosomes so that there are now two nuclei present in the cell. At this point in animal cells, the cytoplasm between the two nuclei constricts, and two cells are formed. Both may retain the ability to divide, or one or both may become specialized and lose their reproductive capacity. In plant cells, the cytoplasm does not constrict to form two new cells; instead, a new cell wall is formed across the cell in the region originally occupied by the equatorial plane of the spindle (Fig. 33.4 and Plate 66*e*).

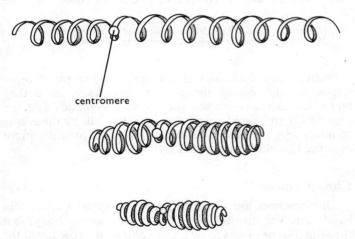

centromere

Fig. 33.3   Diagram illustrating how chromosomes appear to become thicker and shorter during prophase

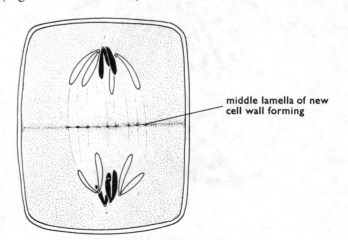

middle lamella of new
cell wall forming

Fig. 33.4   Late anaphase in a plant cell showing how separation of daughter cells differs from animal cells

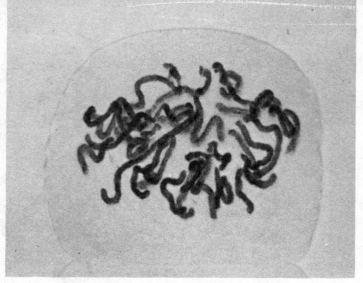

(a) PROPHASE
The chromosomes have become short and thick

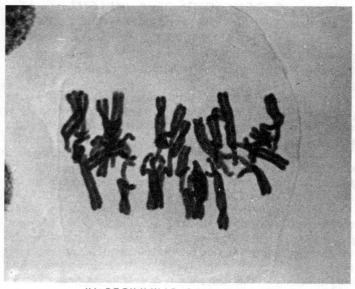

(b) BEGINNING OF ANAPHASE
Each chromosome is seen to consist of two chromatids which are attached to the equator of the spindle and are beginning to separate

(c) ANAPHASE
The chromatids have completely separated. The spindle is not visible in this photograph

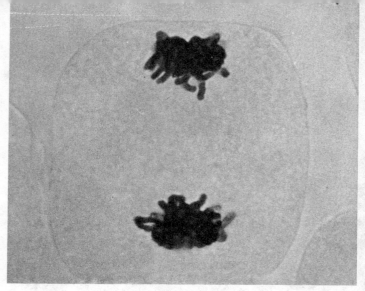

(d) TELOPHASE
The chromosomes are becoming less distinct

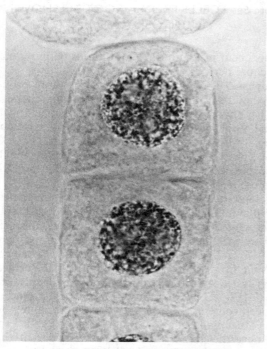

(e) THE END OF CELL DIVISION
The daughter nuclei are separated by a new cell wall

Plate 66. STAGES IN MITOSIS (×1800)

(Photomicrographs of cells from the root tip of Lilium regale from McLeish and Snoad, Looking at Chromosomes, Macmillan, 1958)

From a study of mitosis it seems very likely that the chromosomes are the site of the genetic instructions since they reproduce themselves when they form chromatids and the chromatids are shared equally between the cells by the events of mitosis. A further study of chromosomes provides more evidence that they carry genetic information.

Chromosomes

Chromosomes are so called because they take up certain basic stains very readily (*chromos*=colour, *soma*=body), but they can also be observed by phase contrast microscopy in the unstained nuclei of dividing, living cells. When the cell is not dividing, the chromosomes cannot be seen in the nucleus, even after staining. Nevertheless, it is thought that they persist as fine, invisible threads, isolated patches of which still respond to dyes and show up as flakes or granules of deeply staining material. The chromosomes consist of protein and a substance called *deoxyribonucleic acid* (DNA: *see* p. 186), but the exact relationship between these two components in forming the chromosome is not known.

Counts of chromosomes show that there is a definite number in each cell of any one species of plant or animal, e.g. mouse 40, crayfish 200, rye 14, fruit fly (*Drosophila*) 8, and man 46 (*see* also Fig. 33.5). This confirms our expectation that the chromosomes determine the difference between one species and another. It can also be seen (Plate 68) that the chromosomes exist in pairs, although not actually joined together, each pair having a characteristic length and, during anaphase, a characteristic shape (Plate 66c) governed by the position of the centromere at which the chromatids are pulled apart; e.g. a V shape if the centromere is central, or a √ shape if it is close to one end. In other words, human cell nuclei contain 23 pairs of chromosomes, mouse cells 20 pairs and so on, one member of each pair having been derived from the male and one from the female parent. The members of each pair are called *homologous chromosomes*, and the total number of chromosomes in each cell is the *diploid number*.

Although the constituent chemicals of the cytoplasm of a cell are constantly being broken down and rebuilt from fresh material, the chemicals of the chromosomes remain remarkably stable. Other investigations show that during cell division, no protoplasmic material is shared so exactly as that of the chromosomes in the nucleus. Such evidence points again to the chromosomes as the main source of the chemical information which determines that a cell should become like its parent cell, and that in their development, the cells of the organism will endow the animal or plant with all the characteristics of its species. Although the supporting evidence is very sketchily outlined here, the hereditary material must almost certainly lie on the chromosomes of the nucleus and be passed on to each daughter cell by the process of mitosis. In a similar way the nuclei of the gametes carry a set of chromosomes from the male and female parent, and these chromosomes determine that the zygote grows and develops to an animal or plant of the same species as the parents, reproducing to quite a minute degree, individual characteristics of both parents, e.g. flower colour or blood group.

(World Health)

Plate 67. STUDYING HUMAN CHROMOSOMES
A member of a team of scientists under Professor Jerôme Lejeune at the Institut de Progénèse, Paris, identifies and prepares pictures of human chromosomes

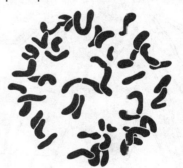

(a) Man [46]

(b) Kangaroo [12]

(c) Domestic fowl [36]

(d) Drosophila [8]

Fig. 33.5 **Chromosomes of different species**   (*From C. C. Hurst*, The Mechanism of Creative Evolution, *Cambridge University Press, 1933*)

Plate 68.   PREPARING A "KARYOGRAM"

(World Health)

The chromosome silhouettes from the photomicrograph on the left are cut out and arranged in order on the right-hand chart

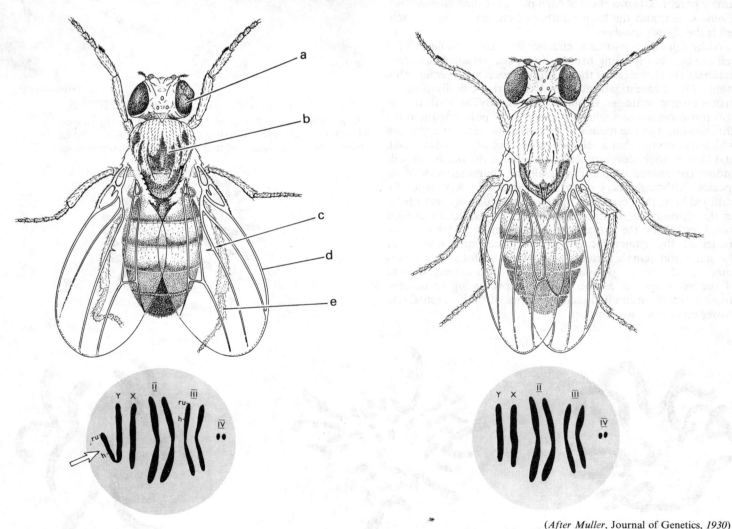

(After Muller, Journal of Genetics, 1930)

The cells of the fruit fly on the left have an extra segment of chromosome no. 3 which has become attached to the Y chromosome. As a result, this fly has (a) mis-shapen eyes, (b) dark patterned thorax, (c) imperfect cross veins, (d) broad wings, (e) incurved hind legs. The normal fly is shown on the right.

Fig. 33.6   Effects of a chromosome mutation in Drosophila

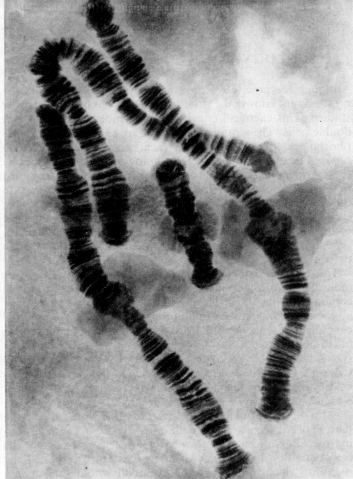

(Courtesy of Professor Wolfgang Beermann, Max Planck Institute, Tübingen, from Sci. Amer., April 1964)

Plate 69. GIANT CHROMOSOMES (×600)
Four giant chromosomes from a cell in the salivary gland of the midge larva, *Chironomus tentans*, showing transverse banding.

The fruit fly, *Drosophila melanogaster*, which has four pairs of chromosomes in its nuclei is, for various reasons, a suitable subject for study. By breeding many hundreds of flies through many generations, geneticists have become very familiar with the detailed anatomy of the fly and the appearance of its chromosomes. It has been observed on many occasions that some unexpected change in the external appearance of a fly is associated with a change in the chromosome pattern (*see* Fig. 33.6). This chromosome aberration must take place at an early stage in the development of the fly for it to affect so many parts of the body, or it may have occurred in one of the gametes from which the zygote was formed.

Cells from the salivary glands of Drosophila and other flies have very large chromosomes called giant chromosomes, on which bands can be seen (Plate 69). The size, shape and position of these bands is quite consistent and characteristic for any pair of chromosomes. If, due to some accident in replication, one or other of the bands is lost, there is a corresponding malformation in the adult fly (Fig. 33.7). Although the bands can be seen on only these rather unusual, giant chromosomes, it is thought that they represent the site of genes, or gene activity on all the chromosomes in the body.

Genes. A gene is a theoretical unit of inheritance, theoretical in the sense that the word was coined long before chromosome structure was investigated in detail or the DNA theory of

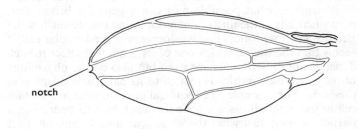

notch

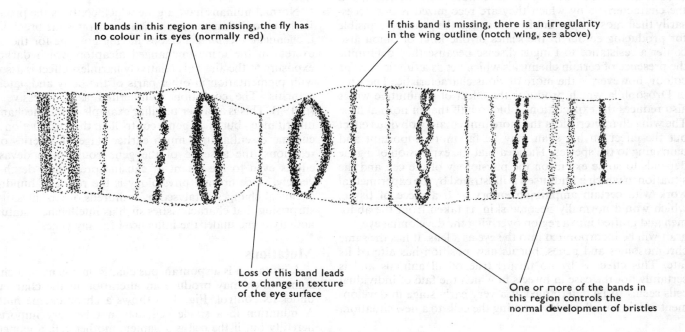

If bands in this region are missing, the fly has no colour in its eyes (normally red)

If this band is missing, there is an irregularity in the wing outline (notch wing, see above)

Loss of this band leads to a change in texture of the eye surface

One or more of the bands in this region controls the normal development of bristles

Fig. 33.7 Part of salivary gland chromosome of Drosophila showing location of four genes

(*From Curt Stern*, Principles of Human Genetics, *2nd edn., W. H. Freeman & Company, 1960. After Slizynska*, Genetics, *1938, 23*)

inheritance put forward. The gene is one of the "words" in the genetic "instructions" (*see* p. 179). For example one gene will specify whether the rabbit is to have black fur or white fur. Another gene will determine whether the fur is long or short. Today the gene is thought to consist of a sequence of bases on the DNA molecule in a chromosome.

Gene function

The picture that emerges from this and other evidence is that the genes which determine the characteristics of the organism are somehow arranged in line down the chromosome. These genes control the production of enzymes which in turn determine what functions go on in a cell, and eventually in the organs and entire organism. If anything happens to a gene it will affect the organism, e.g. in mice there is a gene which determines that the coat will be coloured. If this gene is missing, the mouse will be without pigment; it will be white with pink eyes. In this case, as in many others, more than one gene will in fact play a part in determining the characteristic.

The number of genes in man is not known but it could be about 1000 per chromosome. At mitosis, each chromosome, and therefore each of any of the genes it carries is exactly reproduced.

Two problems arise from this account. Since every cell in the body carries an identical set of chromosomes and since cell structure and function are determined by the genes on the chromosomes, why is not every cell of the body identical? Furthermore, what possible part can a gene for brown eyes play when it is in the nucleus of a cell lining the stomach wall? Briefly, when we follow the development of a particular cell, it seems that the way in which one of its genes will affect the cell depends not only on the gene itself but also on the physiology of the cell, which in turn is related to its particular position in the body. For example, the chemical environment in a certain cell in the scalp allows the gene for black hair to operate in a particular way. Just what the same gene does in another part of the body is not certain, its action may simply be suppressed, but it is known that most genes have more than one effect, and the characteristic by which they are recognized is not necessarily their most important function; e.g. the genes responsible for producing colour in the scales of one kind of onion also confer a resistance to fungus disease because they determine the presence of certain chemicals which act as a fungicide. The colour, however, is the more obvious characteristic. The gene in Drosophila which produces the effect of diminutive wings also reduces the expectation of life to half that of normal flies. The wing characteristic is the more immediately obvious effect but the effect on life span may be far more important and damaging to the species. The idea that the expression of a gene depends to some extent on the physiology of the cell and the situation in which it finds itself is illustrated by the experimental work with certain amphibian embryos. If a piece of tissue, which would normally become skin, is taken from the abdomen and grafted into a region overlying the developing eye, the graft will be incorporated into the eye as a lens. It has the same chromosomes and genes, but its new position has altered its fate. This effect is by no means true of all animals and is certainly not the case in insects in which the fate of individual cells seems to be determined at a very early stage in development and is not affected by moving the cells to a new situation.

How genes work

It was mentioned earlier (p. 183) that chromosomes consist of protein and a nucleic acid, deoxyribonucleic acid (DNA).

Although the precise relationship between the DNA and the protein is not known, the structure of DNA has been intensively studied. This chemical consists of long molecules coiled in a double helix. The strands of the helix are chains of sugars and phosphates, the sugar being a 5-carbon compound, *deoxyribose*. The two helices in a DNA strand are linked together by cross-bridges made by pairs of organic nitrogenous bases joined to the sugar molecules (Figs. 33.8 and 33.9). Although there are only four principal kinds of base in the DNA molecule, *adenine*, *cytosine*, *thymine* and *guanine*, it is thought that the sequence of these bases is the important factor in heredity, and that a gene may consist of a particular sequence of up to 1000 base pairs in a DNA molecule.

The different sequence of bases *along the length* of the DNA molecule seem to act like a code, instructing the cell to make certain proteins, the order of bases indicating the sequence of amino acids to be joined up in order to make the protein. For example, the sequence CAA (cytosine-adenine-adenine) specifies the amino acid *valine*; three thymines in a row, TTT, specify *lysine*, while AAT specifies *leucine*. So the sequence of bases CAA-TTT-AAT would direct the cell to link up the amino acids valine-lysine-leucine to make the appropriate peptide, and in a similar way proteins are formed.

Most of the proteins made are enzymes which direct the pattern of chemical activity in the cell. Thus DNA, by determining the kinds of enzyme formed in the cell, will control the cell's activities. This in turn will effect the nature of the cell, the organ of which it is a part and eventually the organism as a whole. A change in the sequence of bases in the DNA molecule will result in a different order of amino acids and, hence, a different and probably ineffective enzyme. This will usually act adversely on the metabolism of the cell.

A rabbit with coloured fur has a gene which controls the production of the enzyme *tyrosinase*. This enzyme converts *tyrosine*, a colourless amino acid, to *melanin*, a black pigment. An albino rabbit has no gene for tyrosinase production, and consequently no pigment is formed from the tyrosine in its body.

Normal humans have a gene which controls the production in the blood of an enzyme which accelerates the breakdown of a chemical, *alcapton*. Persons having no gene for the enzyme excrete in the urine unchanged alcapton which darkens on exposure to the air. This relatively harmless effect is associated with pigmentation in other parts of the body and, later, with arthritis. The condition is inherited as a recessive factor (p. 194). This is a rather peculiar example of the mechanism of inheritance, but if a gene controlled the production of an enzyme essential in a much earlier stage of a series of vital reactions, the absence of the gene could have devastating effects even to the extent of causing premature death. Conversely, since normal physiology is the result of hundreds of chemical changes catalysed by hundreds of enzymes, it is not surprising that characteristics such as intelligence, stature and activity come under the influence of many genes.

Mutations

A mutation is a spontaneous change in a gene or a chromosome which may produce an alteration in the characteristic under its control. Fig. 33.6 shows a chromosome mutation. A mutation in a single cell may not be very important in heredity but if the cell is a gamete mother cell, a gamete or a zygote, the entire organism arising from this cell may be affected. Since the mutant form of the gene is inherited in the usual way, the mutation may persist in subsequent generations.

On the whole, genes are stable structures because DNA is a stable chemical, but once in a hundred thousand replications or more a gene may mutate. The frequency with which certain genes mutate has been estimated for many abnormalities in man and in experimental animals. Most mutations that produce an observable effect seem to be harmful if not actually lethal. This is not surprising, since any change in a well- but delicately-balanced organism is likely to upset its physiology. Most mutations however, are recessive* and so may not find expression in the heterozygous* form.

In humans, a form of dwarfism arises as a result of a dominant mutation having a frequency of about 1 in 20,000. A fairly frequent form of mental deficiency known as mongolism results from a chromosome mutation in which the ovum carries an extra chromosome so that the child has 47 chromosomes in his cells instead of 46.

Radiation and mutations. The cause of mutation is not known, but exposure to X-radiation, gamma-radiation, ultra-violet light, etc., is known to cause an increase in the mutation rate in experimental animals such as fruit flies and mice. The artificially-induced mutations are the same as those which occur naturally, but the frequency with which they occur is greatly increased.

There is a fairly constant background of radiation on the Earth's surface as a result of cosmic rays. Individuals also receive radiation from X-rays used in medicine, television tubes and luminous watch dials. Workers in atomic power stations and other people handling radioactive materials in industry or research may receive additional radiation. The radioactive fall-out from atomic explosions has increased the background radiation.

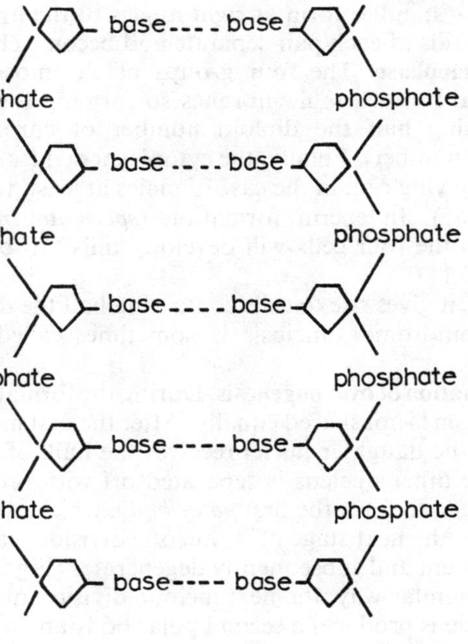

It is of obvious importance to assess the effect of any increase in radiation on the health of individuals and, as a result of mutations in their reproductive cells, the health of their children.

There is, so far, insufficient information to determine the correlation between the radiation dose and the mutation rate in man. The maximum safe dose in respect of direct effects on the individual, e.g. leukaemia, is still a matter of controversy. Nevertheless it is probably safe to say that any increase in the mutation rate is likely to have harmful effects on the population. Consequently, the exposure of individuals to the hazards of radiation is limited, though somewhat arbitrarily, by law.

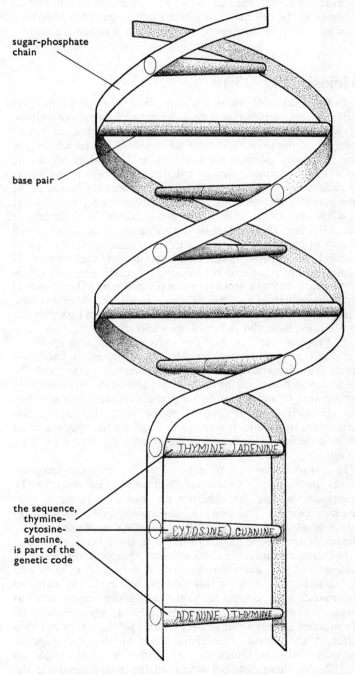

Fig. 33.9 The drawing shows schematically part of a DNA molecule. The lower part is shown uncoiled to emphasize the position of the base pairs

Fig. 33.8 Part of a DNA molecule

($\bigcirc = deoxyribose$)

* See p. 194

Variation

The exact replication of chromosomes and genes and their equal distribution between cells at mitosis produces conformity; the organism breeds true to type, e.g. a herring reproduces herring and not whiting or haddock. Nevertheless, the offspring will differ in many respects from its brothers and sisters and from its parents. It is possible for two black mice to have some white babies as well as black ones. Both your parents could have had brown eyes and yet you could have blue eyes. Sometimes variations arise from gene or chromosome mutation but these are infrequent and usually harmful. The variations of, for example, eye and hair colour result from rearrangement of parental genes and chromosomes in the zygote. This rearrangement is the direct result of the way in which the chromosomes separate at the cell division which leads to gamete formation. This sequence of chromosome separation is called meiosis.

Meiosis (Fig. 33.10)

Cells in the reproductive organs, which are going to form gametes, e.g. sperms and ova, undergo a series of mitotic divisions resulting, in the case of male gametes, in a vast increase of numbers. The final divisions, however, which give rise to mature gametes are not mitotic. Instead of producing cells with 46 chromosomes in man, they form gametes with only 23 chromosomes. When, at fertilization, there is a fusion of the two gametes, the resulting zygote contains the diploid number of 46 chromosomes, and this number is present in all the cells of the offspring. The halving of the chromosome number which occurs at gamete formation, ultimately maintains the diploid number of chromosomes characteristic of the species. If gametes were produced by mitosis, a human egg and sperm would each contain 46 chromosomes and when they fused at fertilization would give rise to a zygote with 92 chromosomes. The gametes from the resulting organism would in turn give rise to offspring with 184 chromosomes and so on.

1. **Prophase.** In meiosis, the chromosomes appear in the nucleus (i.e. they become visible when the cell is fixed and stained or observed by phase contrast microscopy) in much the same way as described for mitosis, but although it is reasonably certain that two chromatids are present in each chromosome, the chromosomes still appear to be single threads. Other differences from mitosis seen at this stage are the appearance of chromomeres and the failure of the chromosomes to shorten by coiling.

In complete contrast to mitosis, the homologous chromosomes now appear to *attract* each other and come to lie alongside so that the centromeres and chromomeres correspond exactly. The pairs of chromosomes so formed are called *bivalents*, e.g. a cell with a normal complement of six chromosomes would have, at this stage, three bivalents.

In this paired state the chromosomes shorten and thicken by coiling, and now each chromosome is seen to consist of two chromatids. As soon as this occurs, however, the pairs of chromatids seem to repel each other and move apart, except at certain regions called *chiasmata* (Plate 70). In these regions the chromatids appear to have broken and joined again but to a different chromatid. The significance of this exchange of sections of chromosomes, or "crossing over", is discussed on p. 192. All these changes occur during prophase while the nuclear membrane is still intact.

2. **Metaphase.** The nuclear membrane disappears, a spindle is formed and the bivalents approach the equatorial region.

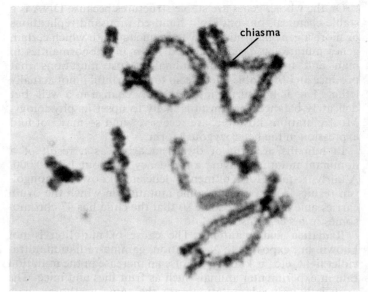

(From colour slide set, Meiosis in Chorthippus brunneus published by Harris Biological Supplies Ltd)

Plate 70.   MEIOSIS IN GRASSHOPPER TESTIS
(late prophase) (× 2000)

3. **Anaphase.** The paired chromatids of each bivalent now continue the separation that began in prophase and move to opposite ends of the spindle in a manner superficially similar to that of chromatids in mitosis. The outcome is that only half the total number of paired chromatids reaches either end of the spindle. Thus although there may originally have been six chromosomes in the nucleus, there are now only three paired chromatids at either end of the spindle.

4. **Second meiotic division.** A nuclear membrane does not usually form round the paired chromatids at this stage. Instead, two new spindles form at right angles to the first one and the chromatids of each pair separate and become chromosomes.

5. **Telophase.** The four groups of chromosomes are now enclosed in nuclear membranes so forming four nuclei, each containing half the diploid number of chromosomes (the *haploid* number). Finally, the cytoplasm divides to separate the nuclei, giving rise, in the case of males at least, to four gametes (Plate 65). In sperm formation (*spermatogenesis*) in most animals the four cells will develop "tails" to become sperms (Fig. 33.11).

Since it gives rise to cells containing half the diploid number of chromosomes, meiosis is sometimes called the *reduction division*.

Formation of ova: oogenesis. During the formation of ova, the cytoplasm is not shared equally. After the first meiotic division, one of the daughter nuclei receives the bulk of the cytoplasm and the other nucleus is separated off with only a vestige of cytoplasm to form the first *polar body*, which, although it may undergo the next stage of its meiotic division, cannot function as an ovum and subsequently degenerates (Fig. 33.11).

In a similar way, the next meiotic division of the remaining egg nucleus produces a second polar body and a mature ovum. In many vertebrates, the first polar body is not formed until after the potential ovum is released from the ovary, and the second polar body, until after the penetration of the sperm in fertilization. In man, when a sperm bearing 23 chromosomes fuses with a 23-chromosome ovum, a 46-chromosome zygote is formed.

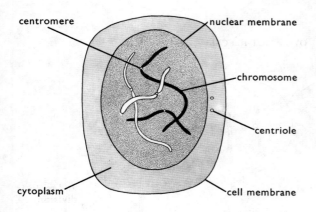

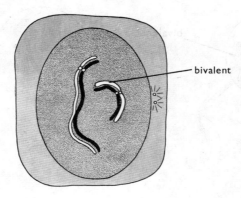

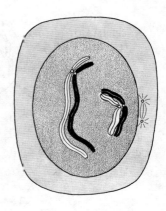

Prophase

(*a*) The diploid number of chromosomes appear

(*b*) Homologous chromosomes pair with each other, shorten and thicken

(*c*) Replication has occurred and the chromatids become visible

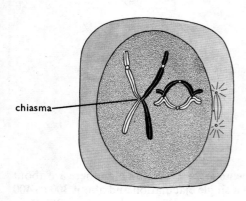

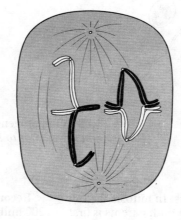

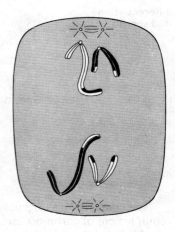

(*d*) Homologous chromosomes move apart except at the chiasmata where the chromatids have exchanged portions

Anaphase

(*e*) A spindle forms and homologous chromosomes move to opposite ends taking the exchanged portions with them

(*f*) Homologous chromosomes separated but not enclosed in nuclear membranes

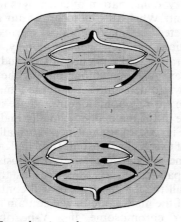

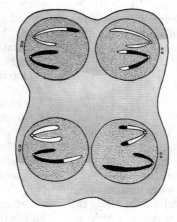

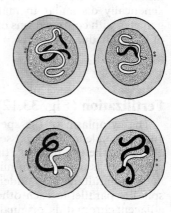

Second meiotic division

(*g*) Spindles form at right angles to the first one and the chromatids separate

Telophase

(*h*) Four nuclei appear, each enclosing the haploid number of chromosomes

(*i*) Cytoplasm divides to form four gametes

Fig. 33.10   Meiosis in a gamete-forming cell

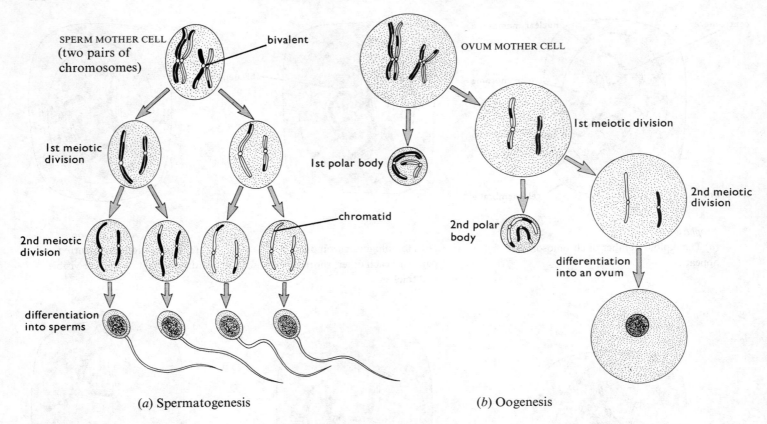

(a) Spermatogenesis   (b) Oogenesis

Fig. 33.11   Gametogenesis
(diagrammatic and not to scale)

New combinations of genes in the gametes. In mitosis the full complement of chromosomes derived from both parents is first doubled and then shared equally between daughter cells. The result of this replication is that the daughter cells receive identical genetic information. In meiosis, on the other hand, the genetic information on the chromosomes is not shared in exactly the same way between all the gametes. If homologous chromosomes were identical in their gene content, this variability of chromosome distribution would have no effect. Since, however, an individual's parents are likely to be genetically dissimilar in many respects, there will be many gametes with combinations of genes quite different from either of the individual's parents.

Fertilization (Fig. 33.12)

The cytoplasm of the sperm fuses with that of the ovum and the male nucleus passes into the ovum, coming to lie alongside the egg nucleus: the zygote is formed, but in many cases there is no fusion of nuclear material at this stage. Each nucleus simultaneously undergoes a mitosis with the axes of the spindles parallel to each other, but at the telophase stage the adjacent chromatids, originally from different parents, become enclosed in the same nuclear membrane, thus restoring the diploid number of chromosomes. These events are followed by the first cleavage, the zygote dividing into two cells. Subsequent mitotic division produces a multicellular organism with the diploid number of chromosomes in all its cells.

Recombination of genes in the zygote. In man there are about 200 million sperms in a single ejaculation and about 300 to 400 eggs produced in a reproductive lifetime. The eggs will have a genetic content differing from the sperms (e.g. wife with curly red hair; husband with straight black hair). When the chromosomes of the sperm and egg combine, the zygote may contain genes for red, black, curly and straight hair. The genes for curly and black are dominant (p. 194) to those for straight and red, so that although the child may carry all 4 genes, his hair will be curly and black—a new combination of characters not represented in either parent. When this child grows up and produces his own gametes, he may produce sperms with 4 alternative combinations of these genes; namely, red and straight, black and straight, red and curly, black and curly.

Determination of sex. In humans, one pair of the smallest chromosomes is known to determine sex. In the female, these two chromosomes are entirely homologous and are called the X chromosomes, while in the male, one is smaller and is called the Y chromosome (Fig. 33.5b). Femaleness normally results from the possession of two X chromosomes and maleness from possession of an X and a Y chromosome. At meiosis, the sex chromosomes are separated in the same way as the others (Fig. 33.13), so that all the female gametes will contain an X chromosome, but half the male gametes will contain an X and half will contain a Y chromosome. If a Y-bearing sperm fertilizes an ovum, the zygote will be XY and give rise to a boy. Fertilization of an ovum by an X-bearing sperm gives an XX zygote which develops to a girl. There should be an equal chance of X or Y sperm meeting an ovum, and therefore equal numbers of boy and girl babies should be born. In fact,

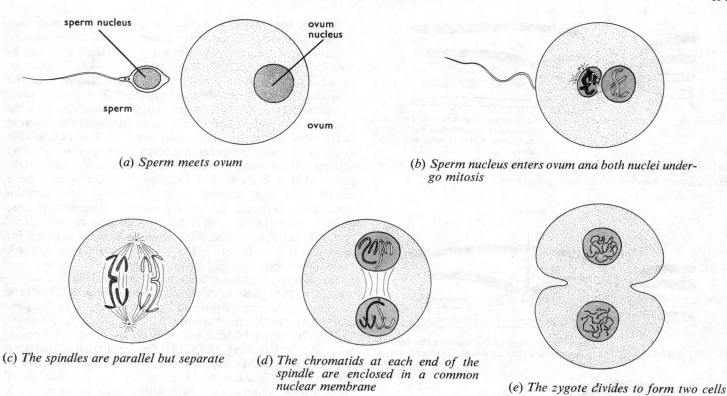

(a) *Sperm meets ovum*

(b) *Sperm nucleus enters ovum and both nuclei undergo mitosis*

(c) *The spindles are parallel but separate*

(d) *The chromatids at each end of the spindle are enclosed in a common nuclear membrane*

(e) *The zygote divides to form two cells*

Fig. 33.12   Fertilization (*polar bodies not shown*)

slightly more boys than girls are born in most parts of the world. The reason for this is not clear, but it also happens that the mortality rate for boy babies and men is slightly higher than for girl babies and women, which tends to restore the balance.

Although the X and Y chromosomes determine sex, it does not necessarily follow that male and female characteristics are determined by genes found only on the sex chromosomes. In man, genes for male and female characters may be scattered fairly evenly throughout all the chromosomes, but the presence of the Y chromosome in an XY zygote may tip the balance in favour of maleness. Femaleness results from the absence of the Y chromosome but this is not the case in all the animals studied; e.g. it is true for the mouse but not for Drosophila.

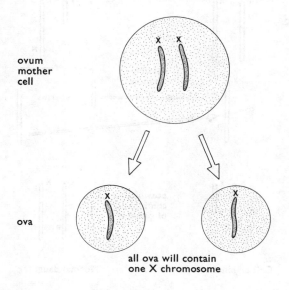

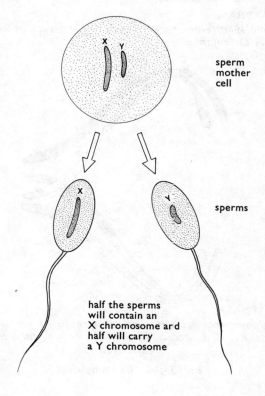

Fig. 33.13   Determination of sex (*diagrammatic only*)

Note (a) Only the X and Y chromosomes are shown.
   (b) The Y chromosome is not smaller than the X in all animals.
   (c) Details of meiosis have been omitted.
   (d) Theoretically, four gametes would be produced, but two are sufficient here to show the distribution of X and Y chromosomes.

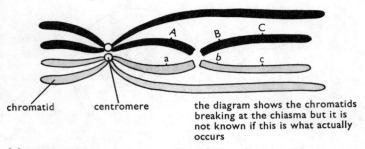

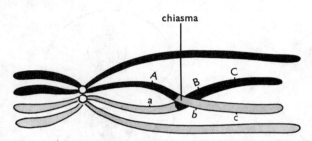

chromatid   centromere   the diagram shows the chromatids
breaking at the chiasma but it is
not known if this is what actually
occurs

(a) PROPHASE
Homologous chromosomes have paired up

chiasma

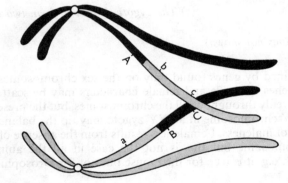

(b) PROPHASE   the terminal portions of the adjacent chromatids
have become attached to the opposite chromatid

(c) METAPHASE
*The homologous chromosomes seem to repel each other
except at the chiasma*

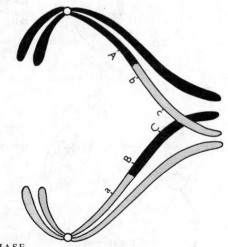

(d) ANAPHASE
*The chromosomes separate, but as a result of crossing over,
the genes A, B, C and a, b, c on the 'inner' chromatids are
rearranged*

Fig. 33.14   Crossing over

Linkage and crossing over. During that early stage of meiosis when homologous chromosomes pair up, the chromatids of maternal and paternal chromosomes exchange portions as mentioned on p. 188 and shown in Fig. 33.14. This leads to variability in the gene combinations in the gametes. In the absence of crossing over, the maternal genes A-B-C on the same chromosome would always appear together no matter how the chromosomes were assorted in meiosis. Similarly the paternal genes a-b-c would always remain together. For example, in Drosophila, since the genes for black body, purple eyes and vestigial wings occur on the same chromosome, one might expect that a black-bodied Drosophila would always have purple eyes and vestigial wings. Crossing over between chromatids, however, gives the possibility of breaking these *linkage groups* as they are called, so that new combinations, ABc, Abc, aBC, abC, aBc, AbC, could arise in the gametes, for example black body with red eyes and normal wings, or black body with purple eyes and normal wings.

Sex linkage. Certain genes which occur on the X chromosome are more likely to affect a male than a female. The gene or genes for a certain form of colour blindness in man are carried on the X chromosome. Normal vision is dominant (p. 194) to colour blindness so that if a colour-blind woman, who must be homozygous (p. 194) for the character, marries a normal man, all their sons but none of their daughters will be colour blind. This can be explained by the fact that the Y chromosome is homologous with only a small section of the X chromosome and the non-homologous part of the X chromosome carries genes which are not represented on Y. It is assumed that the Y chromosome plays no part in the determination of colour vision. Fig. 33.15 shows how this type of sex linkage produces its effect.

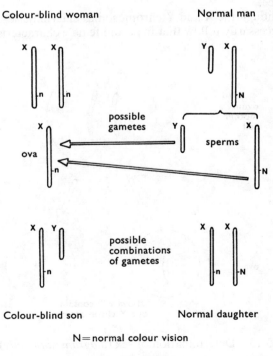

Fig. 33.15   SEX LINKAGE, showing the possible distribution of X and Y chromosomes between the gametes and the chances of combination in the zygotes

The normal daughter is heterozygous (*see* p. 194) for colour blindness and is therefore a "carrier" for the recessive gene. If she marries a normal man the possible combinations of genes in the children are shown by:

Parents:	XN Xn carrier woman	XN Y normal man
Gametes:	XN Xn	XN Y

Possible combinations of gametes:

XN XN normal girl	Xn XN girl carrier	XN Y normal boy	Xn Y colour-blind boy

The theoretical expectations are that all the girls will be normal but half of them will be carriers, half the boys will be normal and half of them colour blind. The types of children expected from the marriage between a woman carrier and a colour-blind man, or a normal woman and a colour-blind man can be worked out in a similar way.

Other X-linked factors are *haemophilia* and brown enamel on the teeth. Haemophilia causes a delay in the clotting time of the blood. Although there are two kinds of sex-linked haemophilia, at least three other clotting disorders are known which are controlled by genes not on the sex chromosomes.

Sexual characteristics such as bass voice, beard and muscular physique in males, mammary glands and wide pelvis in females, are not the result of sex-linked genes but the different expressions of the same genes present in both sexes. Both sexes carry genes controlling the growth of hair, mammary glands and penis but in the physiological environment of maleness or femaleness they have different effects, with the result that e.g. the mammary glands in males are small and functionless; the penis in females is represented by only a small organ, the *clitoris*.

Only a few rare abnormalities are thought to be linked to the Y-chromosome and even these are now open to doubt.

PRACTICAL WORK

1. *Squash preparation of chromosomes using acetic orcein.*
Material. *Allium cepa* (onion) root tips. Support onions over beakers or jars of water using tooth-picks as shown in Fig. 33.16. Keep the onions in darkness for several days until the roots growing into the water are 2–3 cm long. Cut off about 5 mm of the root tips, place them in a watch glass and
(*a*) cover them with 9 drops acetic orcein and 1 drop molar hydrochloric acid;
(*b*) heat the watch glass gently over a very small Bunsen flame till steam rises from the stain, but do not boil;
(*c*) leave the watch glass covered for at least five minutes;
(*d*) place one of the root tips on a clean slide, cover with 45 per cent acetic (ethanoic) acid and cut away all but the terminal 1 mm;
(*e*) cover this root tip with a clean cover-slip and make a squash preparation as described below.
Making the squash preparation. Squash the softened, stained root tips by lightly tapping on the cover-slip with a pencil: hold the pencil vertically and let it slip through the fingers to strike the cover-slip. The root tip will spread out as a pink mass on the slide; the cells will separate and the nuclei, many of them with chromosomes in various stages of mitosis (because the root tip is a region of rapid cell division) can be seen under the high power of the microscope ($\times 400$).

2. *Meiosis*
To see chromosomes in meiosis, the same technique is adopted using material where gamete mother cells are giving rise to gametes, e.g. the developing anthers of bluebell flowers, where pollen grains are being produced.
Dig the bulbs early in the year when the leaves just appear above the ground and dissect out the inflorescences, fixing them straight away in Clarke's fluid for 24 hours. Select flowers of a suitable age, i.e. those higher up on the stalk than the ones in which the anthers are beginning to turn yellow. Dissect out the anthers and stain them in acetic orcein as before.

Preparation of reagents
(i) **Clarke's fluid.** 25 cm³ glacial acetic (ethanoic) acid; 25 cm³ ethanol.
(ii) **Acetic orcein.** 2 g orcein; 100 cm³ glacial acetic (ethanoic) acid. Dilute a small portion with an equal volume distilled water just before use.
(iii) **Molar hydrochloric acid** (i.e. one mole of HCl per litre). Make up 87·3 cm³ of concentrated acid to 1 litre by adding distilled water.

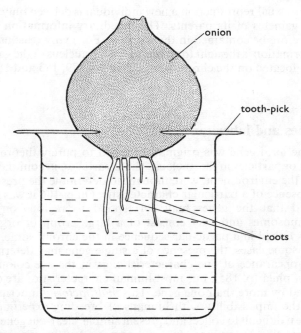

Fig. 33.16   Method of supporting an onion to promote growth of roots

QUESTIONS

1. Sometimes, at meiosis, the bivalent chromosomes fail to separate properly with the result that the gamete so formed contains the diploid number of chromosomes. If such a diploid sperm were to fertilize a normal monoploid ovum, (*a*) what effect would you expect this to have on the zygote and offspring and (*b*) supposing the zygote grew into a normal individual, what might happen when the individual produced gametes by meiosis?
2. A horse and a donkey are related closely enough to be able to reproduce when mated together. The offspring from this mating is a mule and though healthy in all other respects is sterile. Suggest an explanation, to do with chromosomes and meiosis, for this phenomenon.
3. It is possible for a cross between a short-winged, grey-bodied Drosophila and a normal-winged, black-bodied Drosophila, to produce some short-winged, black-bodied offspring. Explain how this could happen (*a*) if the genes for body colour and wing size are on different chromosomes and (*b*) if these genes are on the same chromosome.

34 Heredity and Genetics

FROM its parents an individual inherits the characteristics of the species, e.g. man inherits highly developed cerebral hemispheres, vocal cords and the nervous co-ordination necessary for speech, a characteristic arrangement of the teeth and the ability to stand upright with all its attendant skeletal features. In addition, he inherits certain characteristics peculiar to his parents and not common to the species as a whole, e.g. hair and eye colour, blood group and facial appearance. The study of the method of inheritance of these "characters" is called genetics.

In sexual reproduction a new individual is derived only from the gametes of its parents. The hereditary information must therefore be contained in the gametes. For many reasons, this information is thought to be present in the nucleus of the gamete and located on the chromosomes (see pp. 179, 183 and 185).

Genes and inheritance

The term gene was originally applied to purely theoretical units or particles in the nucleus. These particles, in conjunction with the environment, were thought to determine the presence or absence of a particular characteristic. On p. 186 it was suggested that the genes may correspond to regions on the chromosomes and may consist of a large group of organic bases linked in a particular sequence in the chromosome.

In some cases, the presence of a single gene may determine the appearance of one characteristic, as in the eye colour of Drosophila (p. 185), but most human characteristics are controlled by more than one gene. This *multifactorial* inheritance and the impossibility with humans of breeding experiments, make it difficult to collect and present simple, clear-cut genetical information about man. In order to provide some clear ideas about heredity, simple cases amongst other animals will first be considered.

Single-factor inheritance. If a pure-breeding i.e. homozygous (see below) black mouse is mated with a pure-breeding brown mouse, the offspring will not be intermediate in colour i.e. dark brown or some combination of brown and black, but will all be black. The gene for black fur is said to be *dominant* to that for brown fur because, although each of the baby mice, being the product of fusion of sperm and egg, must carry genes for both blackness and brownness, only that for blackness is expressed in the visible characteristics of the animal. The gene for brown fur is said to be *recessive*. The black babies are called the first filial or F_1 *generation*. If, when they are mature, these F_1 black mice are mated amongst themselves, their offspring, the F_2 *generation* will include both black and brown mice and if the total number for all the F_2 families are added up, the ratio of black to brown babies will be approximately 3 to 1. It must not be assumed, however, that if two black F_1 mice have 4 babies, 3 will be black and one brown. In a mating which produced, say, 8 babies, it would not be at all unusual to find all black, or 5

black to 3 brown etc. The ratio $3:1$ appears only when large numbers of individuals are considered.

The appearance of brown fur in the second generation is proof of the fact that the F_1 black mice carried the recessive gene for brown fur even though it did not find expression in their observable features.

In explanation, it will be assumed that a pure-breeding black mouse carries, on homologous chromosomes (see p. 183), a pair of genes controlling the production of black pigment. The genes are represented in subsequent diagrams (Fig. 34.1 *a* and *b*) by the letters BB, the capital letters signifying dominance.

In the same position on the corresponding chromosomes in brown mice are carried the genes bb for brownness. The genes B and b are called *allelomorphic genes* or *alleles*. During the formation of gametes the process of meiosis (p. 188) will separate the homologous chromosomes, so that the gametes will contain only one gene from each pair. All the sperms from the pure-breeding black parent will carry the factor B and all the eggs from the brown parent will carry the factor b. When the gametes fuse, the zygotes will contain both factors B and b but since B is dominant to b, only the former gene is expressed, i.e. the offspring will all be black.

When, later on, these black F_1 mice produce gametes, the process of meiosis will separate the chromosomes carrying the B and b factors (see Fig. 34.1b) so that half the sperms of the male parent will carry B and half will carry b. Similarly, half the ova from the female will contain B and half b. At fertilization there are equal chances that a B-carrying sperm will fuse with either an egg carrying the B gene or an egg with the b gene so producing either a BB or a Bb zygote. Similarly there are equal chances of a b-carrying sperm fusing with either a B- or a b-carrying ovum to give bB or bb zygotes.

This results in the theoretical expectation of finding, in every four F_2 offspring, one pure-breeding black mouse BB, one pure-breeding brown mouse bb, and two "impure" black mice Bb.

The separation at meiosis of the alleles B and b into different gametes is called *segregation*. The pure-breeding black (BB) and brown (bb) mice are called *homozygous* for coat colour and the "impure" black mice (Bb) are called *heterozygous*. The heterozygous mice will not breed true i.e. if mated with each other their litters are likely to include some brown mice. The homozygous BB mice mated together can produce only black offspring and the bb homozygotes only brown offspring.

Genotype and phenotype. The BB mice and Bb mice will be indistinguishable in their appearance i.e. they will both have black fur and they are thus said to be the same *phenotypes*, in other words they are identical in appearance for a particular characteristic, in this case blackness. Their genetic constitutions, or *genotypes*, however, are different, namely BB and Bb. In short, the black phenotypes have different genotypes.

To distinguish between the black phenotypes, the usual practice is to do a further breeding experiment called a *back-cross*.

SINGLE-FACTOR INHERITANCE

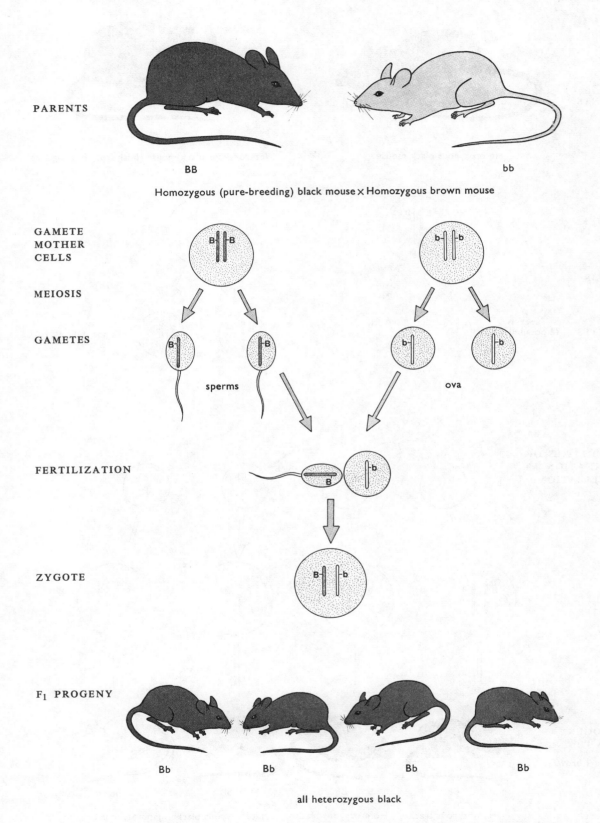

PARENTS

BB

bb

Homozygous (pure-breeding) black mouse × Homozygous brown mouse

GAMETE
MOTHER
CELLS

MEIOSIS

GAMETES

sperms

ova

FERTILIZATION

ZYGOTE

F₁ PROGENY

Bb

Bb

Bb

Bb

all heterozygous black

Fig. 34.1(a) Inheritance of a single factor for coat colour in mice

SINGLE-FACTOR INHERITANCE

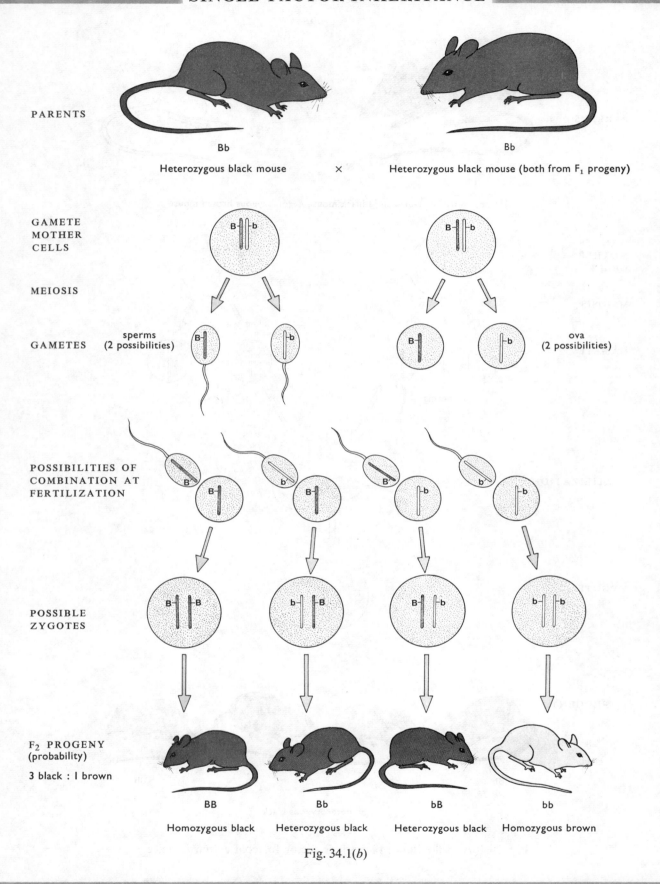

PARENTS

Bb
Heterozygous black mouse

×

Bb
Heterozygous black mouse (both from F₁ progeny)

GAMETE
MOTHER
CELLS

MEIOSIS

GAMETES   sperms
(2 possibilities)

ova
(2 possibilities)

POSSIBILITIES OF
COMBINATION AT
FERTILIZATION

POSSIBLE
ZYGOTES

F₂ PROGENY
(probability)

3 black : I brown

BB
Homozygous black

Bb
Heterozygous black

bB
Heterozygous black

bb
Homozygous brown

Fig. 34.1(b)

The back-cross. To discover their genotypes, the black F_2 phenotypes are each mated with mice of the same genotype as their brown grandparent i.e. the homozygous recessive, bb. Half the gametes from the heterozygous black mice Bb will carry the B gene and half will carry the b gene. The gametes from the homozygous black mouse, BB, will all carry the B gene. Similarly, the gametes of the homozygous recessive brown mouse will all carry the b gene. Thus, when the black parent is heterozygous, one would expect the back-cross to yield approximately equal numbers of black and brown babies in the litters but if the black parent was homozygous, the babies must all be black since they receive the dominant gene for blackness, B, from this parent (Fig. 34.2).

Incomplete dominance (*codominance*). If red Shorthorn cows are mated with white Shorthorn bulls the coats of the calves carry both red and white hairs, giving a *red roan*. Neither the red nor the white factor is dominant.

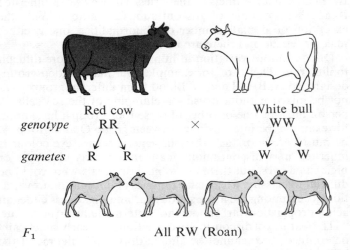

Red cows and bulls when mated together will breed true, i.e. all their offspring will have red coats. White cows and bulls, similarly, are homozygous and will breed true. The F_1 roan cattle, however, are heterozygous and will not breed true; their progeny will include red calves and white calves as well as roans.

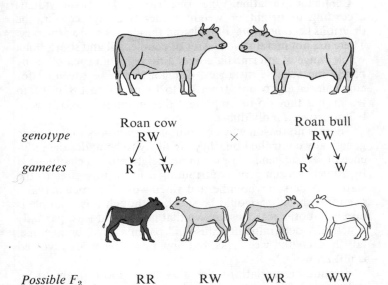

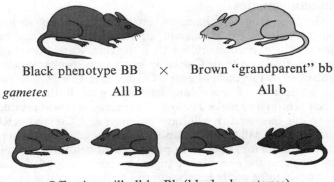

Offspring will all be Bb (black phenotypes)

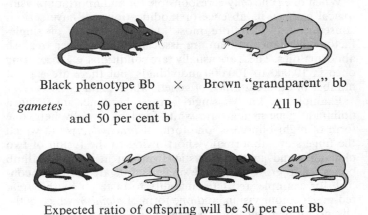

Expected ratio of offspring will be 50 per cent Bb (black) and 50 per cent bb (brown)

Fig. 34.2   The back-cross

The inheritance of the ABO blood groups in man is an instance of incomplete dominance. According to whether their blood will mix without clotting during a transfusion, people are classified into four major blood groups, A, B, AB and O. The blood group is controlled by three genes, A, B and O acting at the corresponding site on homologous chromosomes. A person will inherit two of these genes, one from each parent. Gene O is recessive to both A and B, but A and B are co-dominant, i.e. if a person inherits gene A from one parent and gene B from the other, he will be group AB, neither gene being dominant to the other. It follows that group O people must have the genotype OO while group A persons could be AA or AO and group B individuals BB or BO. The following example shows the possible blood groups of children born to a group A man and a group B woman both of whom are heterozygous for these genes.

Phenotype	group A		group B	
Genotype	AO		BO	
Gametes	A and O		B and O	
F_1 Genotype	AB	AO	OB	OO
Phenotype	group AB	group A	group B	group O

Human genetics

The "one gene–one character" effects described above illustrate very clearly the Mendelian* principles of inheritance, but they are the exceptions rather than the rule. Rarely do single genes control one trait. Colour in sweet peas, for example, is controlled by two pairs of genes, CC and RR. Gene C controls the production of the colour base and gene R the enzyme which acts on it to make a colour. The recessive cc will produce no colour base and rr will have no enzyme, CCrr and ccRR combinations will thus be unable to produce coloured flowers. At least six factors operate to produce coat colour in mice. In man, eight of the chemical changes involved in blood clotting are known to be under genetic control so that several genes are responsible for coagulation; absence of any one of them may lead to a blood-clotting disease such as haemophilia (p. 193).

When one gene only is responsible for an important physiological change, its absence or modification will have serious consequences. Therefore most known instances of single-factor inheritance in man are associated with rather freakish abnormalities. These are usually rare conditions, e.g. occurring once in 10,000 to 100,000 individuals, but there are a great number of different kinds of genetic abnormality.

Examples of known single-factor inheritance involving a dominant gene in man are white forelock, woolly hair, one form of night-blindness, one form of *brachydactyly* in which the fingers are abnormally short owing to the fusion of two phalanges and achondroplastic dwarfism in which the limb bones fail to grow. Recessive single factors are known to control, for example, red hair, inability to taste phenylthiourea, red-green colour vision, and one form of *albinism* which is the absence of pigment from the eyes, hair and skin.

Eye colour is determined to some extent by single genes. The gene for brown eyes is dominant to that for blue eyes. A blue-eyed person will be homozygous, bb, for the characteristic but a brown-eyed person could have the genotype Bb or BB. Thus one would expect blue-eyed parents to have only blue-eyed children but brown-eyed heterozygotes, Bb, could have blue-eyed children in their families. In fact, eye colour is probably controlled by more than one pair of alleles and some blue-eyed people carry genes for eye pigment. Thus it is possible but unusual for a brown-eyed child to be born to blue-eyed parents.

In experimental animals or plants, the type of inheritance and the genetic constitution can often be established by breeding together the progeny of the F_1 generation, or by backcrossing one of the F_1 individuals with the mother or father and producing numbers of offspring large enough to give results that have statistical significance. These methods are obviously not applicable to man and our knowledge of human genetics comes mainly from detailed analyses of the pedigrees of families, particularly those showing abnormal traits such as albinism, from statistical analysis of large numbers of individuals from different families for characteristics such as sex ratio, intelligence, susceptibility to disease, etc., and from individual studies of identical twins (*see* below).

Despite the scarcity of evidence for single-factor inheritance in humans, there is plenty of evidence to suggest genetic control of many physiological, physical and mental characteristics. Body height, eye colour, hair colour and texture, susceptibility to certain diseases, and facial characteristics are all genetically controlled but in a more complex way than that described for mice on p. 195.

* The term is derived from an Austrian monk, Gregor Mendel, who in the 1850s first discovered the type of inheritance described here.

Discontinuous and continuous variation

The individuals within a species of plants or animals are alike in all major respects; indeed, it is these likenesses which determine that they belong to the same species. Nevertheless, even though an organism recognizably belongs to a particular species, it may differ in many minor respects from another individual of the same species. A mouse may be black, brown, white or other colours, the size of its ears and tail may vary but despite these variations, it is still recognizably a mouse.

Discontinuous variation. The variations in coat colour are examples of discontinuous variation because there are no intermediates. If black and brown mice are bred together they will produce black or brown offspring. There are no intermediate colours and no problems arise in deciding in which colour category to place the individuals. It is not possible to arrange the mice in a continuous series of colours ranging from brown to black with almost imperceptible differences of colour between adjacent members of the series. The way sex is inherited is another example of discontinuous variation. With the exception of a small number of abnormalities, one is either male or female and there are no intermediates.

Discontinuous variation in humans is rather more difficult to illustrate. There are, for example, four major blood groups designated A, B, AB and O. Blood from different groups cannot be mixed without causing a clumping of the red cells. A person must be one or other of these four groups; he cannot, for example, be intermediate between group O and group A, he must be one or the other. In general terms, eye colour is inherited in a discontinuous manner; one has blue eyes or pigmented eyes, but there are some individuals who would be difficult to classify. Clear-cut examples of discontinuous variation occur among the more serious variants e.g. one is either an achondroplastic dwarf or one is not; intermediates do not occur.

The features of discontinuous variation are clearly genetically determined; they cannot be altered during the lifetime of the individual. You cannot alter your eye colour by changing your diet. An achondroplastic dwarf cannot grow to full height by eating more food. An albino cannot acquire a suntan because he is unable to produce melanin in his skin. Moreover, the variations are likely to be under the control of a small number of genes. One dominant gene makes you an achondroplastic dwarf; the absence of one gene for making pigment causes albinism.

Continuous variation. When one tries to classify individuals according to height or weight rather than eye colour, the decisions become more difficult and the classes more arbitrary. There are not merely two classes of people, tall and short, but a whole range of intermediate sizes differing from each other by barely measurable distances. Categories can be invented for convenience e.g. people from 1·4 to 1·6 m, 1·6 to 1·8 m, 1·8 to 2·0 m, but they do not represent discontinuous variations of 0·2 m between individuals.

There is no reason why continuous variations should not be genetically controlled but they are likely to be under the influence of several genes. For example, height might be influenced by 20 genes, each gene contributing a few centimetres to the stature. A person who inherited all 20 would be tall whereas a person with only 5 would be short. Although this example is purely hypothetical, it is known that height is at least partially genetically determined because tall parents have, on average, tall children and vice versa, but how many genes are involved is not known.

Continuous variations are also those most likely to be influenced by the environment. A person may inherit genes for tallness but if he is undernourished in his years of growth he will not grow as tall as he might if he had received adequate food. In fact, most continuous variations result from the interaction of the genotype with the environment. A person will grow fat if he eats too much food; he will lose weight if he goes on a diet. This seems to be an entirely environmental effect until one realizes that another person may eat just as much food and yet remain slim because of his different, inherited constitution. Whether one catches a disease or not would appear to be dependent on exposure to the disease germs, an exclusively environmental effect, and yet it is apparent that one may inherit susceptibility or resistance to a disease. If a person with inherited susceptibility to an infectious disease is never exposed to the infection, he will not develop the disease.

Heredity or environment?

It is possible to experiment with plants and animals to discover whether an observed variation is due primarily to the genetic constitution or to environmental differences. A species of plant growing in a valley may have larger leaves and taller stems than individuals of the same species growing on a mountain side. If the two varieties are collected, planted in the same situation and grown through one or two generations and still show the differences of leaves and stem size, one may assume that the differences are genetically controlled. If, however, the two varieties, after growing in the same environment, produce offspring which are indistinguishable, the original variations must have been due solely to the environmental differences.

Similar experiments with man are not possible or desirable. Even when situations occur which resemble the experiment, such as the "uniformity" of an institutional environment for orphaned children, the observations are always susceptible to more than one interpretation. Thus, there is usually a great deal of argument about very little evidence when people discuss whether our intelligence, for example, is predominantly due to the genes we inherit or the conditions of home and school in which we were brought up. One source of evidence in the "nature v nurture" controversy is the study of identical twins.

Identical twins

Twins may be either identical or fraternal. If they are fraternal, they are the result of the simultaneous fertilization of two separate ova by two separate sperms. The resulting zygotes will thus contain sets of chromosomes very different from each other as explained on pp. 190–192, and the twins, although they develop simultaneously in the uterus, will not necessarily be any more alike than if they were brothers or sisters born at different times, e.g. they can differ in sex. Identical twins, on the other hand, result when a single fertilized ovum, usually after a period of cell division, separates into two distinct embryos. The two embryos will thus have in their cells identical sets of chromosomes, since they are derived by mitosis (pp. 179–182) from a single zygote. The twins often share a placenta, although they may be enclosed in separate amnions. Such twins, having the same genotypes, usually resemble each other very closely and are invariably of the same sex, though variations in their position and blood supply while in the uterus may produce differences at birth.

Since the one-egg twins carry identical sets of genetic "instructions", it can be argued that any differences between them are due, not to their genes, but to the effects of their environment. Identical twins, therefore, are a valuable source

of evidence for assessing the relative importance of heredity and environment. For example, the average difference in height of 50 pairs of identical twins reared together was only 1·7 cm while the average difference for the same number of non-identical twins was 4·4 cm. These and other points of comparison are given in Table 34.1.

Table 34.1. Average differences in selected physical characteristics between pairs of twins:

Difference in:	50 pairs of identical twins reared together	50 pairs non-identical twins reared together	19 pairs of identical twins reared apart
Height (cm)	1·7	4·4	1·8
Weight (lb)	4·1	10·0	9·9
Head length (mm)	2·9	6·2	2·2
Head width (mm)	2·8	4·2	2·85

(*From Freeman, Newman and Holzinger*, Twins: A study of heredity and environment, *Univ. Chicago Press, 1937*)

The scores of identical twins in intelligence tests (Table 34.2) show a greater correlation than those of non-identical twins even when the former have been brought up in different environments, but the results show that educational background can make a considerable difference. (The reliability of the data for these results has recently been called into question.)

Table 34.2. Corrected average differences in IQ tests for 50 pairs of identical twins:

Identical, reared together. 50 pairs	Non-identical. 52 pairs	Identical, reared apart. 19 pairs
3·1	8·5	6·0

(*From Freeman, Newman and Holzinger*, Twins: A study of heredity and environment, *Univ. Chicago Press, 1937*)

Detailed histories of identical twins give even more impressive results. There is the case of girl twins, separated soon after birth; one was brought up on a farm and the other in a city, and both became affected with TB at the same age. Identical twin sisters were separated and adopted shortly after birth but both became schizophrenic within two months of each other in their 16th year. Such individual examples are of interest but cannot lead to any far-reaching conclusions about inheritance of human characteristics in general.

Applications of genetics to human problems

(*a*) **Eugenics** is the study of human genetics from the point of view (i) of encouraging breeding for good characteristics to improve human stock and (ii) of eliminating or reducing harmful characteristics. The simplest notions of eugenics as appreciated by the layman have tended to fall into some disrepute since they are based on naïve assumptions and insufficient evidence. It would seem obvious at first sight that if all the congenitally*insane were sterilized or prevented from breeding

* In this context the term *congenital* is taken to mean a hereditary condition rather than an effect resulting from environmental causes during gestation or birth.

in some way, the numbers of congenital idiots in the world would be greatly reduced or the condition completely eliminated. That this is not the case is shown by considering the condition of recessive albinism. Only people with the two recessive genes, aa, will show the characteristics. These number about one in 20,000 of the population. The marriage of two albinos would give rise to all albino children but such a marriage is unlikely. Calculations show that 1 in 70 of the general population must be a carrier for albinism, i.e. have the genetic constitution Aa. Such persons are indistinguishable, at the moment, from normal AA individuals but a marriage between two such people could produce some albino children aa. Of all albinos, 99 per cent arise from such marriages between unwitting carriers and clearly there could be no question of sterilization or voluntary birth control at least until after the first albino child was born and the genetic constitution of the parents so revealed. If no homozygous recessive persons, aa, were to breed, it would take 22 generations to reduce the frequency of a harmful gene from 1 per cent to 0·1 per cent.

Similarly, if every congenitally mental defective were sterilized, the incidence of the disease would fall by only 8 per cent. This fall, however, represents the elimination of suffering for a large number of people; mentally defective

Aa, has a chance of 1 in 420 begetting an albino child if he marries a normal person. The chances for other relations, such as aunts and uncles, can also be calculated.

Genetic advice could be far more accurate if heterozygous affected people could be distinguished from homozygous normal. In some hereditary diseases such a distinction is becoming possible. Heterozygotes for a disease called *phenylketonuria* can be detected with reasonable certainty by a test for the level of *phenylalanine* in their blood.

(c) **Consanguinity.** The study of human genetics enables predictions to be made on the chances of the recombination of two harmful recessive genes in the children of marriages between first cousins.

Fig. 34.3 shows diagrammatically the theoretical pedigree of two cousins, Bill and Jane. Cousin Bill is assumed to be heterozygous for a recessive gene, Nn. (He would be known to be Nn for certain, only if one of his parents was nn.) There is a 1 in 2 chance that the gene came from grandparents B, and in this case there is also a 1 in 4 chance that Jane has inherited the gene. There is thus a chance of 1 in 8 ($\frac{1}{2} \times \frac{1}{4}$) that cousin Jane is also Nn, in which case the chance of an affected child from their marriage is 1 in 4. The overall chances of an affected child if Bill is Nn and marries Jane, are thus 1 in 32 ($\frac{1}{8} \times \frac{1}{4}$).

Common grandparents

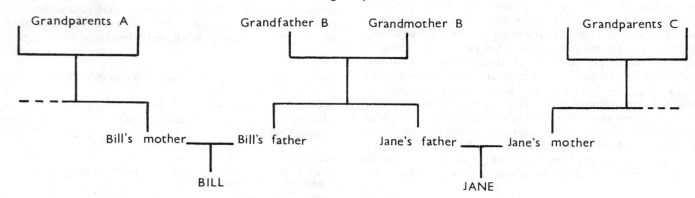

Fig. 34.3   Lineage of first cousins

people are hardly likely to provide a suitable home even if they have normal children. Several Scandinavian countries have passed laws which allow abortions to be induced for eugenic reasons.

The elimination of a harmful dominant trait, Dd or DD, can theoretically be accomplished in one generation provided that the condition appears before the reproductive age.

(b) **Genetic advice.** Even if the more drastic measures of sterilization, etc. are not adopted, advice based on sound knowledge of genetics can sometimes avoid unsatisfactory breeding and several countries have "Heredity Clinics" to give information about inborn defects.

For example, a person suffering from congenital juvenile cataract (a defect of the eye), caused by a dominant gene would be told that, if he married a normal person, half their children might inherit the disease; this can be seen by considering the possible F_1 offspring from the mating of Dd (affected) and dd (normal). An albino (aa) marrying a normal person who might be either AA or Aa would find that the chance of any one of his children being an albino is 1 in 140, while the normal brother of an albino, who might be a carrier,

If the gene is fairly rare in the general population, e.g. occurs once in 100 individuals, the chances of Bill marrying an Nn person from the general population are 1 in 100. The overall chances of an affected child in this case are $1/100 \times 1/4$, i.e. 1 in 400.

Although these considerations would apply equally well to beneficial genes, cousin marriages are not usually encouraged and brother-sister marriages forbidden by law. This does not reduce the total number of homozygous recessives which occur in a population but does reduce the chances of their occurring in a particular family. Consanguinity is bound to occur sooner or later, otherwise we should need to have had an impossibly large number of ancestors.

Intelligence

Intelligence is a product of a person's genetic constitution and the effect of his environment; i.e. he may inherit from his parents the mental equipment for intelligent thought, but this will not produce his full potential of intelligent behaviour unless he is educated.

That part of intelligence which is genetically controlled is almost certainly influenced by a large number of genes and is not susceptible to simple analysis. When a graph or histogram is plotted to show the different numbers of individuals possessing a particular IQ value, the type of picture obtained is that shown in Fig. 34.4, often called a

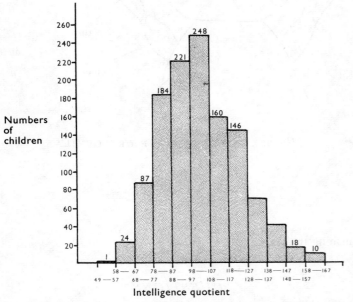

Fig. 34.4 Distribution of IQ rating in a random sample of 1207 Scottish eleven-year-old children

(*From C. O. Carter*, Human Heredity, *Penguin Books, 1962*)

"normality" curve or *curve of normal distribution*. Similar curves are also obtained for factors such as height and skin colour and can be explained on the basis of several genes influencing the characteristic. Instead of the straightforward presence or absence of a condition, such as albinism, there is a continuous variation with every grade of intermediate.

PRACTICAL WORK

Breeding experiments with Drosophila

Drosophila is a small fly which is easy to breed in large numbers in the laboratory. By carrying out controlled cross-breeding experiments with mutant forms and wild types, it is possible to illustrate and investigate some of the principles of heredity.

Sources. Wild type and several mutant strains can be obtained from the usual biological supply firms.

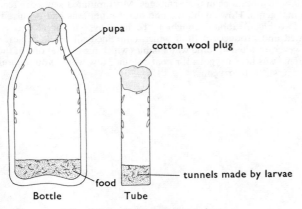

Fig. 34.5 Drosophila cultures

Culture medium. Mix together 50 g maize meal, 15 g agar, 13 g dried yeast, 25 g brown sugar, 800 cm³ water and boil gently for several minutes. Dissolve 2 g "Moldex" (a mould-inhibitor) in 40 cm³ boiling water and add it to the mixture. Pour the mixture into 100×25-mm specimen tubes to a depth of 20 mm, plug the tubes with cotton wool wrapped in butter muslin and sterilize in an autoclave at 10^5 N/m² (15 lb/in²) for 15 min. On the day before introducing the flies, add 3 drops of a suspension of fresh yeast to each tube.

The tubes will hold about one hundred flies for experiments but for maintaining stocks of Drosophila, wide-mouthed bottles, e.g. half-pint milk or cream bottles, should be used and new cultures started every five or six weeks (Fig. 34.5).

Setting up experiments. It is essential that the females used for the breeding experiments have not already mated with the males in the culture bottle. From a flourishing culture with many unhatched pupae, all the flies are shaken into a clean, dry bottle and the culture bottle re-plugged. The flies that emerge from the pupae in the culture bottle during the day will be virgins and can be easily recognized by the unexpanded wings. These flies are etherized, and males and females sorted into separate dry tubes where they can recover from the ether. About three virgin females of a given strain, e.g. wild type, are transferred to a fresh culture tube and six males of a different strain, e.g. vestigial wings, are introduced. The males need not be virgins and can be taken directly from stock cultures.

The flies will mate and lay eggs in the culture medium. Larvae hatch from the eggs, burrow through the food, grow, and in about 10–14 days from laying, pupate on the sides of the tube. The parent flies should be removed after one week. When the F_1 progeny have been emerging from the pupal cases for about 10 days they should be etherized, the different sexes and strains counted and recorded, and the flies killed in alcohol or retained for F_2 experiments. From the results, the ratios of the different strains can be calculated, and interpretations attempted on the lines of the principles of Mendelian inheritance.

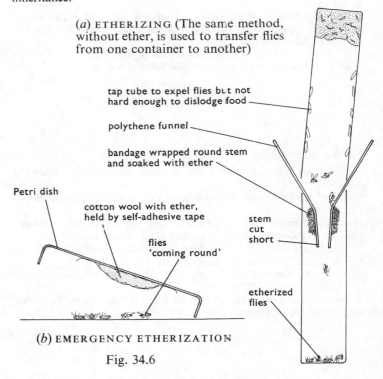

(*a*) ETHERIZING (The same method, without ether, is used to transfer flies from one container to another)

tap tube to expel flies but not hard enough to dislodge food

polythene funnel

bandage wrapped round stem and soaked with ether

Petri dish

cotton wool with ether, held by self-adhesive tape

flies 'coming round'

stem cut short

etherized flies

(*b*) EMERGENCY ETHERIZATION

Fig. 34.6

Etherizing (ether is very inflammable; no naked flame should be allowed while it is in use). The technique is depicted in Fig. 34.6a. The flies should not be exposed to ether for more than one minute. If they begin to recover while being counted, etc. they can be covered for a few seconds by a Petri dish lid carrying a small pad of ether-soaked cotton wool (Fig. 34.6b). If etherized flies are placed directly

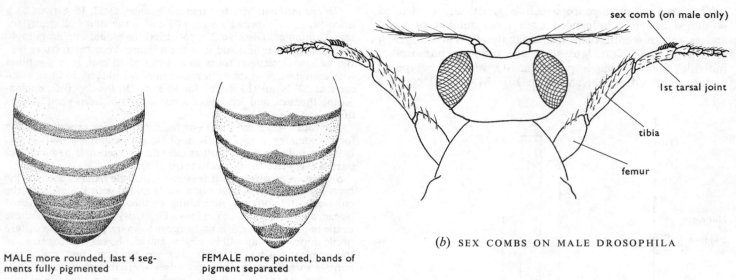

MALE more rounded, last 4 seg-   FEMALE more pointed, bands of
ments fully pigmented   pigment separated

(*a*) DROSOPHILA ABDOMEN: DORSAL ASPECT

These differences are not easy to see on newly hatched flies or ebony-body mutants.

(*b*) SEX COMBS ON MALE DROSOPHILA

Fig. 34.7   Distinguishing the sexes

into a culture tube, the latter should have dry sides and be placed horizontally, otherwise the anaesthetized flies will stick to the food or the glass.

Sexing (*see* Fig. 34.7 *a* and *b*). The presence or absence of sex combs on the fore-legs is the surest guide. A hand lens or binocular microscope is essential.

Suitable crosses. Wild type × vestigial wing; wild type × ebony body; wild type × white eye: each cross should be made in two ways, reversing the sexes, e.g.

        vestigial winged male × wild type female

and   wild type male × vestigial winged female.

Further information

More detailed instructions can be obtained from the following books

Practical Heredity with Drosophila, by G. Haskell (Oliver & Boyd, 1961).

Biological Science: An Inquiry into Life (BSCS Yellow Version, published in Great Britain by Rupert Hart-Davis, 1963):

    Teachers' Manual for Students' Laboratory Guide, p. 193;

    Students' Laboratory Guide, p. 213.

Biology: An Environmental Approach, Man and His Environment (John Murray, 1972) p. 58.

Genetics for Schools (Modern Science Memoir No. 31), by Professor K. Mather (John Murray, 1953).

QUESTIONS

1. Two black guinea pigs are mated together on several occasions and their offspring are invariably black. However, when their black offspring are mated with white guinea pigs, half of the matings result in all black litters and the other half produce litters containing equal numbers of black and white babies.

From these results, deduce the genotypes of the parents and explain the results of the various matings, assuming that colour in this case is determined by a single pair of genes (alleles).

2. The blood groups A, B, AB and O are determined by a pair of allelomorphic genes inherited one from each parent. The gene for group O is recessive to both A and B. Thus a group A person may have the genotype AA or AO. Inheritance of A from one parent and B from the other produces the phenotype AB.

(*a*) What are the possible blood groups likely to be inherited by children born to a group A mother and group B father? Explain your reasoning.

(*b*) A woman of blood group A claims that a man of blood group AB is the father of her child. A blood test reveals that the child's blood group is O. Is it possible that the woman's claim is correct? Could the father have been a group B man? Explain your reasoning.

3. A geneticist wishes to find out the colour of F_1 flowers from a cross between red- and white-flowered insect-pollinated plants such as antirrhinum. Revise, if necessary, the section on pollination (p. 33) and describe how he should conduct his experiments, assuming that pollen can effectively be transferred by means of a dry paintbrush.

4. Individuals of a pure-breeding line of Drosophila are exposed to X-rays to induce mutations in their gametes. Most mutated genes are recessive to normal genes. How could one find out if mutations had occurred?

5. Two black rabbits thought to be homozygous for coat colour were mated and produced a litter which contained all black babies. The F_2, however, resulted in some white babies which meant that one of the grandparents was heterozygous for coat colour. How would you find out which parent was heterozygous?

35 Evolution and the Theory of Natural Selection

THE theory of evolution offers an explanation of how the great variety of present-day animals and plants came into existence. It supposes that life on Earth began in relatively simple forms which over hundreds of millions of years gave rise, by a series of small changes, to a succession of living organisms which became more varied and more complex. Taken to its logical conclusion, the evolutionary theory must suppose that life itself evolved from non-living matter. It must be emphasized that evolution is a theory and not an established fact. In general terms, it is an acceptable hypothesis to account for the existence of the living organisms which we know today.

The arguments for evolution are drawn from the kind of evidence which follows, though much of the evidence is either very incomplete or circumstantial.

1. Reproduction and spontaneous generation

As far as we know, all living organisms are derived by reproduction from pre-existing organisms and do not arise spontaneously from non-living matter. This knowledge weakens any alternative theory to evolution, if it claims that each kind of organism known today arose spontaneously or was created suddenly at different points in time. e.g. that horses were created suddenly one million years ago and have remained the same ever since, reproducing their kind exactly over thousands of generations.

The evolutionary theory would thus assume that when new forms of life appear on the Earth, they have been derived by reproduction from organisms which already exist, e.g. that mammals were derived from reptiles, reptiles from amphibia and amphibia from fish. In general, for vertebrates at least, the fossil record supports this contention. The question which must arise, however, is "Where did the *first* living creatures come from?" To this the biologist is obliged to say that, although spontaneous generation of life from non-living substances is not known to occur today and is thought to be very improbable during the geological period of which we have some knowledge, there was a time, some 500 million years ago or more, when conditions were favourable for such an event or series of events. For example, the atmosphere might have been devoid of oxygen but rich in methane and ammonia which makes feasible the production of amino acids and proteins.

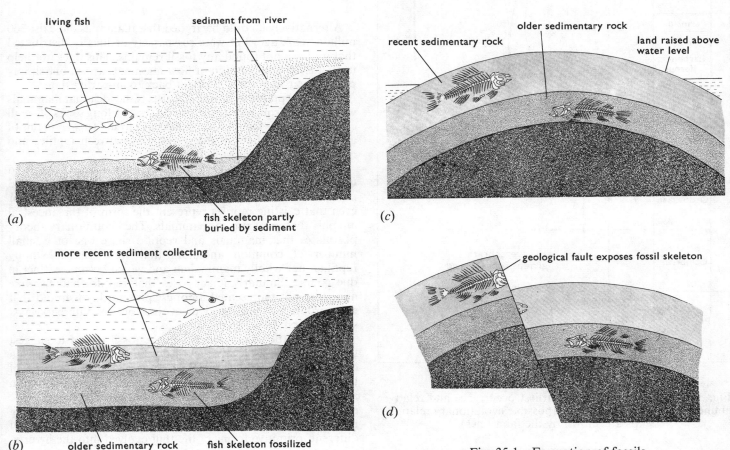

Fig. 35.1   Formation of fossils

2. The fossil record

Sedimentary rocks were formed by the settling down of mineral particles in lakes and oceans. These particles often became cemented together and the layers of sediment were compressed over millions of years to form rock. The dead remains of animals and plants, falling to the bottom of the lakes or seas became incorporated in the sediment and so preserved in a variety of ways as *fossils* (Fig. 35.1). As a result of slow earth movements, many sedimentary rocks were raised above the water and the fossils they contained became accessible for study.

If not too contorted by the earth movements, in an exposed series the lowest sedimentary rocks will be the oldest, and from the fossils in successive layers, the scientist can form some idea of the animals and plants present millions of years ago. When the fossils in various layers are studied it appears (*a*) that many present-day animals and plants are not represented (Fig. 35.2), and (*b*) that a vast number of organisms represented by skeletal remains in the rocks no longer exist today. For example, 300 million years ago there were apparently no mammals, that is, no fossil remains have yet been found; but there were at that time some "armour-plated" fish (Fig. 35.3) which no longer seem to exist.

Such evidence seems to detract from any idea that all the organisms existing today have been reproduced exactly since life began. Even if spontaneous generation or *biogenesis* could have taken place 300 million years ago, it seems unlikely that it would "generate" anything so complex as a mammal in a single operation or even a number of operations in a very short time.

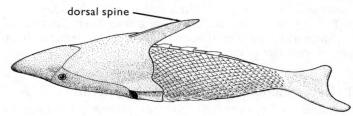

dorsal spine

Fig. 35.3   Pteraspis—one of the extinct, "armour-plated" fish of the Silurian period (a reconstruction from fossil remains)

Alternatively it could be argued that mammals did exist 300 million years ago, but were so sparsely or locally distributed that we have not found any fossil remains. This does not help us to answer the question of how mammals arose, but simply pushes their hypothetical origin back to an earlier date.

More acceptable to most biologists is the idea that mammals were derived from reptile-like ancestors by a long series of small changes, and the fact that scientists have found fossil remains of animals intermediate in many respects between reptiles and mammals lends support to this idea.

It must not be supposed, however, that mammals evolved from reptiles in the sense that present-day reptiles could produce a mammal, even over a very long period of time; or even that existing reptiles represent the form of the ancestral animals that gave rise to mammals. The evolutionary theory postulates that mammals and reptiles share one or a small number of common ancestors that were neither wholly reptilian nor wholly mammalian and which became extinct in due course. Similarly it supposes that reptiles had fish-like ancestors not represented amongst present-day fish and that each group has continued to evolve, but in different ways, since the first divergence.

There is little or no evidence in the fossil record to suggest that any of the large invertebrate groups of animals share a common ancestor. This may be because (*a*) our knowledge of the fossil record does not go back far enough, (*b*) the common ancestors, if they existed, have not been preserved, or (*c*) the events which produced living organisms from non-living matter occurred more than once, initiating a number of different, primitive forms of life that evolved into the invertebrates but share no common, living ancestor.

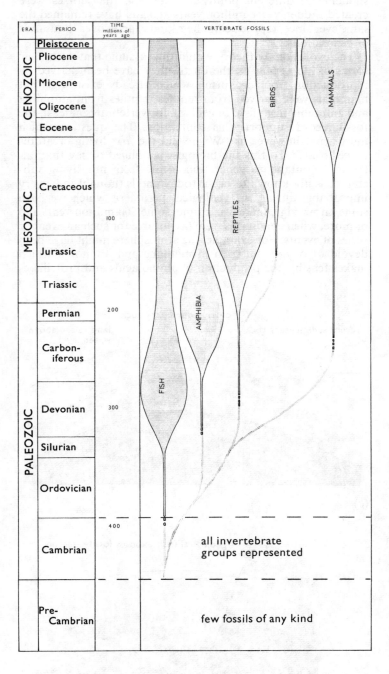

Fig. 35.2   Chart to show the earliest occurrence and relative abundance of fossil vertebrates (possible evolutionary relationships are shown by the faint lines)

(*After Grove and Newell,* Animal Biology, *University Tutorial Press*)

3. Circumstantial evidence

By studying the structure and distribution of modern animals it is possible to point out features which can be interpreted as supporting the theory of evolution. One such study is *comparative anatomy*, and perhaps the most familiar example is the skeleton of mammalian limbs (Fig. 35.4). The limbs of seals, moles, bats and antelope look very different from each other and are adapted to the functions of swimming, digging, flying and running. Despite this difference of appearance and function, they all have basically the same skeletal structure. If these mammals did not originate from a common ancestor but arose independently (and spontaneously?), there seems no convincing reason why the pattern of bones in limbs performing such different functions should be so similar. On the other hand, it does seem reasonable to visualize these limbs as

modifications of the primitive, unspecialized limbs of a common ancestor, the differences coming about during evolution as the limbs became more closely adapted to each special method of progression, while retaining the fundamental pattern of bones and joints.

Many other examples of comparative anatomy could be cited in the vertebrates, including fundamental similarities of the skeletal, circulatory and nervous systems. Also, other lines of circumstantial evidence, such as the similarity between the embryonic forms of different vertebrates, could be discussed. Being circumstantial evidence, i.e. attempting to fit known facts into a pattern of events in the past which is suspected but impossible to reproduce, the evidence is often subject to alternative interpretations and further discussion is suited to more advanced study than this book can offer.

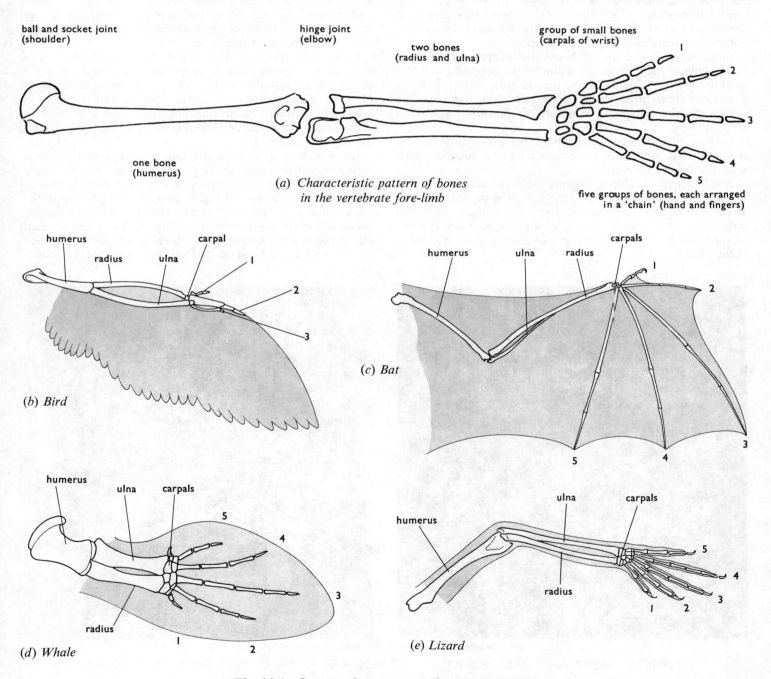

Fig. 35.4   Comparative anatomy of vertebrate limbs

NATURAL SELECTION

In 1858 Charles Darwin and Alfred Russel Wallace put forward a theoretical explanation of how evolution could have taken place and new species arisen. The theory of evolution by natural selection is the one which so far fits most, but not all, of the observed facts and has been strengthened by discoveries since 1858.

The arguments for the theory of natural selection can be summarized as follows:

(*a*) *Observation* 1. The offspring of animals and plants outnumber their parents.

(*b*) *Observation* 2. Despite this tendency to increase, the numbers of any particular species remain more or less constant.

(*c*) *Induction* 1. Since fewer organisms live to maturity than are produced, there must be a "struggle" for survival.

(*d*) *Observation* 3. The individual members within any plant or animal species vary from each other by small differences; some of these differences can be inherited.

(*e*) *Induction* 2. (i) Some of these varieties are better adapted to the environment or mode of life of the organism and will tend to survive longer and leave behind more offspring. If the variations are harmful, the organism possessing them may die before reaching reproductive age and so the variation will not be passed on.

(ii) If an advantageous variation is inherited by an organism, it will also live longer and leave more offspring, some of which may also inherit the variation.

Small but favourable variations may thus accumulate in a population over hundreds of years until the organisms differ so much from their predecessors that they no longer interbreed with them. The "variety" would now be called a new species. Certain points in the argument above need elaboration to make them clear.

(*a*) If a pair of rabbits had 8 offspring which grew up and formed 4 pairs, eventually having 8 offspring per pair, in four generations the number of rabbits stemming from the original pair would be 512, i.e. $2 \rightarrow 8 \rightarrow 32 \rightarrow 128 \rightarrow 512$.

(*c*) An induction is an argument in which a generalization is made as a result of observations of actual events. The potential number for the fourth generation of rabbits is 512. If, however, the population of rabbits is to remain constant, 510 must have died or not been born, leaving only two of this vast potential family still alive.

The "struggle" for survival, however, does not imply actual fighting; the participants may never meet but they could be in competition for food and shelter. Often the competition will not only be between adults, but between eggs or larvae or seeds, i.e. the most prolific stage of the life history of a species at which mortality rate is often high. The "struggle" is often quite passive and may depend on the relative resistance of eggs to adverse conditions, or concealment patterns on the body leading to effective camouflage.

(*d*) Man is a species and the variations between members of this species are obvious at a glance. Variations in other organisms are less obvious but very evident to an experienced observer. If a variation is to play an effective part in evolution it must be heritable. A variation acquired in the lifetime of an organism, e.g. the well-developed muscles of an athlete, is not usually heritable.

(*e*) A species of a normally light-coloured moth, *Biston betularia*, the peppered moth, produces from time to time a black variety. This black variety was first recorded in 1848 in Manchester but by 1895 it had increased to 98 per cent of the population in this district. Observations showed that the light-coloured form was well camouflaged against the lichen-covered tree trunks where it normally rested. The atmospheric pollution of industrial areas, however, reduced or eliminated the lichens

(From the experiments of Dr H. B. Kettlewell, University of Oxford)

Plate 71.   LIGHT AND DARK FORMS OF THE PEPPERED MOTH AT REST ON TREE TRUNKS

(*a*) Soot-covered oak trunk near Birmingham   (*b*) Lichen-covered trunk in unpolluted countryside

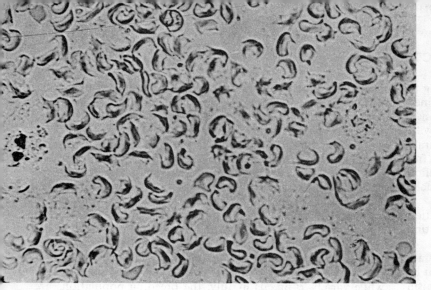

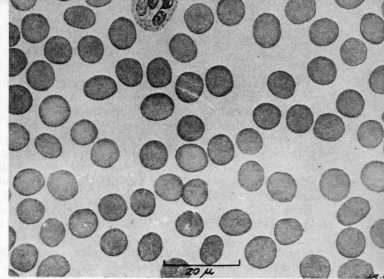

Plate 72.   SICKLE CELL ANAEMIA (× 1000)

(Wellcome Museum of Medical Science)

(a) At low oxygen concentrations the cells become distorted

(b) Normal red cells for comparison

on the tree trunks and also darkened them with deposits of soot, so that the dark forms were better concealed (Plate 71) than the light forms. Better concealment led to fewer moths being eaten by birds, e.g. a Redstart in Birmingham was observed to eat 43 pale forms and only 15 dark forms from equal numbers resting on trees. The "struggle" here is very indirect, being a "struggle" for concealment, but the outcome is that more of the dark forms would survive and lay eggs. The dark colour in many cases is due to a single, dominant gene (p. 194) and so would be inherited by some of the offspring. The dark variety, however, is not yet a new species since it will interbreed with the light form, but this account illustrates how a new species could arise by natural selection.

HERITABLE VARIATION

The sources of heritable variation were not known to Darwin but have since been shown to arise principally in two ways: by mutation and by recombination.

Recombination and natural selection

A mutant gene which is harmful in one genetic constitution (genotype) may be neutral or beneficial in another. Similarly, certain advantageous genes may exist in one population and different beneficial genes in another. If interbreeding takes place between the populations, individuals may arise in which the two beneficial genes recombine, so conferring a selective advantage on the new phenotype. For example, the genes in a wild grass which render it resistant to fungus disease have been combined by cross-breeding with the genes for large grain size in cultivated wheat thus producing a variety with a high yield and good resistance to disease. Although segregation (p. 194) tends to separate these beneficial genes their combined selective advantage could cause them to maintain a constant frequency in a population as happens with sickle cell trait (Plate 72).

Sickle cell anaemia occurs when two recessive mutant genes controlling haemoglobin production, combine in an individual. Such individuals have severe anaemia and generally die before reaching the reproductive age, i.e. natural selection removes the "h" genes from the population. The recessive genes remain in the population, however, in the heterozygous individuals, e.g. HH = normal; hh = sickle cell anaemia; Hh = heterozygous for sickle cell anaemia. Although from one quarter to one half of

the haemoglobin of the heterozygotes is affected, usually less than 1 per cent of their red cells show sickling in low oxygen concentrations. Such heterozygotes are said to have the sickle cell *trait*. When two heterozygotes marry, one would expect on average a quarter of their children to have sickle cell anaemia

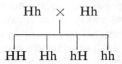

and would die before reaching reproductive age. In this way the h genes would be eliminated by selection from the population since for every four offspring from the HH genotypes there will be only three from the Hh genotypes. In certain African populations, however, as many as 34 per cent of the people carry the recessive gene and there is evidence to suggest that this is because the heterozygotes are more resistant to malaria, i.e. persons with the genotype Hh enjoy a selective advantage in malarious districts while the HH genotypes are reduced by the selective pressure of malaria. Investigations of Negro populations in America show an incidence of only 4 to 5 per cent of the trait compared with the 15 to 20 per cent characteristic of the African population from which the migrants were originally derived. In some non-malarial regions therefore, it looks as if the heterozygotes lose their selective advantage.

Mutation and natural selection

Mutation (p. 186). A mutation is a change in a gene or a chromosome. Most mutations are harmful, e.g. vestigial wings in Drosophila, and so are unlikely to be handed on to the offspring. The mutations which result in advantageous characteristics, however, may be passed on to the offspring and preserved in a population. The dark variety of the peppered moth, mentioned above, results in most cases from the mutation of a single gene which is dominant to the normal gene. Usually gene mutations are recessive and the characters controlled by the gene may not be expressed, e.g. if gene *A* mutates to *a* the organism *Aa* may not noticeably differ from *AA*. If the mutation occurs frequently in the population, however, there is a chance of *a* and *a* coming together at fertilization, so producing a homozygous recessive *aa* in which the new character is fully expressed.

207

Mutation rate. One cannot predict when a gene is going to mutate but the frequency of its occurrence can be determined in some cases; for example in achondroplastic dwarfism it is possibly as high as one mutation in 20,000 compared with one in 100,000 for many genes. The rate of mutation is characteristic of particular genes in particular species but the frequencies are such that in a human ejaculate of, say, 200 million sperms, there are likely to be a considerable number of nuclei bearing gene and chromosome mutations.

Exposure to radio-activity, X-rays and ultraviolet radiation is known to increase the rate of gene and chromosome mutation (*see* p. 187).

Significance of mutation. Since so many mutations are harmful to the individual bearing them and are eliminated by natural selection it might seem that a well-adapted organism with no mutations would be at an advantage. This may be true so long as the environment does not alter and the organism does not move to a different situation where it is subjected to new selection pressures for which it is poorly adapted. If a species of bacteria were incapable of mutation, exposure to streptomycin might eliminate the species. The fact that streptomycin-resistant mutants occur, allows the bacteria to survive such an adverse change in the environment.

The dark mutant of the peppered moth had little selection advantage prior to the Industrial Revolution but with the advent of pollution and the consequent darkening of tree trunks, the mutation enjoyed favourable selection pressure. The elimination of mutants by natural selection in one environment is the price which populations have to pay if they are to retain the ability to adapt to a changing environment. Mutations should not therefore be regarded as "accidents" since they are events which occur normally in all populations and are essential for survival and evolution, though they may well be harmful to the many individuals in which they occur.

Balanced polymorphism

The existence of genetically controlled varieties within a population provides the raw material on which natural selection acts, with one or other variety being favoured or reduced by the selective process. However, in most populations over a short term, the different varieties persist in about the same numbers; for example, although in certain environments the light forms of some moths have a selective advantage, dark mutants still occur with a low but consistent frequency partly as a result of new mutations appearing in the population and partly because the dark forms enjoy some positive selective advantage. It seems that although the dark forms are more easily seen when resting on trees, they are better concealed from predators while in flight, thus selection operates in their favour on certain occasions.

The term "balanced polymorphism" refers to the persistence of such varieties in a population.

Variations in eye and hair colour and the existence of different blood groups are examples of genetically controlled variations in human populations. It is not always possible to discern the selective advantage of one or other variant but this does not mean that the variant is selectively neutral. The purple pigmentation of some onions seems to have little selective advantage until it is demonstrated that the purple variety is also more resistant to fungus disease. This is also an illustration of the fact that the characteristic which we recognize as being controlled by a gene or group of genes is not necessarily the one on which selection is acting.

Isolation and the formation of new species

A mutation or recombination may give rise to a variety with characteristics which enable it to colonize new areas not accessible to the parent stock. In this way the variety may form a breeding population which is isolated from the original population. Further mutations may occur in this isolated population and accumulate until individuals are incapable of interbreeding with the original stock. In this way a variety will have given rise to a new species.

There are many ways in which populations can become isolated. Geographical isolation is an obvious case where rivers, oceans or mountains separate populations. Another cause could be a difference in breeding season or incompatibility of mating behaviour. If variety A breeds only in April while variety B breeds in July, the breeding populations are effectively isolated even if they occupy the same area.

Evolution by natural selection can be visualized as proceeding as follows: mutations and genetic recombinations arise in a population; those which are not so harmful as to be lethal give rise to a balanced polymorphism; environmental change or migration of the population favours certain varieties which leave more offspring which inherit these same variations; isolation allows other favourable genes to accumulate in the population until it differs so much from the parental stock that interbreeding is impossible.

"Survival of the fittest"

This is an expression often used to summarize the theory of natural selection. In this context it must be emphasized that "fit" refers not to health but to the production of a large number of offspring which survive to reproductive age. A person may be physiologically "fit", i.e. healthy and strong, but if he fails to have any children he is less fit in the evolutionary sense than a physically weak and unhealthy individual who nevertheless has a large family. Generally speaking, an organism which is healthy and well adapted to its surroundings will survive for a long time and, reproducing each year of its life, will leave behind many offspring which inherit its advantageous characters. In human populations, natural selection can be said to operate on a characteristic only if it affects reproductive capacity. For example, achondroplastic dwarfs are less fit because they have, on average, only 0·25 children per parent compared with 1·27 children born to their normal brothers or sisters. Fewer dwarfs marry, and childbirth is hazardous for the females. The gene for dwarfism being dominant would therefore tend to disappear from the population were it not maintained by mutation.

From an evolutionary point of view it is better to consider the fitness of a population of organisms than the fitness of an individual. Although, in the short term, fitness is dependent on the number of viable offspring, in the long term it depends also on the potential for change and adaptation in the population. There will be a great deal more variation for selection to act on within a population of organisms than there is in a single creature and its progeny.

Natural selection in man

There is no reason apparent at the moment why natural selection should not have been responsible for the evolution of man from an ape-like ancestor and there is no reason why selection should not still be acting on human populations. The incidence of the sickle cell trait is likely to be the result of selection pressure. Where there is heritable variation and environmental pressure, evolution by natural selection is possible.

However, with the growth and spread of modern medicine, public health measures, education, and other benefits of civilization, natural selection has ceased to play so immediate a role in human populations as it does in other organisms. Weakly babies who would die in "natural" conditions can be saved, and the congenitally blind, deaf, and lame are enabled to live and reproduce. Some people see in this artificial preservation a tendency for mankind to evolve into a race of weaklings. This is to suppose, however, that one day we shall be without the benefits of modern medicine, good food, surgery, etc. It also wrongly supposes that every weakly child will grow into an enfeebled adult and have unhealthy children. Natural selection is a wasteful process and would have eliminated our diabetics, our short sighted, and our Rhesus babies, many of whom may have an important part to play in human affairs.

Man has devoted much thought and effort to minimizing the adverse effects of environmental change by providing himself with, for example, clothing, housing, central heating, air conditioning, an agricultural industry, and protection from predators and competitors. This does not mean that selective pressures have disappeared altogether, but they have changed. Resistance to diseases of modern society, indifference to crowding and noise, ability to withstand the pace of modern life or assimilate a modern diet without developing ulcers, may all have selective advantages if they are controlled to some extent by genes and are manifest before the end of normal reproductive age.

However, in most advanced countries selection will operate largely through parental decisions on how many children they will have, since it may be assumed today that nearly all the children will reach reproductive age. The "fittest" will still be those with the largest families but this will depend on the conscious decision to procreate rather than on the number left after miscarriage, infant mortality, and disease have taken their toll.

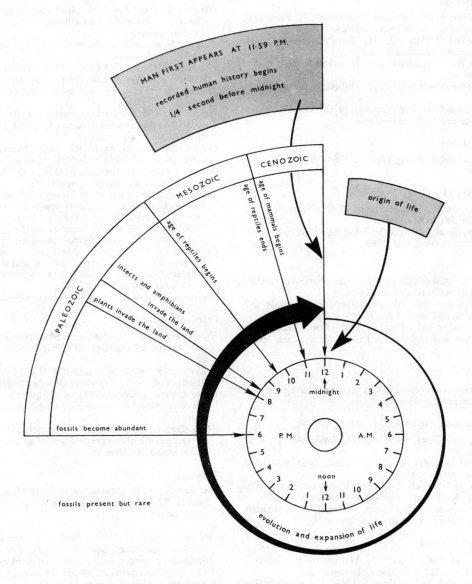

Fig. 35.5   The history of life on a 24-hour scale

Books for further reading

Cells, tissues, plant anatomy (Chapters 3 and 4)

The Cell, by Swanson (Prentice Hall), 1969. A general study of cell structure and function, including activity in cells and cell division.

The Microstructure of Cells, by Hurry (John Murray), 1965. Electron photomicrographs of cells; explanatory notes. Mainly animal cells.

An Atlas of Biological Microstructure, by Dodge (Arnold), 1968. Electron photomicrographs and notes; plant and animal cells.

An Atlas of Histology, by Freeman and Bracegirdle (Heinemann), 1966. Photomicrographs and drawings of animal tissues.

An Atlas of Plant Structure Volume 1, by Bracegirdle and Miles (Heinemann), 1971. Photomicrographs and drawings of plant anatomy.

Plant physiology (Chapters 8–13)

Principles of Plant Physiology, by Bonner and Galston (W. H. Freeman), 1951. Readable, explicit, advanced reference book.

Organization in Plants, by Baron (Arnold), 1963. For "A" level students; general physiology with some practical work.

The Living Plant, by Ray (Holt, Rinehart and Winston), 1963. General physiology, concise, explicit.

Plants and Water, by Sutcliffe (Arnold), 1967. Diffusion, transpiration, stomata, transport.

Plant growth, by Black and Edelman (Heinemann), 1970. Plant hormones, differentiation, flowering, environmental effects.

General botany (Chapters 4–16)

Botany, by Robbins, Weier and Stocking (Wiley), 1964. Well illustrated, advanced reference book.

Bacteria and fungi (Chapters 14 and 15)

The Biology of Fungi, by Ingold (Hutchinson), 1967. Structure and reproduction of the main classes of fungi.

The Biology of Fungi, Bacteria and Viruses, by Stevenson (Arnold), 1967.

Microbes and Man, by Postgate (Penguin), 1969.

Soil (Chapter 16)

The World of the Soil, by Russell (Collins), 1961. The physics, chemistry and biology of the soil; agriculture and landscape.

Life in the Soil, by Jackson and Raw (Arnold), 1966. Soil animals and micro-organisms, including methods of extraction and culture.

Soils and Fertilizers, by Coker (Macdonald), 1971. For readers with horticultural interests. Some straightforward ideas for practical exercises.

Food and diet (Chapter 17)

Human Nutrition, by Mottram (Arnold), 1963. A comprehensive reference book requiring no specialized knowledge.

Mammalian and human physiology (Chapters 17–27)

Biology of the Mammal, by Clegg (Heinemann), 1963. An explicit, readable reference book.

The Living Body, by Best and Taylor (Chapman and Hall), 1952. Advanced reference book on human physiology.

Animal Biology, by Grove and Newell (University Tutorial Press), 1969. General zoology for "A" level studies; including invertebrates.

An Introduction to Human Physiology, by Green (Oxford), 1963. Written for first-year medical, dental and physiotherapy students.

Insects and invertebrates (Chapter 28)

Insect Natural History, by Imms (Collins), 1971. Written for the layman; interesting to read and illustrated with many photographs.

Insect Physiology, by Wigglesworth (Methuen), 1966. A concise useful monograph.

Animals without Backbones Volumes 1 and 2, by Buchsbaum (Penguin), 1971. An advanced but easy to read, well illustrated reference book.

Vertebrates (Chapters 29–31)

The Life of Vertebrates, by Young (Oxford), 1962. An advanced, well illustrated reference book.

Man and the Vertebrates Volumes 1 and 2, by Romer (Penguin), 1970.

The Life and Organization of Birds, by Yapp (Arnold), 1970. Comprehensive reference book of bird biology.

Single-celled organisms (Chapter 32)

The Protozoa, by Vickerman and Cox (John Murray), 1967. Detailed accounts of seven protozoa with many line drawings and photographs.

Genetics and evolution (Chapters 33–35)

Genetics for "O" Level, by Head and Dennis (Oliver and Boyd), 1968. Covers similar ground to Chapters 33–35 but more fully and from a different point of view.

Genes, Chromosomes and Evolution, by Ashton (Longman), 1967. Suitable for "A" level studies.

The Genetic Code, by Asimov (John Murray), 1962. Carefully explained account of DNA, written for the layman.

Elementary Genetics, by George (Macmillan), 1964. Suitable for "A" level studies.

Human Heredity, by Carter (Penguin), 1970. Written for the layman.

Genetics and Man, by Darlington (Penguin), 1966. History and philosophy of ideas about genetics.

Genetics for Schools, by Mather (John Murray), 1953. Concise but very useful survey of genetics for "A" level students.

Applied Genetics, by Paterson (Aldus), 1969. Important aspects of genetics in agriculture and human populations.

Heredity and Development, by Moore (Oxford), 1963. Advanced text which introduces much of the experimental evidence for genetical knowledge.

Facets of Genetics, from *Scientific American* (W. H. Freeman), 1969. Collection of 35 articles from *Scientific American*. Needs more background for understanding than provided by Chapters 33–35.

Cytogenetics, by Swanson, Merz and Young (Prentice Hall), 1967. Advanced book about chromosomes and their role in genetics.

Principles of Genetics, by Sinnott, Dunn and Dobzhansky (McGraw Hill), 1968. Advanced reference book.

Outline of Human Genetics, by Penrose (Heinemann), 1959.

Practical work

Experimental Work in Biology, by Mackean (John Murray), 1971 (*see* p. 1). A collection of experiments under subject headings. Detailed instructions.

A Certificate Course in Practical Biology, by Lee and Martin (Mills and Boon). Volume 1: *Experimental Investigations* (1968). Volume 2: *Plant and General Studies* (1971). A complete collection of experiments and relevant theory in Volume 1. Good drawings of plant specimens in Volume 2.

Elementary Microtechnique, by Peacock (Arnold), 1966. Comprehensive reference book about preparation of specimens for microscopic examination; stains, fixatives etc.

Questions

Past papers may be obtained from the publication offices of the examining boards. The address is usually given in the booklet of regulations and syllabuses.

New Questions in "O" Level Biology, by Head (Oliver & Boyd). Book 1: *The Mammal* (1966). Book 2: *Plants and Invertebrates* (1967). Questions demanding understanding and intelligent application of the principles and scientific aspects of biology rather than simple recall of information. Answers published in separate books.

Objective and Completion Tests (1–20) by Clarke *et al.* (John Murray), 1970. Multiple-choice tests under 20 headings, 30–50 questions in each. Answers provided.

Visual material

Colour photomicrographs by Gene Cox

These high-quality colour photomicrographs consist of six sets, each of twenty slides. There is also a seventh set available (set G) for use with the tropical edition of this book.

The slides are individually mounted and packed in "Viewpacks" (plastic folders with individual pockets for each slide) and each set includes a folder of notes for the teacher.

Set titles

A Plant Anatomy

B Cells, Angiosperm Reproduction, Bacteria and Fungi

C Mammalian Physiology 1

D Mammalian Physiology 2

E Insects

F Vertebrates, Protista and Cytology

Outline drawings for Chapter 28 "Insects"

Seventeen transparencies for the overhead projector. They consist of combinations of drawings from Chapter 28 with labelling overlays.

Available from Audio-Visual Productions, Hocker Hill House, Chepstow, Gwent NP6 5ER.

Outline drawings for human physiology by D. G. Mackean

These eighty-four black and white transparencies for the overhead projector are divided into nine sets (packed in seven wallets) and include notes for the teacher and references to *Introduction to Biology*.

Set titles

A Digestion (8 foils)

B Circulation (12 foils)

C Breathing (6 foils)

D Excretion, Skin (6 foils)

E Reproduction (14 foils)

F Skeleton and movement (10 foils)

G Teeth (5 foils)

H The sensory system (12 foils)

I Co-ordination (11 foils)

Sample material is available to teachers on request from Educational Department JOHN MURRAY 50 Albemarle Street London WIX 4BD.

Reagents

Benedict's solution (1 litre)

Dissolve 173 g sodium citrate crystals and 100 g sodium carbonate crystals in 800 cm³ warm distilled water. Dissolve separately 17·3 g copper sulphate crystals in 200 cm³ cold distilled water. Add the copper sulphate solution to the first solution with constant stirring.

Note. (1) Benedict's solution is preferable to Fehling's solution as it is less caustic and does not deteriorate on keeping.

(2) The red deposit of cuprous oxide that coats the inside of test-tubes used for the sugar tests can be removed with dilute hydrochloric acid.

Iodine solution (1 litre)

Dissolve 10 g iodine and 10 g potassium iodide in 1 litre distilled water by grinding the two solids in a mortar while adding successive portions of the water. The solution should be further diluted for class use, e.g. 5 cm³ in 100 cm³ water.

Hydrogencarbonate indicator

Dissolve 0·2 g thymol blue and 0·1 g cresol red powders in 20 cm³ ethanol. Dissolve 0·84 g "Analar" sodium hydrogencarbonate in 900 cm³ distilled water. Add the alcoholic solution to the hydrogencarbonate solution and make the volume up to 1 litre with distilled water.

Shortly before use, dilute the appropriate amount of this solution 10 times, i.e. add 9 times its own volume of distilled water.

To bring the solution into equilibrium with atmospheric air, bubble air from outside the laboratory through the diluted indicator using a filter pump or aquarium pump. After about 10 minutes, the dye should be red.

Acknowledgements

I am indebted to all the people who have provided photographic material, allowed me to reproduce or adapt drawings, and to quote or use data from their experiments. They are acknowledged individually with the captions to the relevant illustrations.

I am grateful to the Joint Matriculation Board and the Cambridge, London, and Oxford examining boards for permission to reproduce some of their O level examination questions.

I am particularly grateful to Mr W. G. Goldstraw, Mr S. W. Hurry, Dr R. G. Pearson, Dr C. H. Rice, and Dr C. O. Carter for carefully reading the manuscript as a whole or in part, and for their corrections and valuable suggestions; to Mr A. E. Ellis, Miss H. G. Q. Rowett, and Mr K. Thomas for their constructive criticisms of the first edition; to Mr A. E. Pound for his helpful comments on the second edition and Mr E. Holden for suggestions for the fourth edition; and to Mr Denys Baker for his advice in the early stages of making the drawings.

Unless otherwise stated, most of the drawings in this book are the copyright of the author whose permission should be sought before they are reproduced or adapted in other publications.   DGM

Examination questions

The following questions are selected from the long answer section of the O level Biology papers set between 1970 and 1972 by four of the major examining boards in England. [O=Oxford, L=London, C=Cambridge, J=Joint Matriculation Board (JMB)]

The Flowering Plant

1. Make a large labelled diagram of a vertical section of a buttercup flower. In tabular form, compare the parts of the flower shown in your diagram with the corresponding parts of a named zygomorphic* flower, such as the pea, so as to bring out the differences between them. (*No* diagram of the zygomorphic flower is required.) [O]

2. Make a large, fully labelled drawing to show the external features of a twig of a *named* tree as seen in winter. *List three* distinctive features of the tree you have chosen, and describe how its seeds are dispersed. *Briefly* compare the functions of its bark with the functions of the skin of a mammal. [L]

3. (a) Distinguish between pollination and fertilization.
(b) Describe the events that take place in the ovary of a *named* insect-pollinated flower after fertilization until the fruit is formed.
(c) Give a labelled diagram of the fruit formed.
(d) How is seed dispersal carried out in this plant? [C]

4. (a) What are (i) the advantages, (ii) the disadvantages, of vegetative reproduction as compared with sexual reproduction in the flowering plant?
(b) By reference to *two named* examples, explain the difference between perennation and vegetative reproduction.
(c) By means of large labelled diagrams, describe how in *named* flowering plants the following structures are used for vegetative reproduction: (i) stem tubers, (ii) runners. [C]

Plant Physiology

5. (a) Describe a klinostat and how it may be used.
(b) Describe an experiment which shows that geotropic curvature occurs only in that part of the root which is increasing in length. [O]

6. What are the functions of the vascular tissues ("veins") in an herbaceous dicotyledon such as a buttercup. Describe an experiment that you could perform to demonstrate *one* of the functions you mention.
  With the *aid* of diagrams, compare the arrangement of vascular tissue in the stem and root of the herbaceous dicotyledon and briefly suggest a reason for the difference in arrangement. [L]

7. Describe carefully and concisely how you would investigate whether
(a) photosynthesizing leaves change in dry weight,
(b) green shoots give off carbon dioxide in the dark. [J]

8. (a) What is *transpiration* and what is its importance?
(b) What are the physical factors of the environment that affect the rate of transpiration?
(c) It is sometimes said that transpiration is unavoidable in the plant. What does this mean?
(d) Describe *one* experiment by which you could compare the rate of transpiration of a *potted* plant in the dark and in the light [C]

9. You are provided with some peas or beans that have been soaked in water and have just started to germinate. Describe with full experimental details how you would
(a) show that the peas or beans are still living (saying that they will consume oxygen or that they will grow is *not* the answer required);
(b) show that if they are to continue to grow they will require a supply of oxygen;
(c) find the percentage dry weight of a batch of ten of these seeds. At what stage would the dry weight of seedlings start to increase? [C]

Bacteria, Protista, Fungi

10. What do you understand by (a) bacteria, (b) viruses?
Write a brief account of *either* the role of bacteria in the soil, *or* the role of viruses in disease. [L]

11. Make labelled diagrams to show the structure of *Amoeba* and *Spirogyra*. Explain, by reference to these two organisms, the essential differences between animals and plants. What features do they both show which are peculiar to living things? [L]

12. Some diseases of plants and animals are due to the presence of parasites.
(a) (i) What is a parasite? (ii) How does it differ from a saprophyte? (iii) Describe *three* ways in which saprophytes are directly or indirectly beneficial to man.
(b) Name a disease caused by (i) bacteria, (ii) viruses, (iii) fungi. For each disease describe how the organism reaches the animal or plant in which it causes disease and how the disease may be prevented or controlled. [C]

13. Processes which are characteristic of living organisms include respiration, reproduction, response to stimuli, nutrition, growth, and excretion.
(a) Describe how *each* of the above processes is carried out in *Amoeba* (or *Paramecium*).
(b) Explain how *Amoeba* (or *Paramecium*) can get rid of excess water from its body.
(c) Give *four* reasons why *Mucor* (or *Rhizopus* or *Penicillium*) is not considered to be an animal. [C]

Soil

14. Describe the composition and structure of fertile soil. In what ways may a farmer improve the quality of his land? [L]

15. (a) Describe experiments which would enable you to test the truth of the following hypotheses: (i) different soils have different capillarities, (ii) different soils have different rates of drainage. You should include diagrams of apparatus, the method of setting up the experiments, and an indication of the method of measuring your results.
(b) Give your reasons why (i) poor drainage, (ii) excess of drainage, hinder the growth of plants. [C]

16. (a) Explain fully two ways in which bacteria improve the fertility of soil for plants.
(b) The soil contains insoluble organic compounds. Of what importance are these compounds to the bacteria?
(c) Beginning with a sterile nutritive medium and a sample of soil how would you culture soil bacteria? [J]

Animal Physiology

17. What is a gland? Name (a) a ductless gland, (b) a gland which possesses a duct, (c) an organ which has both types of gland. What are the functions of each? [O]

18. Make a large labelled diagram of the respiratory organs of a mammal and describe the way in which they function. What happens to the waste products of respiration? Briefly compare the respiration of a green plant with that of a mammal. [O]

19. Explain concisely what is meant by (a) osmoregulation and (b) excretion. Describe in outline how these two processes occur in mammals and *one* other *named* animal. [L]

20. In relation to animal nutrition, explain the following terms briefly: ingestion; digestion; absorption; assimilation; egestion.
  Illustrate your answer by reference to carbohydrate in the diet. [L]

21. Give two differences between the mode of operation of nervous and chemical co-ordination in a mammal. For each of the following explain how co-ordination is achieved:
(a) The change in the size of the pupil in a human eye when light intensity increases;
(b) control of blood sugar level in a mammal;
(c) a positive phototropic response in a plant shoot. [J]

22. (a) Give *three* differences in structure between red blood cells and white blood cells.
(b) Describe *two* ways in which the blood of a mammal protects the body from disease.

(c) There may be an increase in the number of red blood cells in the circulatory systems of members of a mountaineering expedition of several weeks duration. Explain, in detail, how this change would benefit the mountaineers.

(d) Give three differences between the composition of the blood in the pulmonary vein and that in the portal vein connecting the intestine to the liver. [J]

23. If you run as quickly as possible in a race you may finish feeling hot, sweaty, breathing heavily, and with your pulse beating more quickly than usual.

(a) Describe the processes which have brought about each of these changes.

(b) What is the benefit to you of each of these changes?

(c) If you feel hot, what is the source of this heat, and where is it released? [C]

24. (a) A man decides to bend his arm so that his fingers touch the shoulder. Describe with the help of diagrams how the forearm is bent and then straightened again as a result of this decision.

(b) (i) The body of an animal may be co-ordinated through the nervous system or by means of hormones. Name a hormone and describe its functions within the body of man. (ii) What are the differences between nervous control and control by hormones? [C]

Insects

25. Give a concise, illustrated account of the external features of the adult stage of any *named* insect. How does this insect feed and breathe? Very briefly compare this method of breathing with that of a mammal. [L]

26. Briefly describe the appearance of a *named* adult insect with special reference to its adaptation to life on land. How does its young form (larva or nymph) differ from the adult in habits and appearance? [L]

27. By means of large *fully labelled diagrams only* illustrate the structure of:
(a) a named butterfly as seen in side view at rest;
(b) the head of its caterpillar as seen from the front. Write an account of its life history. [O]

28. (a) (i) Describe the life history of a *named* insect vector of disease. (ii) How has our knowledge of the life history influenced the methods used to control the insect?
(b) What harmful results may be produced by the use of poisonous sprays for controlling insect pests? [C]

Vertebrates

29. Make a large labelled diagram of a named fish. Describe how it swims and reproduces. [O]

30. Make labelled diagrams to show the external appearance of a *named* amphibian larva at two different stages in its development, giving one diagram for each stage. How is this larva adapted to life in water? Briefly *list* the changes it undergoes during metamorphosis which adapt it to life on land. [L]

31. (a) With reference to shape, body structure, and feathers, show how birds are adapted for various forms of flight and landing.
(b) In what ways do the wings of an insect (i) resemble, and (ii) differ from those of a bird? [C]

Genetics

32. When true breeding coloured flowers are crossed with true breeding white ones of the same species the offspring (F$_1$ generation) all produce coloured flowers. If only one pair of genetical factors is involved what types of flower would you expect if this F$_1$ generation is (a) self-pollinated and the resulting seeds sown; (b) cross-pollinated with white flowers and the resulting seeds sown; (c) propagated vegetatively? Give reasons for your statements. [O]

33. The fruit fly (*Drosophila*) is a small insect, one variety of which has a light-coloured body and the other a black body colour. A cross was made between a pure breeding *Drosophila* with a light-coloured body and one with black body colour. The resulting offspring all had light-coloured bodies and when these were interbred their offspring consisted of 380 light body coloured flies and 116 flies with black bodies. Assuming that body colour is controlled by one pair of alleles give a genetical explanation of these results. Include in your account the following terms: gene, dominant, recessive, homozygous, heterozygous, genotype, and phenotype. [J]

34. When Mendel crossed pure breeding pea plants, which had smooth coated seeds, with plants whose seeds had wrinkled coats, all the offspring had smooth coated seeds. When he selfed† plants of this F$_1$ generation he obtained 822 plants with smooth coated seeds and 299 plants whose seed coats were wrinkled. Explain these results.

How would you establish the genotype of one of the plants with smooth coated seeds of the F$_2$ generation? [J]

General Principles

35. Explain the importance of respiration to all living organisms. Distinguish carefully between aerobic and anaerobic respiration. Describe an experiment which demonstrates anaerobic respiration. [L]

36. Describe some of the ways in which (a) light and (b) temperature affect the lives of plants and animals. [L]

37. Describe how the living cells of (a) a green plant and (b) a *named* mammal obtain their supplies of glucose. Briefly explain what may happen to the glucose once it has reached the cells of the green plant and the mammal. [L]

38. A molecule of carbon dioxide enters the air spaces of a leaf of a cabbage. Trace the possible fate of the carbon in such a molecule until it is eliminated as carbon dioxide in the expired air of a mammal. [L]

39. (a) It can be said that all animal nutrition is directly or indirectly dependent on the utilization of the sun's energy by green plants. Explain this statement using as your examples (i) man keeping cows for meat, and (ii) any aquatic carnivore such as a shark.
(b) Name *three* vitamin deficiency diseases; for each, name the deficient vitamin and a source from which it is normally obtained.
(c) Describe briefly how one example of vitamin deficiency is recognized and successfully treated. [J]

40. Give a concise account of the importance of air, water, and light to (a) animals and (b) plants. [O]

41. What is meant by *irritability* in living organisms? Illustrate this phenomenon by reference to the behaviour of (a) *Amoeba*, (b) a germinating seed, (c) a potted plant growing in a window, (d) a mammal. [O]

42. Describe some of the ways in which man has changed his environment. [L]

43. By reference to man, illustrate the distinctive characteristics of (a) a living organism, (b) an animal as distinct from a plant, (c) a mammal. [O]

44. How are the following adapted for the function stated:
(a) the skin for protection and heat conservation in the mammal,
(b) the ileum for absorbing digested food in the mammal,
(c) the root of the plant for absorbing water and mineral salts,
(d) the primary root of the seedling for growing downwards in the soil? [C]

45. Make labelled diagrams of a named unicellular (acellular) animal, and a cell from the palisade layer of the leaf of a green plant. List (a) three similarities in structure, (b) three differences in structure, (c) three differences in function, between the two cells you have drawn. [J]

46. The main requirements of organs for gaseous exchange in animals and plants are as follows: (i) a source of oxygen, (ii) a large surface area for exchange of gases, (iii) a medium for transport of gases. Describe how these requirements are met in the following: an insect, a bony fish, a frog and the leaf of a plant. (Details of breathing mechanisms are *not* required.) [C]

---

* Zygomorphic: the flower is not symmetrical, e.g. lupin, white deadnettle.
† Selfed: self-pollinated

Glossary

The explanations of terms given in this glossary are not meant to be formal definitions, but reminders, and are restricted to the context in which these terms are used in the book. The numbers in brackets are page references.

abdomen (106) the part of the body below the diaphragm which contains stomach, intestines, kidneys, liver, etc; in insects it refers to the third region of the body which has no limbs or appendages

abscission (28) the shedding of leaves

accommodation (132) changing the shape and hence the focal length of the eye lens to focus on near or distant objects

adipose tissue (91) fatty tissue in mammals

adrenal gland (144) one of the endocrine glands, situated above the kidneys; it produces the hormone adrenaline

aerobic respiration (47) the chemical process which uses oxygen to release energy from food

albinism (198) absence of colour in skin, hair, and eyes; an inherited condition

algae (6) simple green plants, single celled or many celled but not organized into root, stem, and leaves

alimentary canal (86) the tube running through an animal from mouth to anus; digestion and absorption of food takes place inside it

alleles (194) genes controlling the same characteristic (e.g. hair colour) but producing different effects (e.g. black or red), and occupying corresponding positions on homologous chromosomes

amino acids (82) chemicals produced by the digestion of proteins or assembled to make proteins

amnion (115) a fluid-filled bag or sac surrounding an embryo

amphibia (6) frogs, toads, newts, etc; they can move and breathe equally well on land or in water

ampulla (135) a bulge in each of the semicircular canals; contains sensory organs

amylase (85) a starch-digesting enzyme

anaerobic respiration (49) the chemical process which releases energy from food but does not need oxygen to do so

anaphase (181) a late stage in mitosis or meiosis when the chromosomes or chromatids are separating

androecium (30) the male part of a flower; the stamens

anther (30) the upper part of a stamen; contains the pollen

antibodies (93) chemicals made in the blood which combat bacteria or their poisons

antigen (94) a foreign organism or chemical in the blood which stimulates the system to make antibodies against it

antitoxin (73) an antibody which neutralizes bacterial poisons

aorta (95) the main artery from the heart to the body

aqueous humour (130) the fluid in the front part of the eye

arteriole (97) a branch of an artery

arthropod (146) an animal with an outer cuticle and jointed legs (e.g. crabs, lobsters, insects)

asexual reproduction (22) production of new individuals by a single organism (e.g. budding) without involving gametes

atrium (98) the upper chamber of the heart receiving blood from the veins

auto-radiograph (54) a photographic image produced by a radioactive source

auxin (44) a plant hormone which may affect rates of growth in roots and shoots

axon (138) part of a nerve cell; the fibre carrying impulses away from the cell body

back-cross (194) mating an organism with the same genetic type as its grand-parents; one method of investigating the genetic make-up of an organism

bile (84) a green fluid made in the liver and delivered to the small intestine where it assists digestion of fats

bivalent (188) a pair of homologous chromosomes, closely associated at prophase of meiosis

bract (32) a leaf borne on a flower stalk

bronchiole (102) a branch of a bronchus

bronchus (102) either one of the two air-pipes branching from the windpipe and going to the lungs

caecum (73) a blindly ending sac at the junction of the small and large intestine

calyx (30) the ring of sepals on the outside of a flower

cambium (16) a layer of cells inside a plant stem; by their division the cambial cells increase the thickness of the stem

capillary (97) the smallest type of blood vessel in the circulatory system; its walls are only one cell thick

carpals (120) the wrist bones

carpel (30) a component of the female part of the flower which contains the potential seeds

central nervous system (139) the brain and spinal cord

centriole (181) a small body in an animal cell which divides first at cell division

centromere (180) that part of a chromosome which becomes attached to the spindle at mitosis or meiosis

centrum (122) the solid, cylindrical part of a vertebra

cerebellum (142) part of the brain; an outgrowth of the roof of the hind-brain

cerebral cortex (142) a layer of grey matter on the outside of the cerebral hemispheres; many connections between nerve cells are possible here

cervix (112) the narrow passage connecting the top of the vagina with the uterus (womb)

chemoreceptor (149) a sense organ which responds to chemicals; produces sensations of smell or taste

chiasma (188) region of close contact between homologous chromosomes at meiosis; when first formed, a chiasma indicates where exchange of chromatid portions is taking place

chlorophyll (50) a green chemical present in all green parts of plants; can trap sunlight and use its energy for promoting chemical changes

chloroplast (18) a cytoplasmic body in a plant cell; it contains the chlorophyll

choroid (130) a layer of cells and blood vessels in the eye, between the retina and sclerotic

chromatid (180) when a chromosome replicates, it produces two identical chromatids which are separated at cell division, one going to each new cell

chromosome (182) a long, rod-like structure in a nucleus; it appears at cell division and is thought to carry the genes

chrysalis (157) the pupa of a moth or caterpillar

cilia (13) hair-like cytoplasmic projections from a cell; they can flick and so produce movement

ciliary body (130) part of the eye between the lens and the choroid which contains the ciliary muscle

ciliary muscle (130) the muscle in the eye which alters the shape and focal length of the lens during accommodation

clavicle (119) the "collar-bone"; it joins the shoulder to the breast-bone

cleavage (165) cell division in a fertilized egg

clinostat (43) an apparatus which rotates plants to make the effects of light and gravity equal on all sides

coccus (72) a bacterium, spherical in shape

cochlea (134) a coiled tube in the inner ear; contains sensory nerve endings which respond to sound vibrations

cocoon (158) a case formed round the outside of the pupa of an insect

codominance (197) where both alleles are equally expressed in the phenotype, i.e. neither of two contrasting allelomorphic genes is dominant

coleoptile (41) the protective casing outside the shoot of a cereal seedling

collecting tubule (107) the final part of the kidney tubule where the water content of the blood is adjusted

colon (86) the first part of the large intestine; water is absorbed here

compensation point (51) the time when photosynthesis is as rapid as respiration; carbohydrates are produced as fast as they are used

cone (131) a light-sensitive cell in the retina of the eye; it is colour sensitive

conjugation (130) one form of sexual reproduction in which individual cells join and exchange gametes

conjunctiva (130) the transparent skin over the front part of the eyeball

consanguinity (200) intermarriage of close relations having a relatively high proportion of genes in common

consumer (58) an organism assigned a position in a food chain because it feeds on other organisms or their products

contraception (118) prevention of fertilization by "artificial" means

co-ordination (137) the linking together of systems to make them work effectively together

corm (24) a swollen, underground stem; it stores food and produces new plants from its buds

cornea (130) the transparent part of the sclera in the front of the eyeball

cornified layer (109) the outer layer of the skin consisting of dead cells

corolla (30) a ring of petals on a flower

corpus luteum (114) the structure formed from the follicle of an ovary in a mammal after an ovum has been released; it produces the hormone progesterone

cortisone (144) the hormone produced from the outer layer (cortex) of the adrenal gland

cotyledon (39) a specialized leaf, often swollen with food reserves, making up part of the embryo in a seed

covert (168) one of the feathers of a bird contributing to the general covering of the body

Cowper's gland (113) part of the male mammalian reproductive system; it adds a secretion to the sperms at ejaculation

crista (135) the sensory part of an ampulla in the semicircular canal

crossing over (192) occurs when chromatids at meiosis exchange portions with their homologous partners

crustacea (6) a group of animals with an external cuticle and jointed legs (e.g. crabs and lobsters)

cupula (135) a structure in the ampulla of a semicircular canal which is deflected during rotatory movements

cuticle of a leaf (18) a thin, non-cellular layer outside the epidermis

cuticle of an insect (146) the hard casing covering the body and limbs

cuticular lens (148) that part of an insect or crustacean cuticle which covers the compound eye and helps to refract light

cytoplasm (10) the semi-fluid matter, other than nucleus, in a cell; it contains enzymes and other substances which maintain life in the cell

deamination (92) removal of the nitrogenous part of an amino acid prior to converting it to glycogen

dendrite (138) one of many fibres in a nerve cell conducting impulses towards the cell body

dendron (138) a single, long fibre of a nerve cell conducting impulses to the cell body

denitrifying bacteria (59) bacteria in the soil which convert compounds of nitrogen into gaseous nitrogen

dentine (125) the layer of bone-like tissue beneath the enamel of a tooth

deoxyribonucleic acid (DNA) (186) the chemical in the chromosomes which controls activities in the cell; it is thought to constitute the genes

dermis (109) the deepest layer of the skin

detoxication (92) the rendering harmless by the liver of poisonous chemicals in the blood

diabetes (144) the disease resulting from an inability to control the level of sugar in the blood due to inadequate production of insulin by the pancreas

dialysis tubing (63) cellophane tubing with semi-permeable properties used for experiments on osmosis

diaphragm (86) the sheet of muscular tissue separating the thorax and abdomen

diatoms (57) single-celled green plants constituting a large part of the phytoplankton

dicotyledon (6) a plant having two cotyledons in its seed

diffusion (62) movement of molecules of a gas or dissolved solute from a region of high concentration to one of low concentration

diploid (183) the number of chromosomes in the body cells of most animals and plants; there are two of each type of chromosome

discontinuous variation (198) relatively large differences between members of the same species, without intermediates (e.g. black or white mice)

dominant (194) the gene, of a pair of alleles, which is expressed in the phenotype

dormancy (42) a "resting" stage in which no growth or movement is observable from outside

dorsal root (140) the spinal nerve bringing sensory impulses into the spinal cord

Drosophila (183) the small fruit fly used for breeding experiments

duodenum (88) the first part of the small intestine, opening from the stomach

dwarfism (187) normal size head and trunk but very short limbs (one type); an inherited condition

ecdysis (119) moulting or shedding the cuticle in insects, usually accompanied by rapid increase in size

echinoderms (6) a group of invertebrate animals such as starfish

ectoplasm (176) the outer layer of clear cytoplasm in a single-celled animal such as Amoeba

egestion (86) the passing out of undigested food residues and other matter from a food vacuole or the intestine

electron micrograph (13) a high power photograph taken by an electron microscope

electron microscope (12) a microscope which gives enormous magnifications by using electron beams instead of light

embryo (112) the stage during which an organism develops from a fertilized egg to an independently functioning individual

enamel (125) the outer layer of a tooth

endocrine gland (144) a gland which produces chemicals called "hormones", which are released directly into the blood circulation

endolymph (134) the fluid in the semicircular canals and the cochlea of the ear

endoplasm (176) the inner, granular, fluid cytoplasm of a single-celled animal such as Amoeba

endoplasmic reticulum (13) a series of flattened tubes running through the cytoplasm of a cell

endoskeleton (119) a skeleton inside the organism (e.g. that of man and other vertebrates)

endosperm (40) a food store inside a seed

endothelium (96) the layer of cells lining the inside of blood vessels

enzyme (85) a chemical made in the protoplasm of cells which speeds up the rate of certain chemical reactions

epicotyl (39) that part of the shoot of a seedling between the cotyledons and the plumule

epidermis: human (109) the outer layer of the skin

plant (14) the layer of cells on the outside of a structure such as a stem or leaf

epididymis (113) a long coiled tube between the testis and the sperm duct; sperms are stored here

epigeal (41) the type of germination in which the cotyledons are brought above the ground

epiglottis (86) a flap of cartilage behind the tongue, which helps direct food away from the windpipe and into the gullet

epithelium (89) a layer of cells in an animal, lining the inside of certain organs

erosion (80) loss of topsoil due to action of wind or rain

erythrocyte (93) a red cell in the blood

etiolation (44) the effect of growing a shoot in the absence of light

eugenics (199) the application of genetics to eliminate or reduce hereditary diseases and improve human qualities

Euglena (176) a single-celled green plant

Eustachian tube (87) the tube running from the middle ear to the throat

eutrophication (60) over-production of microscopic plants in lakes and rivers due to excess of mineral salts reaching them

evolution (203) the production of new species of organisms from existing organisms by a series of small changes over a long period

excretion (106) the getting rid of the waste products of chemical reactions in the cells of the body

exoskeleton (146) a skeleton outside the body of an organism (e.g. the cuticle of an insect)

F_1 generation (194) the offspring from the mating of two individuals

F_2 generation (198) the offspring resulting from mating two individuals of the F_1 generation with each other

faeces (89) the undigested material plus bacteria, etc., left in the colon after food has been digested and absorbed

fatty acid (88) one of the chemicals produced when a fat is digested

femur (88) the upper bone of the hind limb

fermentation (49) the breakdown of food material by yeast or bacteria to produce energy plus carbon dioxide and in some cases, alcohol

fertilization (35) the combining of the nuclei of male and female reproductive cells (gametes) to form a zygote

fibre (138) a long strand of cytoplasm which grows out from the cell body of a nerve cell and can conduct impulses to or from it

fibrin (94) the protein fibres which form a network across a wound and cause a blood clot to form

fibrinogen (91) the protein in the blood plasma which, when blood vessels are damaged, turns into fibrin and forms a blood clot

fibula (119) the smaller of the two bones in the lower leg

field capacity (78) the maximum amount of water which the soil can hold against gravity

filament (30) the stalk of a stamen in a flower

flagellate (176) a single-celled organism propelled by means of a flagellum

flagellum (176) a long filament of cytoplasm which projects from a flagellate and whose lashing movements propel it through the water

foetus (115) the later stages of an animal's embryo when all the organs are present

follicle (*see* Graafian follicle or hair follicle)

fossil (203) the remains of a plant or animal preserved in sedimentary rock

fovea (131) the small area in the retina of the eye having the greatest concentration of sensory cells, hence giving it the most accurate vision

fruit (30) the fertilized ovary of a flower

fungi (75) a group of plants lacking chlorophyll; their bodies are made of hyphae rather than cells and they feed as saprophytes or parasites

gall bladder (86) a small sac in or near the liver which stores bile

gamete (30) a reproductive cell such as a sperm or ovum

ganglion (139) a group of cell bodies and sometimes synapses, at one point in a nerve

gene (185) a "particle" in the nucleus which determines the presence or absence of certain characteristics in organisms

genetics (185) the study of the way in which characters are inherited

genotype (174) the genetic constitution of an individual, i.e. all the different genes present, whether expressed or not

geotropism (43) a change in the direction of growth of a root or shoot as a result of the direction in which gravity is acting

germination (39) the development of the embryo in a seed into an independent plant

gestation (116) the period of time between fertilization and birth

glomerulus (107) a coiled "knot" of capillaries in the cortex of the kidney; blood serum is filtered out

glycogen (90) an insoluble carbohydrate similar to starch but stored in the liver and muscles of mammals

Graafian follicle (114) the region in a mammal's ovary which contains a maturing egg

granular layer (109) the living inner layer of cells in the epidermis of the skin

grey matter (142) areas of brain and spinal cord occupied principally by the cell bodies of nerve cells

guard cells (19) the cells in the epidermis of a leaf or stem, on either side of the stoma which control its aperture

guttation (66) the exudation of drops of water from leaves in certain conditions

gynaecium (30) the female part of a flower, consisting of carpels

haemocoel (147) the space in an insect between the body wall and internal organs; it is filled with blood

haemoglobin (93) the red, iron-containing pigment in the red blood cells; it can combine with oxygen

haemophilia (193) an inherited disease in which the time needed for blood to clot is greatly increased

haploid (188) containing a single set of chromosomes in which each chromosome is represented only once; usually applies to the nucleus of gametes

hepatic portal vein (89) the vessels carrying the blood from all parts of the alimentary canal to the liver

heterozygous (194) carrying a pair of contrasted genes for any one character; will not breed true for this character

hibernation (111) prolonged phase of inactivity during adverse winter conditions with, in mammals, a corresponding fall in body temperature

holophytic (50) a type of nutrition involving the building up of food from simple substances; characteristic of plants

holozoic (50) taking in complex substances for food and breaking them down to simpler substances by digestion; characteristic of animals

homeostasis (108) the regulation of the composition, within narrow limits, of the body fluids and internal conditions

homoiothermic (110) body temperature above that of the surroundings and maintained at a constant level

homologous chromosomes (183) corresponding chromosomes of same shape and size derived from each parental gamete

homozygous (194) possessing a pair of identical genes controlling a given character; will breed true for this character

hormones (144) chemicals produced by endocrine glands and released into the circulatory system; they control the rate of various bodily activities.

humerus (119) the upper bone of the fore-limb

humus (78) the finely divided organic matter incorporated into soil crumbs

hydrotropism (44) change in direction of growth in roots supposedly as a result of a one-sided stimulus of water

hyphae (75) the microscopic living threads which make up the structure of a fungus

hypocotyl (39) that part of a seedling between the cotyledon and radicle; its elongation brings the cotyledons above the soil

hypopharynx (153) a component of insects' mouth parts

hypothalamus (108) the region of the brain just above the pituitary gland and which controls the activities of this gland

ileum (86) the major part of the small intestine

imago (156) a fully formed adult insect

incisor tooth (120) one of the teeth in the front of the jaw

inclusion (13) a non-living particle in the cytoplasm of a cell

incus (134) the second of the chain of small bones in the middle ear

indoleacetic acid (IAA) (44) a plant growth substance

inflorescence (30) a group of flowers on the same stalk

ingestion (86) the taking in of food

inoculation (95) injection with a harmless form of a disease so that the system will make antibodies

insulin (144) the hormone produced by the pancreas; controls sugar metabolism

integument (35) the outer coat of a developing seed

intercostal muscles (104) the muscles between the ribs which play a part in the breathing movements

internode (14) the length of stem between two adjacent leaves

isotope (53) one form of an element which can be distinguished from the other forms because of its radioactivity or some other property

karyogram (184) a chart showing the shapes and numbers of the chromosomes in a cell of an individual

labium (152) a component of an insect's mouth parts

labrum (152) a component of an insect's mouth parts

lachrymal duct (87) the tube running from the eye to the nasal cavity

lacteal (89) the tube in the centre of a villus into which pass the products of fat digestion in the intestine

lactic acid (49) an intermediate product of the breakdown of glucose during respiration

lactose (90) a sugar present in milk

lamina (14) the flat part of a leaf

larva (155) the stage in the life cycle of certain insects which hatches from the egg

larynx (86) the upper part of the windpipe which communicates with the pharynx

latent heat (70) the heat energy needed to evaporate a liquid

lateral line (159) a sense organ on each side of a fish which responds to vibrations in the water

leguminous plants (59) plants of the pea family; their roots contain nitrogen-fixing bacteria

lenticel (16) a gap in the bark of a twig through which exchange of oxygen and carbon dioxide can occur

leucocyte (93) a white blood cell

ligaments (120) the tough, fibrous strands holding bones in place at a joint

linkage (192) the occurrence of genes on the same chromosome so that they tend to stay together during inheritance

lipase (85) an enzyme which digests fats

loam (79) a type of soil containing both sand and clay with adequate humus

lymph (100) a fluid in the body, derived from blood plasma and returned to the circulation via the lymphatic system

lymphatic (98) a vessel which returns lymph from the tissues to the circulatory system

lymphocyte (100) a type of white cell which makes antibodies

malleus (134) the first of the chain of small bones in the middle ear

Malpighian layer (109) the deepest layer of epidermal cells in the skin; these cells reproduce and replace the cells above them

maltose (87) a sugar produced by the digestion of starch (e.g. in germinating seeds)

mandibles: bird (168) extensions of the skull and lower jaw to form the beak
insect (147) the first pair of mouth parts on the head

maxillae (152) form the second pair of insects' mouth parts

meiosis (188) the form of nuclear division which takes place when cells divide to produce gametes and the chromosome number is halved

Meissner's corpuscle (128) a sensory nerve ending in the skin which responds to touch

melanin (109) a black pigment in the hair and skin of animals

menstruation (118) the breakdown of the lining of the uterus which occurs when an egg has not been fertilized

mesophyll (17) the palisade and spongy layers of cells in a leaf

metabolism (50) all the chemical changes going on in the cells of an organism

metacarpals (120) the bones of the hand

metamorphosis (144) the series of changes by which a larval form of an animal becomes an adult

metaphase (181) a stage in mitosis or meiosis when the chromosomes are arranged on the equator of the spindle

metatarsals (119) the bones of the foot

micropyle (35) a small hole in an ovule or seed of a plant

micropyle (156) a small hole in an insect's egg

midrib (20) the main vein and supporting tissue running down the middle of a leaf

mitochondrion (13) a cell organelle concerned with respiration

mitosis (180) the events in a nucleus at cell division

molar tooth (125) a tooth at the back of the jaw

monocotyledon (6) a plant having only one cotyledon in its seed

monoculture (61) the agricultural practice of growing large numbers of a single species of plant together

monoploid (188) *see* haploid

mosaic image (149) the kind of "picture" likely to be produced by an insect's compound eye

motor fibre or **neurone** (138) a nerve fibre or cell which carries impulses from the central nervous system and causes some action at the other end (e.g. muscle contraction or enzyme secretion)

Mucor (75) a family of mould fungi

mucus (88) a sticky fluid produced by animals to lubricate and protect delicate surfaces

multifactorial inheritance (194) where many genes are involved in controlling a characteristic

multipolar neurone (138) a nerve cell in the central nervous system with many fibres making synapses with adjacent neurones

mutation (186) a spontaneous change in a gene or chromosome which usually produces an observable effect in the organism concerned

mycelium (75) the mass of hyphae which make up a fungus

natural selection (206) a theory which tries to explain how evolution could have occurred

neural spine (122) the bony projection from the dorsal surface of a vertebra

neurone (138) a nerve cell

nictitating membrane (168) a third, transparent eyelid in birds

nitrifying bacteria (59) bacteria whose activities increase the amount of nitrate in the soil

nodules (59) swellings on the roots of leguminous plants which contain nitrogen-fixing bacteria

nucleic acid (186) a chemical present in nuclei which is thought to control the activities of the cell and the pattern of inheritance

nucleolus (13) a small structure within a nucleus

nucleoplasm (10) the protoplasm of the nucleus

nucleus (10) the structure in a cell which determines the shape and controls the activities of a cell

nymph (152) an immature stage in the life cycle of certain insects

ocellus (150) a simple eye of an insect

oesophagus (86) the gullet, a tube conveying food from the mouth to the stomach

oestradiol, oestrogen, oestrone (144) female sex hormones

ommatidium (149) one of the many light receptors in an insect's compound eye

omnivorous (126) having a diet equally composed of animal and plant material

oogenesis (188) the series of cell divisions in the ovary which give rise to egg cells

operculum (159) the bony plate covering the gills on each side of a fish's head

orbit (120) the cavity in the skull which houses the eye

organ (12) a group of tissues working together to do a particular job

organelle (13) a protoplasmic structure with a particular function in a cell

osmoregulation (67) the control of the quantity of water entering and leaving the cells of an organism

osmosis (63) the movement of water from a weaker to a stronger solution

osmotic pressure (63) the pressure built up in a system as a result of intake of water by osmosis

ossicles (135) a chain of three tiny bones in the middle ear, which transmit vibrations from the ear drum to the middle ear

otoliths (135) chalky granules in the utriculus which helps it respond to changes in posture

oviduct (112) the tube which conveys eggs from the ovary to the uterus or to the outside world

ovipositor (147) an appendage on the last abdominal segment of an insect which helps it to place its eggs below the surface

ovulation (114) the release of mature eggs from the ovary, ready to be fertilized

ovule (30) the part of the ovary of a flower which contains the female gamete and will become the seed

ovum (112) the unfertilized egg of an animal

Pacinian corpuscle (128) a sensory receptor in the skin which responds to pressure

palisade cells (18) the layer of cells on the upper surface of a leaf just below the epidermis

pancreas (86) a gland beneath the stomach which secretes digestive enzymes into the small intestine

Paramecium (178) a single-celled, ciliated animal

parasite (154) an animal or plant which derives its food from another organism without necessarily killing it

patella (119) the bone of the knee-cap

pepsin (87) a protein-digesting enzyme produced in the stomach

peptide (87) a chemical consisting of a chain of amino acids and resulting from the partial digestion of a protein

perennial (22) a plant which survives successive winters

pericarp (30) the outer coat of the ovary of a flower; it becomes the fruit wall

perilymph (134) the fluid in the inner ear

periosteum (121) the fibrous tissue covering bones

peristalsis (86) the muscular contractions which move food down the gullet and along the alimentary canal

petiole (17) a flower stalk

phagocyte (93) a white cell which can ingest foreign particles

phalanges (120) the bones of the fingers or toes

pharynx (86) the area at the back of the mouth cavity leading into the nasal cavity, gullet, and windpipe

phenotype (194) the observable characteristics of an organism

photosynthesis (50) the production of food by green plants from carbon dioxide, water, and salts using light as a source of energy to drive the chemical reactions involved

phototropism (43) a change in the direction of growth of a shoot in response to one-sided illumination

phylum (6) a major group of animals or plants

phytoplankton (57) microscopic green plants living in the surface waters of the sea, lakes, etc.

pituitary body (144) an outgrowth from the floor of the brain which acts as an endocrine gland producing many hormones

placenta, mammal (116) the organ in the uterus which enables food, oxygen, and waste products to pass between the embryo and its mother

plankton (57) the microscopic plants and animals living in the surface layers of natural waters

plasma (98) the liquid component of the blood

plasmolysis (64) partial collapse of a cell as a result of withdrawal of water by osmosis

platelets (93) small bodies in the blood which play a part in clot formation

pleural fluid (104) the fluid secreted by the pleural membrane into the pleural cavity

pleural membrane (104) the membrane lining the outside of the lungs and the inside of the thorax

plumule (39) the leafy part of the embryonic shoot in a seed

poikilothermic (159) body temperature almost the same as the surroundings and varying with the latter

polar body (188) a functionless female gamete produced when an egg mother cell undergoes meiosis

polymorphism (208) a variation which is always present between individuals in a population

potometer (70) an apparatus for measuring water uptake in a shoot

primary feathers (168) the large flight feathers on a bird's wing inserted in the skin over the "hand" bones

proboscis (152) projecting, sucking mouth parts in an insect

producer (58) organisms, mainly plants, assigned a position in a food chain because they make their own food

progesterone (115) the female sex hormone produced by the corpus luteum in the ovary after ovulation

prolegs (157) appendages on some of the abdominal segments of a caterpillar

prophase (181) the earliest stage in mitosis or meiosis; the chromosomes appear and shorten

proprioceptor (129) an internal sense organ responding to changes usually in muscles

prostate gland (113) an accessory male sex organ in mammals; adds a secretion to the sperms at ejaculation

protein (82) the class of foodstuffs which provide the raw material for making the cytoplasm of cells

proteinase (85) an enzyme which breaks down proteins

protista (178) a collective name for single-celled animals and plants

protoplasm (10) the living material in cells; cytoplasm and nucleoplasm

protozoa (176) single-celled animals such as Amoeba

pseudopodium (176) a temporary protrusion of cytoplasm extended by an Amoeba during locomotion

pseudotracheae (154) the channels in the proboscis of a housefly or blowfly which carry saliva and digested food

puberty (114) the stage when men and women become capable of sexual reproduction

pulmonary artery and vein (95) the blood vessels from the heart carrying blood to or from the lungs

pupa (155) the inactive stage in the life cycle of certain insects when the organs of the adult are being formed

pyloric sphincter (86) a ring of muscle which can close the exit of the stomach

radicle (39) the embryo root in a seed

radius (119) one of the bones of the lower fore-limb

receptacle (30) the expanded end of a flower stalk bearing the parts of the flower, petals, etc.

recessive (194) gene which, in the presence of its contrasting allele, is not expressed

recombination (190) a combination of genes in the offspring which was not present in either of the parents

rectum (86) the last part of the large intestine

reduction division (188) *see* meiosis

reflex (139) an automatic response to a stimulus which is not under conscious control

reflex arc (140) the nervous pathway conducting the impulses which result in a reflex action

renal artery or vein (95) the vessels from aorta or vena cava conducting blood to or from the kidneys

rennin (87) a digestive enzyme in the stomach which coagulates the protein in milk

replication (181) making a new gene or chromosome exactly like the original

retina (131) the layer of light-sensitive cells inside the back of the eye

rhabdom (148) a microscopic transparent rod in the centre of one of the components (ommatidia) of an insect's compound eye

rhizome (24) a horizontal, underground stem swollen with stored food and capable of producing new independent plants from its buds

ribosome (13) a cell organelle concerned with protein synthesis

rickets (84) soft, easily distorted bones in children lacking vitamin D

roan (197) a coat colour in cattle resulting from incomplete dominance of two genes for colour

rod (131) a light-sensitive cell in the retina of the eye; it responds to weak light but not to colour differences

roughage (84) the indigestible portion of food; usually cellulose from plants

sacculus (134) a sense organ of the inner ear which may respond to changes of posture

sacrum (120) the part of the spinal column to which the pelvic girdle is joined

saprophytic (50) a form of nutrition in which dead organic matter is digested externally and the products absorbed (e.g. in fungi)

scapula (120) the shoulder blade

scion (26) a bud or shoot which is grafted onto another plant's stem

sclera (130) the tough tissue on the outside of the eyeball

scrotum (112) the sac of skin which carries the testes outside the body cavity

scurvy (83) a disease in which the skin and blood vessels are susceptible to damage and infection due to lack of vitamin C

sebaceous gland (109) a gland which secretes an oily substance on to the skin at the top of a hair follicle

secretin (144) a hormone produced by the lining of the duodenum when the acid contents of the stomach reach it; it stimulates the pancreas to produce enzymes

segment (157) a division of the body into sections visible in certain invertebrates such as worms and insects

segregation (194) the separation of allelomorphic genes into different gametes at meiosis

selectively-permeable membrane (63) a membrane which permits water to pass through it more readily than dissolved substances. Sometimes called a **semi-permeable membrane**

semicircular canals (134) sensory organs consisting of fluid-filled tubes in the inner ear which respond to rotatory movements

semi-lunar valves (98) pocket-like valves in the main arteries leaving the heart which prevent return of blood to the ventricles

sepal (30) a leaf-like structure on the outside of a flower

serum (93) blood plasma from which fibrinogen has been removed

sieve plate (13) the perforated cross-wall which allows communication between two sieve tube cells

sieve tube (16) a row of cells which transport food in the vein of a plant

silt (78) very fine mineral particles, smaller than sand grains, in the soil

special senses (129) the eyes, ears, nose, and tongue which are specialized for detecting certain types of stimulus

specialization (12) the development of a structure or process to do one particular job

species (6) a group of animals or plants possessing a great many features in common and capable of breeding with each other

sphincter (87) a circle of muscle in a tube or duct; its contraction reduces the diameter of the tube or closes it altogether

spikelet (32) a group of flowers in the inflorescence of a grass

spindle (180) a structure in a cell which appears at mitosis or meiosis and plays a part in separating the chromosomes or chromatids

spiracle (146) an opening in an insect's cuticle which admits air to the tracheal system

Spirillum (72) a family of bacteria with a spiral shape

Spirogyra (174) a filamentous green alga occurring in fresh water

spleen (93) an organ in the abdomen, near the stomach; it makes white cells, destroys worn-out red cells, and removes foreign particles from the blood

spongy layer (18) the lower layer of cells in the mesophyll of a leaf blade

sporangium (75) a small capsule containing reproductive spores produced by fungi and other types of plant

spores (72) a reproductive cell of a fungus and certain other types of plant which can grow to produce a new individual

stamens (30) the male reproductive organs in a flower; each stamen consists of a stalk (filament) and an anther

stapes (134) the last of the chain of bones in the middle ear; fits into the oval window

Staphylococcus (72) a family of bacteria with spherical shape whose cell division produces clumps rather than chains

sterilization: bacterial (74) the destruction of bacteria by heat, chemicals, or radiation
   genetic (200) rendering a person incapable of breeding

sternum (104) the breast-bone; it joins the front end of the ribs

stigma (30) the part of the carpel in a flower which receives the pollen

stoma (19) the opening in the epidermis of a leaf or stem through which oxygen, carbon dioxide, and water can pass

Streptococcus (72) a family of spherical bacteria whose cell division produces chains of individuals

style (30) the part of the carpel of a flower between the ovary and the stigma

sucrose (51) a sugar produced mainly from sugar-cane or sugar-beet

suspensory ligament (130) the fibres running from the edge of the lens in the eye to the ciliary body

synapse (138) the junction between one nerve cell and the next across which the nerve impulse has to pass

synovial membrane (120) the lining of the capsule of a moveable joint in the skeleton; it secretes the synovial fluid which lubricates the joint

system (12) a series of organs working together to a certain purpose (e.g. the circulatory system)

tarsals (119) the bones of the ankle

telophase (181) the final stage of mitosis or meiosis when the chromosomes become less visible and the nuclear membrane re-forms

temporalis muscle (123) a muscle from the lower jaw to the skull which closes the jaws during chewing

tendon (99) the bundle of tough fibres attaching a muscle to a bone

terminal bud (14) the bud at the end of a shoot or branch

territory (162) an area defended by an individual against other members of the same species

testa (39) the outer coat of a seed

testis (112) the male reproductive organ of an animal; it produces sperms

testosterone (144) a male sex hormone produced by the testis

thoracic duct (100) a large lymphatic duct emptying the lymph from most of the body into the circulatory system near the heart

thorax: man (102) the upper part of the trunk above the diaphragm which contains the heart and lungs
   insect (146) the three middle segments of the body carrying limbs and wings

thyroid (144) an endocrine gland in the neck; it produces the hormone thyroxine

thyroxine (144) the hormone produced by the thyroid gland; it controls rates of metabolism

tibia (119) the larger of the two bones in the lower hind limb

tissue (12) a group of cells similar in structure and function (e.g. muscle)

tissue respiration (47) the chemical breakdown of food in cells to provide energy for living activities

toxin (73) a poison produced by bacteria

trachea: man (86) the windpipe; it conducts air from the mouth and nose to the bronchi and lungs
   insect (146) a component in the system of tubes which carries air from the spiracles to all parts of the body

tracheole (146) a branch of a tracheal tube in an insect

translocation (68) the transport in plants of food and other dissolved substances

transpiration (69) the evaporation of water from the shoot, particularly the leaves, of a flowering plant

tropisms (43) a change in the direction of growth of a root or shoot in response to the direction of an external stimulus

trypsin (88) a protein-digesting enzyme secreted by the pancreas into the small intestine

tuber (24) a food store producing a swelling in a root or stem

turgid (64) the condition in a plant cell when the vacuole is pressing outwards strongly on the cell wall

ulna (120) one of the two bones in the lower forearm; it runs from elbow to wrist

umbilical cord (116) the cord containing blood vessels which conduct embryonic blood to and from the placenta

urea (91) a nitrogen-containing chemical formed in the liver from excess amino acids; it is an excretory product

ureter (106) the tube conducting urine from the kidney to the bladder

urethra (106) the tube conducting urine from the bladder to the outside world

urine (108) a mixture of water, salts, urea, etc. removed from the blood by the kidneys

uriniferous tubules (106) microscopic tubes in the kidney which help to extract unwanted substances from the blood

uterus (112) the part of the female reproductive system in which the embryo develops

utriculus (134) a sensory organ in the inner ear; it responds to changes in posture

vaccine (95) a preparation of dead, inactive, or harmless bacteria or viruses which, when introduced to the body, cause it to produce antibodies

vacuole (10) the fluid-filled cavity in the centre of a plant cell or the droplets of fluid in the cytoplasm of animal cells

vascular bundle (16) groups of vessels, sieve tubes, and strengthening fibres which conduct water and food through a plant

vector (154) an animal which transmits a disease-causing organism from one plant or animal to another

vena cava (95) the large vein returning blood from the body to the right atrium of the heart

ventilation (47) a method of exchanging the air or water in contact with a respiratory organ

ventral root (140) a nerve carrying motor fibres from the spinal cord to the body

ventricles (95) the lower, more muscular chambers of the heart which pump blood into the arteries

venule (97) a blood vessel emptying into a vein

vertebra (122) one of the bones of the spinal column

vertebral column (122) the "backbone" or spine

vertebrate (6) an animal possessing a vertebral column

vessel (16) a water-conducting tube of a plant made up of dead cells joined end to end

villus (89) one of thousands of finger-like protrusions from the internal surface of the small intestine

virus (74) a sub-microscopic particle living in cells of plants and animals and causing disease

vitamin (83) complex chemicals which must be present in the diet for normal health but have no energy value

vitreous humour (130) the jelly-like fluid in the main part of the eye

viviparous (162) bringing forth live young, as distinct from laying eggs

voluntary muscles (122) muscles under conscious control (in man)

white cells (93) cells in the blood, lacking haemoglobin but possessing a nucleus and, in some cases, powers of independent movement

white matter (142) the area of the brain and spinal cord consisting of nerve fibres as distinct from cell bodies

wilting (64) excessive loss of water from a plant leading eventually to a collapse of the leaves and stem

X chromosome (190) the chromosome which, in the absence of a Y chromosome determines that certain animals will be female

xerophthalmia (84) a disorder of the skin affecting the conjunctiva of the eye resulting from a deficiency of vitamin A

xylem (16) the tissue in the vascular bundle of a plant composed of vessels and supporting cells

Y chromosome (190) the chromosome whose presence determines that certain animals will be males

zona pellucida (115) a structureless "shell" round the ovum of a mammal

zooplankton (57) microscopic animals living in the surface waters of oceans, lakes, rivers, etc.

zygospores (174) the zygote produced by sexual reproduction of some algae and fungi which can resist adverse conditions

zygote (34) the cell produced when a male and female gamete fuse; it can grow into a new individual

zymase (77) a starch-digesting enzyme produced by yeast

Index

(bold figures indicate where subject is defined or given main treatment)

abdomen 106
  in insects 146, 147, 156–8
ABO blood groups 197, 198
abscission 28, 29
absorption of food 85, 87, **89**, 90
accessory food factor 83
accommodation of the eye 132
acetic (ethanoic) acid 77
acetic orcein 193
achondroplastic dwarfism 198, 208
acid soil 79
active transport 63, 69, 100
adaptation:
  beaks and feet 173
  birds to flight 170
  fruits and seeds 38
  insect mouth parts 152–4
  leaf to photosynthesis 18, **50**
  limbs 205
  teeth 126, 127
adenine 186, 187
ADH (anti-diuretic hormone) 108, 144
adipose tissue 90, **91**, 110
adrenal gland 113, **144**, 145
adrenaline 144, 145
adventitious root **20**, 23–6, 41
aerobic respiration **47**, 49
after-birth 116
agar 74, 80
agglutinin 94
aggregates (soil) 78
agriculture 60, 61, 80
air bladder 160
air passage 102
air space:
  leaf 17–19, 50, 51
  soil 78, 79
albinism 198, 200
albumen 165, 170, 172
albumin 93
alcapton 186
alcohol 49, 50, 77
alcoholic fermentation 77
algae 6, 54, 57, 60, 74, 176
alimentary canal 126
  fish 161
  insect 147
  man 85–90, 94
allantois 172
allele 194
allelomorphic gene 194
alveolus (pl. alveoli) 102, 103
amino acid **82**, **83**, 88, 89, 93, 94, 98, 106,
  116, 186, 203
  utilization of 90–2
ammonia 59, 106
ammonium compounds 108
  as fertilizers 60
  in nitrogen cycle 59
ammonium sulphate 79
amnion 115, 116, 172, 199
Amoeba 176, 177
amphibia 6, 8, 112, 163, 204
ampulla 135
amylase 85, 87, 90
anaemia (sickle-cell) 207
anaerobic respiration 47, **49**, 50, 73, 77
anal fin 160
anaphase 180, **181**, 182, 183, 188, 189, 192
androecium 30
animal characteristics 9
animal classification 6–8
ankle 120
annual plants 22
Anopheles 154
antagonistic hormones 145
antagonistic muscles 122, 124, 150
antenna 146, 148, 152, 153, 156–8
anther **30**, 31, 32, **33**, 34, 193
antibiotic 74
antibodies 93–5, 100
antidiuretic hormone (ADH) 108, 144, 145
antigen 94
antiseptic 74
antitoxin 73, 94, 95
anus 85, 86, 89, 113, 159
aorta 95, 96, 98, 99, 106
Apanteles 158
aphid 50, 152
appendix 73, 86, 88, 89, 126
apple 36, 37
aqueous humour 130–2
Arachnid 6, 7
arteriole 97, 99, 106, 108–11
artery 96, **97**, 98, 101, 117
arthropod 146
artificial fertilizer 60, 61, 78
artificial propagation 26
ascorbic acid 84
asexual reproduction 22, 75, 177

association centre 141, **143**
association neurone 140
atrium:
  heart 95, 98–100
  tadpole 166
autoclave 74
autoradiograph 54
auxin 44
axil 14
axillary bud 27
axon **138**, 139

bacillus 72
back-cross 194, 197, 198
bacteria 49, **72–4**, 89, 94, 95, 103, 109, 110,
  155
  in nitrogen cycle 59, 60
  in soil 78, 80
baking 77
balance of nature 57–9, **60**, 61
balance, sense of 129, 135
balanced polymorphism 208
ball and socket joint 120, **121**
barb and barbule 168, 169
bark 17, 69
base (organic) 186, 187
base pair 186, 187
bastard wing 169, 170
beak 168
  adaptations of 173
Benedict's solution (reagent) 85, 92, 211
beri-beri 83, 84
BHC (insecticide) 155
biceps 122, 123, 140, 141
bicuspid valve 98, 99
biennial plant 22
bile 84, 86, 88, 90–2
bile duct 86, 88, 92
bile pigments 106
bile salts 88, 91
binary fission 177
biodegradable 73
biogenesis 204
birds 6, 8, 112, **168–73**, 204
  seed dispersal by 38
birth 116
birth control 118
bitter (taste) 129, 136
bivalent (chromosome) 188
bladder 106–**108**, 112, 113
blind spot 130, 131, 136
blinking 130, 139
blood **93–101**, 103, 108
  circulation of 95–100
  composition of 93
  functions of **93**, **94**
blood cells 93, 101
blood groups 197, 198, 208
blood pressure 98, 101
blood smear 101
blood sugar 91
blowfly 154
body temperature 159
bolus 86, 87
bone 119–21
bone marrow 93, 121
bottling 73
Bowman's capsule 106, **107**, 108
brachydactyly 198
bract 32, 38
brain 128, 129, 131, 134, 137–41, **142**, **143**
breathing:
  fish 161
  frog 164
  insect 146, 147, 155
  man 101–5
  tadpole 166
brewing 77
bristles (sensory) 148, 154
bronchiole 102
bronchus 86, **102**
brooding 172
bubonic plague 154
bud **27**, 28
  dormant 27, 29
  grafting of 26
  lateral (axillary) 14, 22–5, **27**
  terminal 14, 23, 24, **27**, 29
bud scales 27–9
budding (yeast) 77
bulb 14, 22, **23**, 193
buttercup:
  flower of **30**, 31, 33
  pollination of 33
butterfly 153, 156–8
buzzard 173

caecum 73, 86, 88, 89, 126, 178

calcium 119, 125
  in diet 83, 85
  in plant nutrition 54, 55
calorific value 83
calyx **30**, 32
cambium 14, 15, **16**, 17, 26, 179
camouflage 206
canine tooth 126, 127
canning 73
canopy (tree) 25, 61, 80
capillary 89, 96, **97**, 98–103, 106, 107, 109–
  11, 116, 117, 125, 161, 164, 168, 172
capillary attraction 78
capsule:
  joint 120, 121
  plant 37, 38
carbohydrate 47, 50, 51, 54, **82**, 83, 85, 87, 88,
  90
carbon cycle **58**, 59
carbon dioxide 77, 101–3, 106–8, 116, 164,
  172, 174
  diffusion of 62–4, 103
  in carbon cycle 58, 59
  in photosynthesis 50–4, 56, 57
  in respiration 47–50, 53
  transport of 93, 94, 97, 98
carbonic anhydrase 103
carnassial tooth 126
carnivore 125, 126
carnivorous 57, 162, 164
carotene 84
carotid artery 96
carpal 120, 121, 169, 204
carpel **30**, 31, 36
carrier:
  in active transport 69
  genetical 193, 200
cartilage 102, 120–2
catalyst 85
caterpillar 156–8
caudal fin 159
cell **10–13**, 98–100
cell body 137–42
cell division **11**, 115, 116, 165, 174, 177, **180**,
  181, 182, 188, 189
cell membrane 11, 12, 179, 180, 189
cell sap 61, 64–8
cell wall **10**, 11, 12, 44, 51, 54, 64, 66, 72,
  174–6
cellulose 9, 51, 54, 73, 82, 84, 89, 126, 127,
  178
cement 125–7
censer mechanism 38
central nervous system 139
centriole 179, 180, **181**, 189
centromere 180, 181, 183, 188, 189, 192
centrum 122
cerebellum 142, 143
cerebral cortex 142
cerebral hemisphere 142, 143
cerebrum 143
cervix 112, 115–18
chaffinch 168
chemical fertilizer 60, 78
chemical sense:
  insects 148
  man 129, 136
chemo-receptor 149
chewing 87, 126, 127
chewing muscle 123
chiasma 188, 189, 192
chitin 75
chlorination 74
chlorophyll 9, 18, 41, 44, 50–2, 54
chloroplast 12, **18**, 19, 50, 51, 54, 56, 174–6
cholera 72, 73, 155
choroid 130
chromatid 180–3, 188, 189, 191, 192
chromosome 179–81, **182–185**, 186–94, 199
  giant 185
chromosome mutation 184, 185, 187
chrysalis 157
chyme 87, 88
cilia 102, 103, 177
ciliary body 130, 132
ciliary muscle 130, 132
ciliate 177
ciliated cell 114
circulation of blood 94, 95, 168
circulatory system:
  embryonic 116, 117
  fish 161
  insect 147
  man 95, 96
citrus fruit 83
Clarke's fluid 193
clasper 157, 158
classification **6**, 7, 8
clavicle 119, 123, 171
clay 78, 80
Clear Lake 61

cleavage 165
clinostat 43
clitoris 193
Clostridium 72
clot (blood) 94
clotting (blood) 84, 93, 94, 153, 193, 198
cobalt chloride 71
cocci 72
cochlea 134, 135
cocoon 158
cod-liver oil 84
codominance 197
coelenterate 6, 7
cohesion theory 70
"cold-blooded" 159
coleoptile 41
  and growth hormone 44, 45
coleorhiza 41
collecting tubule 107
colon 86, 88–90
colour blindness 192, 193
colour vision 132, 150, 158, 198
combustion 59
common ancestor 204, 205
comparative anatomy 205
compensation point 51
competition 25, 38
complete metamorphosis 156
composite flowers (Compositae) 32
compound eyes 146, 147, **148–150**, 153, 155,
  158
conditioned reflex 140, 141, 143
conditioning 141
conditions for germination 42, 43
cone 131, 132
congenital defects 199, 200
coniferous trees 6
conjugation 77, 175
conjunctiva 130
consanguinity 200
consciousness 143
conservation (soil) 81
consumer 58
continuous variation 198, 199, 201
contraception 118
contraceptive 118
contractile root 24
contractile vacuole 176–8
contraction (muscle) 122–4
control (experimental) **42**, 44, 47–50, 52–4,
  66, 67, 69, 71, 72, 74, 80, 92, 168
control of insect pests 155
controlled experiment 42, 52
converging lens 133
cooking 73
co-ordination **137–145**, 177, 178
copulation 115, 118
coracoid 170, 171
cork 28
corm 14, 22, **24**
cornea 84, 130, 131
cornified layer 109, 110, 129
corolla 30
coronary artery 98
corpus callosum 143
corpus luteum 114, 115, 118, 144
correction of sight defects 133
cortex:
  adrenal 144, 145
  brain 139, 142
  kidney 106–8
  root 20
  stem 14, 15, **16**, 67
cortisone 144, 145
cotyledon 36, **39**, 40, 41, 46, 47
coughing 103
courtship:
  bird 170
  stickleback 162
cousin marriage 200
covert 168
Cowper's gland 113, 115
coxa 148, 158
cremaster 157, 158
cretinism 144
crista 135
crocus corm 24
crop (insect) 147
crop rotation 59, **60**, 81
cross pollination 33
crossing over 188, **192**
crown (tooth) 125
crumb (soil) 78
crumb structure 60, 78, 79
crustacea 6, 7, 57, 119
crystalline cone 148, 149
crystalline lens 130
culture:
  bacteria 74
  Drosophila 201
culture solutions 55, 56

cupula 135
curlew 173
cuticle:
  leaf 18, 70
  insect 146, 148, 150, 156
cuticular lens 148–50
cuttings 26
cyst (Amoeba) 177
cytoplasm **10**, 11–13, 51, 64, 65, 72, 75, 77, 82, 83, 93, 125, 138, 165, 174–81, 183, 188, 189
  semi-permeability of 64–6
cytosine 186, 187

daisy 32
daffodil bulb 23
dandelion:
  flower of 32
  fruit of 38
dark reaction 54
Darwin 206
DDT (insecticide) 61, 155
deamination 90–**92**
decay 58, 59
deciduous 6, 28, 70
decomposer 58
deficiency disease 84
deforestation 61, 80, 81
dendrite 137, **138**
dendron 138, 139
denitrifying bacteria 59
dentine 125–7
deoxygenated blood 103, 106, 117
deoxyribonucleic acid (DNA) 183, **186, 187**
deoxyribose 186, 187
depressor muscle (insect) 151
dermis 109, 110, 128, 129
desmid 175
destarching a leaf **51**, 52
detoxication 92
development:
  bird 172
  frog 165–7
  man 116
diabetes 144
dialysis tubing 63
diaphragm 86, 88, 96, 103–6
diastole 99
diatoms 57, 176
dicotyledon 6, 39
diet 91
differential permeability 63
diffusion 50, 51, 56, **62**, 63, 69, 94, 98, 100, 103, 105, 161, 164, 178
diffusion gradient **62**, 63, 94, 103, 105, 107
digestion 75, **85–88**, 89, 90, 178
digestive juice 87, 90
diploid **183**, 189, 190
Diplococcus 72
direct flight muscles 151
direction of sound 134, 136
discontinuous variation 198
disease 94, 95
  insects as carriers of 154, 155
dispersal of fruits and seeds 38
distance judgement 133
diverging lens 133
DNA (deoxyribonucleic acid) 183, **186**, 187
dominant 187, 190, **194**, 197, 198, 200, 207
dormancy 42
dormant 76, 167, 174
dormant bud 27, 29
dorsal fin 159, 160
dorsal root 140, 141
dorsal root ganglion 140, 141
double image 132, 136
down feather 168, **169**
droplet infection 73, 74
Drosophila 183–6, 192, 201, 202
  culture 201
  sexing 202
dry weight 47, 56
duodenum 86, 87, **88**, 90, 144
dwarfism 187, 198, 208
dysentery 155, 178

ear 134, 135, 142
ear bones 134
ear drum 134, 163
ecdysis 119, 146, 147, 156, 157
echinoderm 6
ectoplasm 176, 177
egestion 9, 86
egg:
  bird 168, 170, **172**
  fish 162
  frog 164, 165
  insect 156, 158
egg cell 114
egg laying 170

egg membrane 115, 165, 172
egg shell 170, 172
Eijkman 83
elbow joint 120, 121
electron micrograph 13
electron microscope 12
elements 53–6, 59, 60
  essential 54–6
  trace 55
elevator muscle 151
embryo:
  bird 172
  frog 165
  man 112, 116, 117
  plant 36, **39**, 40, 41
emulsification of fat 89–91
enamel 125–7
endocrine glands 144–5
endocrine system 137, **144, 145**
endolymph 134, 135
endoplasm 176, 177
endoplasmic reticulum 13
endoskeleton 119
endosperm 40, 41, 46, 47
endothelium 96, 97
energy 77, 82, 83, 90, 92, 101, 107
  from respiration 47, 49, 54, 59
  from sunlight 50, 51, 58
  sources of 58
enterozoic protozoa 177
environment and heredity 199, 200
enzymes 41, 47, 49–51, 73, 75–7, 82, 91, 103, 108, 114, 115, 127, 130, 144, 146, 154, 177, 178, 186, 198
  digestive **85**, 86–8, 90
  experiments on 92
epicotyl **39**, 40, 41
epidermis:
  human 109–11, 128
  plant **14**, 15–17, **18**, 19–21, 67
epididymis 113, 114
epigeal germination 40, **41**
epiglottis 86, 87, 103
epithelium 89, 101, 103, 129
erectile tissue 113
erosion 60, 61, **80**, 81
erythrocyte **93**, 94
essential amino acid 83
ethanol 49, 85
etherizing 201
etiolation 44
eugenics 199, 200
Euglena 176, 178
Eustachian tube 87, 134
eutrophication **60**, 61
evergreen 28, 70
evolution 203–9
excretion 9, **106–8**, 178
excretory organ 106
excretory product 106–8
exhaling 104
exoskeleton 146, 150
expiration 104, 105
explosive fruits 38
extensor muscle 120, 122, 123, 150, 163
external fertilization 112, 162, 165
external gill 166, 168
eye:
  colour of 131, 198
  fish 159
  frog 163
  man 129, **130–133**, 142
  muscles of 133, 143
eyeball 130, 142, 163
eyelid 130, 168

F_1 generation **194**, 195, 196, 198, 201
F_2 generation 195, 196, 201
faeces 89, 155
"false fruit" 36
fat 47, 61, 82, **83**, 85, 100, 111, 172
  digestion of 87, 88
  storage of 91
  test for 85
  utilization of 90, 92
fat cell 110
fat-soluble vitamin 83, 92
fatty acid 88–90
fear 144
feather 168, **169**, 170, 171
feed-back 145
feeding 9
  Amoeba 177
  birds 172, 173
  fish 162
  frog 164, 166, 167
  insects 152–4
  Paramecium 177
  tadpoles 166–8
Fehling's solution 211
femoral artery 96

femur:
  bird 170, 171
  insects 148, 153, 158
  mammals and man 119–21, 124
fermentation **49**, 50, 77
fern 6
fertilization 188, 190, **191**, 195, 196
  birds 170
  definition of 34
  external 112
  fish 162
  flowering plants 34, **35**, 36
  frog 165
  internal 112
  man 112, 115, 118
  Spirogyra 174, 175
fertilizers 60, 61
fibre (nerve cell) 109, 128, 129, **138**, 139, 141, 142
fibrin 94
fibrinogen 91–5, 107
fibrous root system 20
fibula 119, 120, 171
field capacity 78
filament (stamen) **30**, 31–4
filter feeding (fish) 162
filtration (kidney) 107
fins 160
finger 120, 121, 123, 128, 169
fish 6, 8, 67, **159–162**, 204
fitness 208
fixation of nitrogen 59
flaccid cells 64
flagellate 176, 178
flagellum 72, 176
flapping flight 170, 171
flatworm 6, 7
flavour 129
flea 154
flexor muscle 122, 123, 150
flies 74
flight:
  birds 170, 171
  insects 151, 158
flight feather 168, **169**, 171
flight muscle (insect) 147
flocculate 79
floret 32
flower 30–7
flowering plants:
  classification of 6
  nutrition of 50–7
  sexual reproduction of 30–9
  structure of 14–21
focal length 132
foetus 115, 117
follicle (ovary) 115, 118
follicle-stimulating hormone (FSH) 145
food 47, 49, 73, 74, **82**, 83–5, 101, 116, 117, 172, 173
  absorption of 89
  digestion of 85–8
  preservation of 73
  production by plants of 50–7
  storage of 90, 91
  tests on 46, 85
  transport of 93, 97, 98
  utilization of 90, 91
food chain **57**, **58**, 61, 176
food vacuole 176–8
food web 58
foot 120
fore-brain 142, 143
fossil 203, 204
fovea 130, **131**, 132, 133
fraternal twins 117, 199
french bean 40, 41
frog 163–8
frog spawn 165
frontal lobe 142, 143
fructose 90
fruit 30, 31, 34, 36, 37
fruit dispersal 38
fruit fly (Drosophila) 183–5, 187, 201, 202
fruit formation 36, 37
FSH (follicle-stimulating hormone) 145
fuel 59
fungi 6, 49, **75**, 76, 77

galactose 90
gall bladder 86, 88, 92
gamete 30, 34, 112, 175, 186, 188–90, 194–6
gamma radiation 187
ganglion 139, 140
gaseous exchange 47, 51, 53, 102, 103, **105**
  bird's egg 172
  fish 161
  frog 164
gastric glands 90
gastric juice 87, 90

gene **185, 186**, 187, 188, 190, 194, 197, 198, 200, 201, 207
gene function 186
gene recombination 190, 192
general sensory system 128
genetic advice (counselling) 200
genetic code 187
genetics 179, 185–202
genotype 179, 194, 197, 199
geographical isolation 208
geological era or period 204
geotropism 43, 46
germination 39–43
gestation 116, 117
giant chromosome 185
gill bar 161, 162
gill filament 161, 162
gill raker 161, 162
gills:
  fish 159, 161
  tadpole 166, 168
girdle (skeleton) 122
gizzard (insect) 147
gland:
  digestive 85, 87–90
  endocrine 144
  mucous 164
gliding flight 170, 171
globulin 93
glomerular filtrate 107, 108
glomerulus 106, **107**, 108
glucose 41, 47, 49, 50, 53, 82, 88, 89, **90**, 91, 93, 94, 98, 100, 107, 108, 116, 145
  metabolism of 90, 91
  test for 85
glycerol 88–90
glycogen 75, 77, 82, 83, **90**, **91**, 92, 111, 144, 145
Gowland Hopkins 83
Graafian follicle 114, 115, 118, 144
grafting 26
granular layer 109, 110, 129
granulocyte 93
grass
  flower of 32
  pollination of 33, 34
gravity, response of root and shoot 43, 44, 46
greenfly 152
grey matter 139, 140, **142**
growth 9, 82, 83, 116, 119, 126, 144, 145, 159, 167, 179
  bacteria 74
  bud 27, 28
  bulbs, corms and rhizomes 23–5
  insects 146, 156
  minerals needed for 55, 56
  ovary in plants 36
  plant cells 64
  pollen tube 35, 39
  root 20, 46
  secondary 17
  seedling 40, 41
  twigs 29
growth curvature 44–6
growth hormone 144
growth-promoting substance 44, 45
growth responses of seedlings 43–6
grubs 156
guanine 186, 187
guard cell 18, **19**, **64**
gull 173
gullet 86, 88
gulley erosion 80, 81
gum 125, 127
guttation 65
gynaecium **30**, 31

haemocoel 147
haemoglobin 92, **93**, 94, 101, 103, 164, 207
haemophilia 193, 198
hair 109, 110
hair colour 198, 208
hair follicle 109, 110
hair plexus 128
half flower 30, 31, 36, 37
hand 121
haploid 188, 189
harmful bacteria 73, 93, 155
hatching:
  bird 172
  caterpillar 156
  tadpole 166
hay infusion 178
hazel catkin 33–5
head (insect) 146, 147, 152, 153, 156, 157
hearing 129, **134**, 142, 143, 159
heart 96, **98, 99**, 102, 103, 161
heart beat 98, 99
heart rate 98, 101, 144
"heat" (on "heat") 144

heat cramp 110
heat loss 110, 111, 117
heat receptor 128, 129, 140
heat stagnation 111
heat-stroke 111
heavy soil 78, 79
height (inheritance) 198, 199, 201
hepatic portal vein 89, 91, 95, 97
herbaceous plants 6
herbivore 89, **126**, 127
heredity 194–202
    and environment 199, 200
hermaphrodite 112
heterozygous 187, 193, **194**, 195–7, 200, 207
hibernation 111, 158, 167
hilum 39, 40
hind-brain 142
hind-gut 147
hinge joint 120–2, 127
hip joint 120, 121
holophytic 50
holozoic 50, 176
homeostasis 91, 92, 94, 103, **108**, **145**
homoiothermic 6, 110, 111, 168
homologous chromosomes 180, **183**, 188–90, 192, 194
homozygous **194**, 195–7, 200, 207
"honey-guides" 30, 31
hooked fruits 38
hormone 91–4, 107, 108, 114, 115, 117, 118, **144**, **145**, 164
    of plants 44
horse-chestnut twig 29
host 50
housefly 154, 155
human genetics 198–200
humerus 119–21, 123, 169–71, 205
humidity (transpiration) 70
humus 60, 65, **78**, 79, 80
hydrochloric acid 87, 89, 90
hydrogen 54, 73
hydrogencarbonate indicator 53, 211
hydrogen sulphide 73
hydrotropism 44
hypha 75–7
hypocotyl **39**, 40
hypopharynx 153
hypothalamus 108, 143

IAA (indoleacetic acid) 44
ichneumon fly 158
identical twins 117, 199
ileum 86, 88–91, 94, 100
image 130–3
    double 132, 136
    formation of 131
    inversion of 131, 136
    mosaic 149, 150
imago 156, 158
immunity 73, 94, 95
immunization 74
impulse (nervous) 128, 129, 131, 134, 135, 138–41
incisor tooth 120, 125–7
inclusion 13
incomplete dominance 197
incomplete metamorphosis 156
incubation 170, 172
incus 134
indirect flight muscles 151
indoleacetic acid (IAA) 44
infection 73, 74, 93
    prevention of 94, 95
inflorescence **30**, 32, 33
influenza 74
ingestion 86
inhaling 104
inheritance of characters 194
inner ear 134, 159
inoculation 95
insecticide 61, 155
insects 6, 7, 119, **146–158**
    pollination by 30, **33**, 34, 35
inspiration 104, 105
instinct 143
insulin 144, 145
integument 35, 36, 39
intelligence 143
    inheritance of 199, **200**, 201
interaction (hormonal) 145
intercellular spaces 50, 51, 62, 66, 69
intercostal muscle 104, 105
internal environment 92, 108, 145
internal fertilization 112, 162, 170
internal gill 166
internal respiration 47
internode **14**, 23, 25
intervertebral disc 122
intestine 97, 167

inversion of image 136
invertebrate **6**, 7
iodine:
    in diet 83
    in starch test 51, 52, 85, 92
iodine solution 211
iris (eye) 130–2
iron 83, 85, 117
    in blood 93
    storage in liver of 92
irritability 9
isolation 74, 208
isotope 53, 54
isotopic labelling 53, 54

jaw 119, 120, 123, 125–7
    insects 152, 156
    muscles of 127
joints **120, 121**, 122, 123
    insect 150
joule 82

karyogram 184
keel 170, 171
keratin 110
kidney 67, 91, 94–6, **106**, 107, 108, 112, 113, 144
knee jerk 139, 140
knee joint 120, 121

labium 152–5
labour 116
labrum 153, 155
lachrymal duct 87, 130
lacteal 89, 100
lactic acid 49, 91, 110
Lake Erie 60
lamina 17
large intestine 86, 98
large white butterfly 156–8
larva 155–8, 201
laryngeal cartilage 86, 87
larynx 86, 87, 102, 103
latent heat 70, 111
lateral bud 22–5, 27, 29
lateral line 159
lateral root 14, 16, 20, 21, 40, 41
leaf:
    adaptation to function 18
    evaporation from 69, 71
    function of 18
    modification in bulbs and corms 23, 24
    photosynthesis in 50
    structure of **17**, 18–20
    water movement in 66, 69
leaf fall **28**, 29, 70
leaf scar 29
leaping (frog) 163
leguminous plants 59, 60
lens:
    human eye 130–2
    insect eye 148, 149
    spectacles 133
lenticel 14, **16**, 29
leucine 186
leucocytes 73, **93**, 97
lever action of bones 122, 123, 134
life history (life cycle):
    bird 170, 172
    butterfly 156–8
    fish 162
    frog 170, 172
    insects 156–8
    mosquito 155
ligament 120, 121, 160, 169
    suspensory 130, 132
light:
    reaction in photosynthesis of 54
    response of shoots to 43–5
light intensity:
    effect on eye of 130–2
    effect on photosynthesis of 56, 57
lightning 60
light-sensitive cells 131, 132, 148–50
light soil 79
limbs 205
lime 79
lime (plant) 38
lime water 48, 49, 105
limiting factor 56, 57
Lind 83
linkage 192
linkage group 192
lipase 85, 90
liver 86, 88–90, **91, 92**, 93–5, 97, 106, 108, 144
liverwort 6
loam 79
locomotion:

Amoeba 176
bird 170, 171
ciliates 177
Euglena 176
fish 160
frog 163
insect 150, 151
    mammal 119, 122, **124**
    Paramecium 177
    protista 178
    tadpole 167
locust 147
long sight 133
lung 89, 94–6, 101, **102**, 103–6, 164, 167
    capacity of 103, 105
lung-fish 160
lupin 31, 34, 36, 38
lymph 100
lymph gland 93, 100
lymph node 94, 100
lymphatic 98, 99, 100, 109
lymphatic duct 100
lymphatic system 89, 94, 98, **100**
lymphocyte 93, 100
lysin 94
lysine 186

maggot 156
magnesium:
    in diet 83, 85
    in plant nutrition 54, 55
maize:
    fruit and germination of 40, 41
    in diet 84
malaria 154, 155, 207
malarial parasite 178
malleus 134
Malpighian layer 109–11, 179
maltose 77, 87, 88, 90
mammal 6, 8, 204
    in seed dispersal 38
mammary gland 97, 112, 117, 193
mandible:
    bird 168, 171
    insects 147, 152, 153, 156
manure 60, 79
margin (leaf) 17
marrow (bone) 93, 121
mass flow hypothesis 68, 69
masseter muscle 123
mating:
    birds 170
    frogs 164, 165
maxilla 152, 153
measles 95
median fins 160
medulla:
    adrenal 144
    kidney 106, 107
medulla oblongata (brain) 142, 143
meiosis 181, **188**, **189**, 190, 193–6
Meissner's corpuscle 109, **128**
melanin 109, 186
membrane:
    cell 11, 12, 179, 180, 189
    semi-permeable 63, 64
memory 143
Mendel 198
Mendelian genetics 198, 201
menstrual period 118
menstruation 118
mesenteric artery 95, 96
mesophyll 17, 19, 50
metabolism **50**, 53, 91, 111
metacarpals 120, 121
metamorphosis 144
    frog 167, 168
    insect 156–8
    metaphase 180, **181**, 188, 192
metatarsals 119, 120
metatarsus 168, 171
micro-organisms 105
    in soil 80
micropyle:
    in insect egg 156
    in seed 35, 39, 40
mid-brain 142, 143
middle ear 134
middle lamella 10
mid-gut 147
milk 83, **85**, 87, 117
milk teeth 126
mineral salts 73
    in diet 82, **83**, 85
    in plant nutrition 54–6
    in soil 78, 79
    uptake and movement in plants of 69, 70

mitochondrion 13
mitosis 179, **180, 181**, 183, 186, 188, 190, 191, 193, 199
molar tooth 120, 125–7
molecule 62, 63, 82, 85
mollusc 6, 7
mongolism 187
monocotyledon 6, 39
monoculture 61
monoploid 188
mosaic image 149, 150
mosquito 153–5
moss 6
moth 206
motor area 141–3
motor end-plate 138
motor fibre 138–41
mould 75, 77
moulting 146, 156
mouth 86
    digestion in 87, 90
    fish 159, 161, 162
    frog 164
mouth parts of insect **152**, 153–5
movement 9, 119, 120, 122, 124, 163, 167 (see also locomotion)
Mucor 75, 76
mucous gland 164, 166
mucus 88, 89, 94, 102, 103, 129
multifactorial inheritance 194
multipolar neuron 138, 139
muscle 86, 89–91, 96–100, 103–5, 108, 111–14, 116, 119, 120, **122**, 123, 124, 128–30, 138, 140, 141, 160, 163, 166, 168, 170
    ciliary 132
    extensor 120, 122, 123
    eye 130
    flexor 122, 123
    hair 109–11
    insects 150, 151
    muscle attachment 119, **120**, 121, 126, 150
mushrooms 75
mutation 184, **186**, 187, 188, 207
    chromosome 184, 185, 187
    rate of 208
mycelium 75, 76
myxomatosis 58

nasal cavity 87, 129, 130
natural selection 206–9
navel 117
nectar 30, 33, 34, 77, 153, 158
nectary **30**, 31, 33, 34
nerve 109, **139**, 140, 141
    auditory 134
    optic 130, 131
nerve cell 12, **138**, 139, 141
nerve ending 109, 110, 125, **128**, 134, 135, 138, 140
nerve fibre 109, 125, 128, 129, 131, **138**, 139, 141, 142, 148, 149
nerve impulse 128, 129, 131, 134, 135, 138–41
nervous system 137–43
nest:
    bird 170
    stickleback 162
neural spine 122
neurone **138**, 139
newt 163
nictitating membrane 168
nitrate 55, 56, 59, 60, 61, 73, 78
nitrifying bacteria 59
Nitrobacter 59
nitrogen 54, 82, 83, 103, 106
    in plant nutrition 59, 60
nitrogen cycle 56, 58, **59**, 60
nitrogen-fixing bacteria 59, 60
nitrogenous waste 98, 106–8, 116
Nitrosomonas 59
node 14
nodules (root) 59, 60
normal distribution 201
nose 103, 129
nostril:
    bird 168
    fish 159, 161
    frog 163, 164
notch wing 185
nuclear membrane 179–81, 188–91
nucleic acid 186
nucleolus 13, 179, 181
nucleoplasm 10
nucleus **10**, 11–13, 75, 77, 93, 112, 114, 115, 138, 165, 174–7, 179–83, 188–90, 193
nutrition:
    Amoeba 177
    animal 50
    holophytic 50
    holozoic 50

Paramecium 177
parasitic 50
plant 50–7
protista 178
saprophytic 50
Spirogyra 174
types of 50
nymph 152, 156

oblique muscle 130
ocelli 150
oesophagus 86–8
oestradiol 144
oestrogen 114, 118, 144, 145
oestrone 144
olfactory lobe 142
olfactory pits 148
ommatidium 149
omnivorous 126
oogenesis 188
operculum:
    fish 159, 161, 162
    tadpole 166, 167
opsonin 94
optic lobe 142
optic nerve 130, 131, 142
optimum temperature (germination) 43
orbit 120, 123, 126, 129, 130
organ 12
organelle 13
organic base 186, 187
organism 12
osmoregulation 67, 108, 177, 178
osmosis 51, 63, 64–7, 98, 108
    in animals 67
    in plants 64–6, 68, 69
osmotic potential 51, 63, 64, 65, 67, 68, 73, 98, 108, 145
ossicle 134
otolith 135
outer ear 134
oval window 134
ovary:
    flowering plant 30, 31, 32, 34–8
    insect 147
    man 112–15, 144, 145
over-cooling 111
over-grazing 80, 81
over-heating 110, 111
oviduct 112–16, 170
ovipositor 147, 158
ovulation 114, 118
ovule 30, 31, 32, 34–7, 39
ovum 112, 114, 115, 117, 118, 178, 188, 190, 191, 195, 196
ovum mother cell 191
owl 173
oxygen 47, 50, 60, 68, 77, 78, 101–3, 105, 107, 116, 117, 147, 161, 164, 166, 172
    diffusion of 62, 63
    from photosynthesis 50–4, 56
    need for germination of 42
    transport in blood of 93, 94, 97, 98, 100
    uptake in respiration 47–9
oxygen debt 49
oxygenated blood 98, 103, 106, 117
oxy-haemoglobin 93, 94

Pacinian corpuscle 109, 128, 129
pain 109, 125, 128, 129
pain receptor 140, 141
paired fins 160
palate 86, 87, 127
palisade cells or layer, 17, 18, 19, 50, 51
    photosynthesis in 51
palp 152, 153, 155
pancreas 86, 88, 90, 144, 145
pancreatic duct 86, 88
pancreatic juice 88, 90
parachute fruits and seeds 38
parallax 133
Paramecium 178
parasite 154, 158
parasitic fungi 75, 76
parasitic nutrition 50
parental care:
    bird 170, 172, 173
    fish 162
    man 117
patella 119–21
Pavlov 140, 141
peat 79
pectoral fin 159, 160, 162
pectoral girdle 123
pectoral muscles 170
pectoralis major and minor 170
peg and socket joint 150
pellagra 84
pelvic fin 159, 160

pelvic girdle 112, 113, 119–22, 123, 124, 163
pelvis (kidney) 106, 107
pelvis (skeleton) 119–21, 124, 170, 171
penicillin 74, 76
Penicillium 76
penis 113, 115, 118, 193
peppered moth 206–8
pepsin 87, 88, 90
    experiment on 92
pepsinogen 88, 89
peptide 87, 88, 90, 186
perennial plants 22
pericarp 30, 35–8, 41
perilymph 134
periosteum 121, 122
peristalsis 86, 87, 89
permanent tooth 126
permeability (soil) 80
pesticides 61
petals 30, 31–4, 36, 37
petiole 16, 17, 18, 28
pH 108
    experiment on 92
    in enzyme reactions 86–8
pH indicator 53
phagocyte 93, 94
phalanges 120, 121, 171
pharynx 86, 87, 102
phenotype 194, 197
phenyl alanine 200
phenylketonuria 200
phloem 14, 15, 16, 17, 20, 51, 68, 69, 152
photosynthesis 18, 19, 27, 39, 41, 44, 48, 50–57, 64, 68–70, 73, 174, 176
    limiting factors in 56
    rate of 56, 57
phototropism 43, 44
phosphate 60, 73, 186, 187
phosphorus:
    in diet 83
    in plant nutrition 54, 55
phylum 6
Phytophthora infestans 76
phytoplankton 57
pigment cell 164, 168
pinna 134
pitch (sound) 134
pitching 160
pith 14, 15, 16, 21
pituitary body (gland) 108, 142, 143, 144, 145, 164
placenta:
    flowering plant 35, 36
    mammal 116, 117, 199
placental villi 115, 116
plankton 176
plants
    characteristics of 9
    classification of 6
    disease in 76
    hormone in 44
plasma 89, 91, 93, 94, 98–100, 103
plasma proteins 91–3, 98, 100
plasmolysis 64, 67, 73
platelets 93, 94
pleural fluid and membrane 104, 105
Pleurococcus 175
plumule 36, 39, 40, 41
poikilothermic 6, 110, 159, 163
polar body 188
poliomyelitis 74, 155
pollen 30, 33–5, 39
pollen grain 193
pollen sac 33, 34
pollen tube 35, 39
pollination 30, 33, 34, 35
    insect 33–5
    wind 33, 35
polymorphism 208
pons 143
poppy 37, 38
population 118
pore 109
potassium:
    in diet 83
    in plant nutrition 54, 55
    radioactive 69
potato blight 60, 75, 76
potato tuber 24, 25
potometer 70, 71
pregnancy 83, 116
pregnancy test 144
premolar tooth 126, 127
pressure sensitivity 109, 128
primaries (feathers) 168–71
primary feathers 173
proboscis 152–5, 157, 158
producer 58
progesterone 115, 116, 118, 144, 145
proleg 157
prophase 179, 180, 181, 182, 188, 189, 192

proprioceptors 124, 129, 148
prostate gland 113–15
protein 82, 93, 94, 97, 98, 100, 106–8, 172, 173, 183, 186, 203
    digestion of 87, 88, 90
    in diet 82, 83, 85
    metabolism of 91, 144
    synthesis in plants of 54
    test for 85
proteinase 85
protista 178
protoplasm 10, 54, 73, 75, 76, 85, 90, 106
protozoa 176
pseudopodium 176, 177
pseudotrachea 154, 155
puberty 114, 117, 144
pulmonary artery or vein 95–9, 102, 103
pulp (tooth) 125, 127
pulse 101
pupa 155–8
pupation 157, 158, 201
pupil (eye) 130, 132
pure-breeding 194, 195
pus 94
pygostyle 171
pyloric sphincter 86–8
pyramid (kidney) 106, 107
pyrogallic acid 42

quarantine 74
quill 169

rabbit 58, 89, 115, 119, 120, 124
    skeleton of 120
radiation 187
radicle 36, 39, 40, 41, 46
    response to stimuli of 43, 44
radioactive isotope 53, 54, 68, 69, 73
radius (bone) 119–21, 123, 169, 171, 205
raspberry 36
ray 17
reabsorption (kidney) 107, 108
reagents, formulae for 211
receptacle (flower) 30, 31, 37, 38
recessive 186, 187, 193, 194, 197, 198, 200, 207
recombination 190, 192, 207
rectum 86, 88, 89, 112, 113
rectus muscle 130
red cell (blood) 92, 93, 94, 97, 99, 101, 103, 154, 207
red hair 198
reduction division 188
reflex 86, 103, 108, 111, 128–30, 132, 135, 139, 140, 143
    conditioned 140, 141, 143
reflex arc 140, 141
refrigeration 73
regulation:
    blood sugar 91
    body temperature 110, 111
    water in the blood 108
relay neurone 140, 141
renal artery or vein 95–7, 106–7
renal tubules 106–8
rennin 87, 90
replication 181, 185, 188, 189
reproduction 9, 203
    Amoeba 177
    bird 170
    fish 162
    flowering plant 30–7
    frog 164–7
    fungi 75
    mammal 112–18
    protista 178
    Spirogyra 174, 175
    yeast 77
reproductive cells 112
reproductive organs 112, 113, 144
reptile 6, 8, 112, 204
residual air 103
resistance to disease 73, 94, 95, 109
respiration 9, 47–50, 51, 53, 54, 58, 69, 73, 77, 78, 101, 107, 144, 178
    aerobic 47, 49
    anaerobic 47, 49
    demonstration of 47–9
retina 131–3
retinal cell 149
rhabdom 148, 149
rhizome 14, 23, 24, 25
Rhizopus 76
rib 97, 104, 105, 119, 120, 122, 171
rib muscles 103, 104
ribosome 13
rice (diet) 83, 84
rickets 84
rill erosion 80
roan 197

rod (eye) 131
rolling 160
root 14
    adventitious 20, 23–6, 41
    contractile 24
    functions of 20
    lateral 14, 16, 20, 21, 40, 41
    response to gravity by 43, 44, 46
    structure of 20, 21
    uptake of salts by 69
    uptake of water by 65
root (tooth) 125–7
root cap 40, 41
root hair, 20, 21, 40, 41, 65
root nodule 59, 60
root pressure 66
root systems 20
root tip 193
root tuber 24, 25
rose, flower and fruit 36, 37
rotation of crops 59, 60
roughage 82, 84, 89
runner 14
ryegrass 32

sacculus 134, 135, 159
sacrum 120
saliva 87, 90, 141, 154, 164
    experiment with 92
salivary gland 86, 87, 90, 141, 147, 154, 156, 185
salivation 140, 141
salt taste 129, 136
salts 98, 106–8, 116
    in diet 82, 83
    in soil 78
    movement in xylem of 69, 70
    uptake by roots of 69, 70
sand 78, 80
saprophytic 50, 75, 176
scale leaves 23–5
scales 125, 159, 168
scapula 120, 123, 171
scion 26
sclera 130, 132
scrotum 112, 113
scurvy 83, 84
sebaceous gland 109, 110
secondary feathers (secondaries) 168–70, 173
secondary sexual characters 117, 118, 144, 193
secondary thickening 17
secretin 144
sediment 203, 204
sedimentary rock 203, 204
seeds 37, 39–46
    conditions for germination of 42, 43
    dispersal of 38
    dormancy of 42
    dry weight of 47
    formation of 36
    production of carbon dioxide by 48
    production of heat by 49
    structure of 39, 40
    uptake of oxygen by 48
segment (insect) 157
segmentation (insect) 146, 147
segregation of genes 194, 207
selective advantage 207, 208
selective permeability 63, 64, 177
selective reabsorption 107, 108
self-pollination 33, 34
semicircular canal 134, 135, 142
semi-lunar valve 98, 99
seminal fluid 165
seminal vesicle 113, 114
seminiferous tubule 113, 114
sensation 129
sensitivity 9, 43, 136, 178
sensory cell 131, 132, 134, 135, 148
sensory fibre 138–42, 148
sensory nerve ending 109, 110, 125, 128, 134, 135, 138, 140, 159
sensory neurone 138
sensory organs 115, 124, 125, 128–36
sensory receptor 138, 140
sensory system 128–36, 148–50
sepal 30, 31, 32, 36, 37
serum 93, 95, 107
setae 158
sewage disposal 73, 74
sex chromosome 190, 191
sex comb (Drosophila) 202
sex determination 190, 191
sex hormone 144, 145
sex linkage 192, 193
sexual reproduction:
    bird 170
    fish 162
    flowering plants 30–9
    frog 164, 165

man 112–18
  Spirogyra 174, 175
shaft (bone) 121
shaft (feather) 169
sheet erosion 80
shell (egg) 170, 172
shell membrane 172
shivering 111
shoot 14
  response to stimuli by 43
short sight 133
shorthorn 197
shoulder blade 123
shoulder joint 120
shrub 6
sickle cell anaemia 207
sieve plate 15
sieve tube 15, **16**, 19, 51, 68, 69, 152
sight 129, **130–3**, 142, 143
  in insects 149
  long and short 133
silk 156–8
silt 78, 81
simple eyes 150, 156
single-celled animals and plants 6, 54, 57, 60, 63, 108, **174–8**
single factor inheritance 194, 195, 198
sinus 117, 143
skeleton 119–24, 205
  bird 169–71
skin 73, 74, 84, 94, 108, **109–11**, 125, 128, 129
  bird 168
  fish 159
  frog 163, 164
  semi-permeability of 67
skin senses 135, 136
skull 119, 120, **123**, 126, 127, 143
  bird 171
  dog 126
  rabbit 120
  sheep 127
sleeping sickness 154
slimming 91
small intestine 86, 87, 95, 96
smallpox 74
smell (sense of) **129**, 142
  in fish 159
  in insects 148, 149, 158
sneezing 139
snowdrop bulb 22, 23
soda lime 48, 52
sodium 83
sodium bicarbonate 52, 53, 88, 93, 94, 103, 130
sodium chloride 93, 107, 110, 130
soft palate 86, 87
soil 60, 61, **65**, **78–80**, 81
  bacteria in 59, 60, 79
  conservation of 81
  erosion of 60, 61, 79–81
sour 129, 136
spawning 162
special senses 129
specialization 12
  of teeth 125–7
species **6**, 206–8
sperm 112, 114, 115, 118, 165, 170, 178, 188, 190, 191, 195, 196, 208
sperm duct 113, 114
sperm mother cell 190, 191
spermatic cord 113
spermatogenesis 188
sphincter, pyloric 86–8
sphincter muscle 87, 106, 108
spider 6
spikelet (grass) 32
spinal column 104, 120, **122**, 124, 141, 170
spinal cord 122, 124, 128, 137, 138, **139–42**, 143
spinal nerve 137, 140–2
spinal reflex **140**, 142
spindle 180–2, 188–91
spindle fibre 180, 181
spine 104, 120, 122–4, 160
spinneret 156
spiracle 146, 156, 157, 166, 167
Spirillum 72
Spirogyra 174, 175
spleen 93, 94
spongy layer 17, **18**, 19, 50
spontaneous generation 203, 204
sporangium 75, 76
spore:
  bacterial 72–4
  fungal 75–7
squash preparation (chromosomes) 193
stamen **30**, 31–4, 36, 37
stapes 134
Staphylococcus 72
starch 41, 49, 51, 82, 87, 174
  in digestion 87, 88, 90, 92

production in photosynthesis of 51, 52
  test for 51, 52, 85
stem **14–17**, 21–3
  conduction in 68–70
  functions of 14
  strength of 17, 20, 21
  structure of 14–16
  underground 24, 25
stereoscopic vision 132, 133
sterilization:
  bacterial 74
  genetic 200
sternum 104, 119, 120, 170, 171
stickleback 159, 162
stigma **30**, 31–7, 39
stimulus 128, 129, 131, 139–41
  on roots and shoots 43–5
stock (grafting) 26
  digestion in 87, 88, 90
stoma (pl. stomata) 14, 16–**19**, 21, 50, 51, 56, **64**, 69, 70, 72, 76
  mechanism of 64
stomach 86, 88, 95, 97
storage (food, etc.) 90–2
storage organs 22–6
strawberry 36, 37
Streptococcus 72, 93
streptomycin 76
stretch receptors 124, **128**–30, 140, 148
struggle for survival 206, 207
style (flower) **30**, 31, 32, 35–7
subclavian artery or vein 96, 100
sub-cutaneous fat 109, 110
succulent fruit 38
succus entericus 90
suckling 117
sucrase 90
sucrose 51, 54, 68
  test for 85
sugar 186
  metabolism of 144, 145
sulphate 55
sulphur:
  in diet 82, 83
  in plant nutrition 54, 55
sulphur dioxide 73
sunlight:
  and vitamin D 84
  as source of energy 58
  in photosynthesis 50–4
sunstroke 111
superphosphate 60
surface area:
  and heat loss 111
  and volume 111, 178
  feathers 168, 170
  gills 161
  intestine 89
  lung 103
survival of the fittest 208
survival, struggle for 206
suspensory ligament 130, 132
suture 120, 123
swallowing 86, 87, 134
sweat 110, 111
sweat duct 109, 110
sweat gland 109, 110
sweating 111
sweet (taste) 129, 136
swim bladder 160–2
swimming:
  fish 160
  frog 163
  tadpole 167
sycamore fruit 38
synapse 138–41
synovial fluid 120, 121
synovial joint 120
system (definition) 12

tactile bristles 148
tadpole 144, 165–8
tail:
  fish 160
  tadpole 166, 168
tail feathers 168, 170, 171
talons 173
tap root 14, 16, **20**
tarsals 119, 120
tarsus 148, 153, 158
taste **129**, 136, 148
taste bud **129**, 141
tear duct 130
tear gland 130
teeth 125–7
telophase 180, **181**, 182, 188–90
temperature (body) 92, 94, 108, 159
  control (regulation) of 94, 110, 111

temperature, effect on
  bacteria 73, 74
  enzyme action 86, 92
  germination 42, 43
  photosynthesis 56
  transpiration 70
temporalis muscle 123, 126, 127
tendon 99, 122, 129, 140, 170
terminal bud 14, 23, 24, 27, 29
terramycin 74
territory 162
testa 36, **39**, 40, 41
testis 112–14
testosterone 144
tetanus 73, 95
thiamine 84
thirst 108
thoracic duct 100
thorax:
  insect 146, 147, 151, 156–8
  man 102, 104, 105, 119, 123
thymine 186, 187
thyroid 83, **144**, **145**
thyroid stimulating hormone (TSH) 145
thyroxine 144, 145, 168
tibia:
  mammal 119–21, 124
  insect 148, 153, 158
tibio-tarsus 171
tidal air 103
tissue 12
tissue fluid 93, 94, 98–100, 108
tissue respiration 47, 101, 103
toad 163–5
tomato flower and fruit 36
tone (muscle) 122
tongue 86, 87, 129, 136, 164
touch 109, 128, 129, 135, 136, 178
  in insects 148
toxin 73, 94
trace element 55
trachea:
  man 86, 87, 102
  insect 146, 147, 153
tracheole 146, 147
translocation 68
transpiration 68, **69**, 70, 71
  rate of 70, 71
transpiration stream 69
transverse process 122
Treponema 72
triceps muscle 122, 123
tricuspid valve 99
trochanter 148, 158
tropism 43, 44, 46
true leg 156, 157
trypsin 88, 90
tsetse fly 154
TSH (thyroid stimulating hormone) 145
tuber **24**, 25
turgid 64, 65
turgor 19, **64**, 69, 70
turgor pressure 64, 66, 68, 69
twigs 27–9
twins 117, 199
tympanal organs 148
typhoid 72, 73, 155
tyrosine 186
tyrosinase 186

ulna 120, 121, 123, 169, 171, 205
ultra-violet light 84, 109, 187
umbilical cord 116, 117, 172
unicellular organisms 174–8
urea 91–3, 106–8, 110
ureter 106–8, 112, 113
urethra 106, 108, 112–14
uric acid 106, 108
urine 87, 91, 92, 107, **108**, 144
uterus 112–18, 145
utriculus 134, 135, 159

vaccination 95
vaccine 95
vacuole **10**, 11–13, 44, 51, 64–7, 75, 77, 174
  contractile 176–8
  food 176–8
vagina 112, 113, 115–18
valine 186
valve 95, 97–101, 161, 164
vane (feather) 169, 171
variation 188, 192, 199, 206–8
  continuous 198, 199
  discontinuous 198
  heritable 207
variegated leaf 52
variety 206–8
vascular bundle 14, 15, **16**, 17, 18, 20, 29, 34, 72

vaso-constriction 111
vaso-dilation 110, 111
vector 154
vegetative reproduction 22–6
  advantages of 25
vein (blood system) 89, 96, **97**, 98, 101, 117
vein (of leaf) 17–20, 50, 66
vena cava 95, 98–100, 106
ventilation (breathing) 47
  fish 161
  frog 164
  insect 147
  man 101, 103–**105**
ventral fin 159, 160
ventral root 140, 141
ventricle 95, 96, 98, 99, 142, 143
venule 97, 99, 109
vertebra 87, 119, 120, **122**, 123, 143, 160, 170, 171
vertebral column 122, 160, 161
vertebrate 6, 8
vessel (plants) 15, **16**, 17, 19, 21, 28, 65, 66, 68, 69, 72
Vibrio 72
vibrissae 110
villus 89
vinegar 77
virus 74, 94, 95
vision **131**, 133
vitamins 73, 77, 82, **83**, 84, 85
  storage of 92
vitreous humour 130–2
viviparous 162
vocal cord 103
voice 103
voluntary muscle 122
vulva 112, 113

walking (insect) 150, 156, 157
Wallace 206
"warm-blooded" 159, 168
washing 74
water 93, 106, 107, 110
  absorption in colon of 89
  balance of 108
  diffusion of 63
  evaporation of 70, 71
  for photosynthesis 50, 51, 54
  from respiration 47, 49
  in diet 82, 84
  in germination 42
  in osmosis 63
  in soil 78, 79
  movement in leaves of 66
  movement in plants of 65, 66, 69, 70
  purification of 74
  reabsorption in kidneys of 107, 108
  response of roots to 44
  uptake of 70
water cultures 55
water potential 65, 66
water sac (amnion) 116, 172
water-soluble vitamin 83
white cells (blood) 12, 73, **93**, 94, 97, 100
white deadnettle:
  flower of 31
  pollination of 34
  stem T.S. of 15
white matter (nervous system) 139, 140, **142**
wilting 64, 70
wind dispersal 38
wind erosion 60, 78, 80
wind pollination 33–5
windpipe 86, 102
wing movements in insects 151
wing, of bird 168, **169**, **170**, 171
winged fruits 38
womb 112
wood 17
World Health Organization 155
worms 6, 7
wrist 120, 121

X chromosome 190–2
xerophthalmia 84
X-radiation 187
xylem 14–**16**, 17, 20, 65, 66, 68, 69

Y chromosome 190–2
yawing 160
yeast 49, 50, **77**
yellow fever 154
yolk 156, 165, 166, 172
yolk sac 115, 172

zona pellucida 115
zooplankton 57, 162
zygospore 174, 175
zygote 34, 112, 115, 174, 175, 179, 186, 188, 190, 191, 194–6, 199
zymase 77